Microbial Enhanced Oil Recovery

Microbial Enhanced Oil Recovery

J.E. Zajic
D.G. Cooper
T.R. Jack
N. Kosaric

PennWell Books
PennWell Publishing Company
Tulsa, Oklahoma

PennWell Publishing Company
1421 South Sheridan Road/P.O. Box 1260
Tulsa, Oklahoma 74101

Library of Congress cataloging in publication data
Main entry under title:

Microbial enhanced oil recovery.
Bibliography
1. Secondary recovery of oil—Congresses.
2. Industrial microbiology—Congresses. I. Zajic,
J. E. (James E.), 1928-
TN871.M446 1983 622'.3382 82-18049
ISBN 0-87814-212-6

Printed in the United States of America

1 2 3 4 5 87 86 85 84 83

The book is dedicated to Hemisphere Licensing Corp., its subsidiary, Petroleum Sciences Inc. of Houston, and to all of those researchers who are advancing the frontiers of technology in microbial enhanced oil recovery.

Contents

Acknowledgments

I wish to thank the technical and administrative staffs of Hemisphere Licensing Corp. and Petroleum Sciences Inc. for assisting and supporting the publication of this monograph and for their continuing efforts to expand the frontiers of enhanced oil recovery technology.

A very special thanks to Neil Rogen, president of Hemisphere Licensing Corp., who has taken the technological gamble in support of a worldwide effort to commercialize microbial enhanced oil recovery.

1 PHYSICAL SIMULATION OF MICROBIOLOGICALLY ENHANCED OIL RECOVERY

B. Bubela

Baas-Becking Institute, Australia

To devise a successful microbiological method to operate in an oil reservoir, it is necessary to consider in some detail the basic principles of geobiology operating in such an environment.

Geobiology is a science investigating the mutual interaction of biological and abiological parameters taking place within the biosphere of the earth's crust and to a depth of at least 2,000 m. It is evident from the above definition that soil biology forms an integral part of geobiology. Geobiology is not limited to the activities of microorganisms; but for the purpose of an enhanced recovery technique, only microbiological activity will be considered herein.

The biological and abiological factors in a geobiological process are interdependent. Factors such as pH, Eh, temperature, substrate availability, permeability, porosity, and mineralogy of the rock influence such functions as the mean generation times of organisms, their morphology, their nutritional requirements, and production of metabolites.

On the other hand the biological activity can modify the chemicophysical properties of the environment. For example, biological production of carbon dioxide may cause solubilization of carbonate rocks, thus increasing their permeability and porosity. Conversely, should carbon dioxide levels decrease because of an abiological activity in the interstitial fluids, processes resulting in carbonate reprecipitation, cementation, and a considerable decrease in permeability and porosity of the reservoir rock may result. Hydrogen sulfide produced from pore water sulfates by sulfate-reducing organisms will cause precipitation of sulfides of iron and heavy metals. This will again affect the permeability and porosity of sediment and at the same time will counteract the potentially toxic effect of the metals. These examples indicate just a few of the interactions taking place during geobiological processes.

The investigation of problems associated with various branches of geology, sedimentology, and geobiology is usually based on two approaches: (*i*) field work or (*ii*) laboratory investigation.

The field approach has the distinct advantage of direct observation of natural systems. However, field investigations frequently are hindered by the complexity of the systems being studied, and the investigator is handicapped by lack of experimental control of physical and chemical parameters. Furthermore, although some very slow processes may operate long enough in their original environment for measurable results to be obtained, for other processes the operating conditions may be very different at the time of observation. On the other hand, just as the field approach can be limited by the complexity of systems, laboratory investigations may be limited by oversimplification.

In laboratory investigations, the fact that individual processes do not operate independently but are highly interactive with other processes should not be overlooked. A typical example would be to study changes in nutritional requirements of an organism due to an increase in temperature, and then to study independently their requirements in the presence of heavy metals. Valid as they are, such experimental results do not offer much assistance from the point of view of geobiology. Unfortunately, these parameters are not additive in their effects on biological activity but are interacting. An example of such an interaction is the effect of copper and magnesium or copper and manganese on the activity of some bacteria.[1,2,3] In certain instances each of the individual elements may be inhibitory to the organism; but if presented in pairs, antagonistic couples are formed, nullifying to some extent the detrimental effect of the individual metals. Similarly, the effect of biologically produced carbon dioxide on a reservoir rock may be reduced if methanogenic organisms utilizing CO_2 are present in the environment. This loss might be offset by improved pressure in the reservoir.

One of the primary advantages of laboratory experiments is that the starting conditions of processes being studied are known. This is frequently essential for subsequent evaluation of the systems. In addition, in many cases the size of an experiment may be a limiting factor and frequent sampling over a long time may be necessary, which is more easily done in the laboratory. It is true that the walls of a small container may affect processes associated with surface tension, microbiological activity, and permeability. On the other hand, such experiments are usually readily monitored and the answers to many basic questions are relatively easy to obtain.

Finally, one of the most important advantages of laboratory investigations is the experimental freedom of the operator, lacking in most cases of field investigations.

After consideration of all of the advantages and disadvantages of the above approaches, a simulating system was formulated.

The simulating system represents a laboratory attempt to move closer to the field environment. This allows some of the nonessential complexity of field studies to be eliminated while it simultaneously facilitates experimental control. Ideally, there should be continuity from the simple laboratory system through the experimentally manageable simulating system to the naturally complex field environment.

In the design of a simulating system, the questions to be answered should be carefully defined. It is advisable to limit the complexity of the system as much as possible, and the number of parameters to be monitored simultaneously should be similarly limited. Therefore, the selection of these parameters is of the greatest importance. With proper planning, the problems of continuous sampling and side-wall effects can be minimized, if not eliminated.

Requirements for specific minerals and fluids in quantities of kilograms of even tons can create logistic problems, especially when uniformity of chemical composition and mineralogy is important. In cases where fully automatic monitoring and automatic control are desirable, the instrumentation may become very complex. The same would apply to a field study, of course.

The systems developed herein vary from simple units having a capacity of several milliliters to rather sophisticated units with thousands of liters of capacity and electronic control. The latter are capable of simulating hydrological and mineralogical situations occurring in nature, climatic and temperature variations, redox potentials, and biological activities pertinent to the problem under investigation.[4,5,6] Because simulation of geobiological processes has proven successful in the investigation of mineralization processes in sedimentary environments, a similar line of attack is being used in the present investigation of microbiologically enchanced oil recovery.[7–15]

The basic requirements for microbiologically enhanced oil recovery are similar to those used in abiological evaluations. They are:

(*i*) reduction of the interfacial tension of the rock-water-oil system,

(*ii*) improvement in the mobility ratio of the fluids involved, and

(*iii*) increase in the permeability of the reservoir rock.

Before such changes can be achieved by biological means, parameters influencing the biological activity under reservoir conditions must be evaluated. The significance of biological activity on the permeability of sediments and on the directional component of the flow of fluids through the sediments has been discussed.[16,17] Similarly, the development of a strongly negative redox potential due to the biological activity in the sediments and its effect on the distribution of the microbiological population has been shown in simulated systems, thus demonstrating the suitability of applying simulating studies to the problem of microbiologically enhanced oil recovery.[17]

The first approach in simulation was to devise a simple system to observe visually and evaluate quantitatively biological activity and the movement of oil through a porous, permeable material that had been oil- or water-prewetted.

Figure 1 shows a water-wetted simulated reservoir of approximately 500 md permeability and porosity of 40% with oil saturation of 75%. The individual pictures show the states (*i*) where water sweep is moving into the simulated system, (*ii*) where the water broke through the oil and no more oil was being recovered even on additional water passage, (*iii–iv*) additional oil recovery obtained by passing through the system a medium in which organisms had been growing for several weeks, (*v*) termination by fluid breakthrough, and (*vi*) the last stage where lysed organisms were introduced into the system, furthering the recovery. Control experiments with water and with fresh and sterile bacterial medium showed that no significant changes in oil recovery occurred when these substances were used as the sweeping fluid. Figure 2 summarizes the results of this experiment quantitatively. The first increase in oil recovery was due to a decrease of the oil-water interfacial tension effected by the production of a surfactant by the organisms. The final increase was due to the improvement of the mobility ratio because of the increased viscosity of the aqueous phase. The interfacial tension of the medium plus organisms-oil was decreased to 7 mN/m and the viscosity of the aqueous phase after the bacterial lysis was approximately 5 cp. A considerable number of experiments with a variety of organisms, grown under different conditions on different substrates and

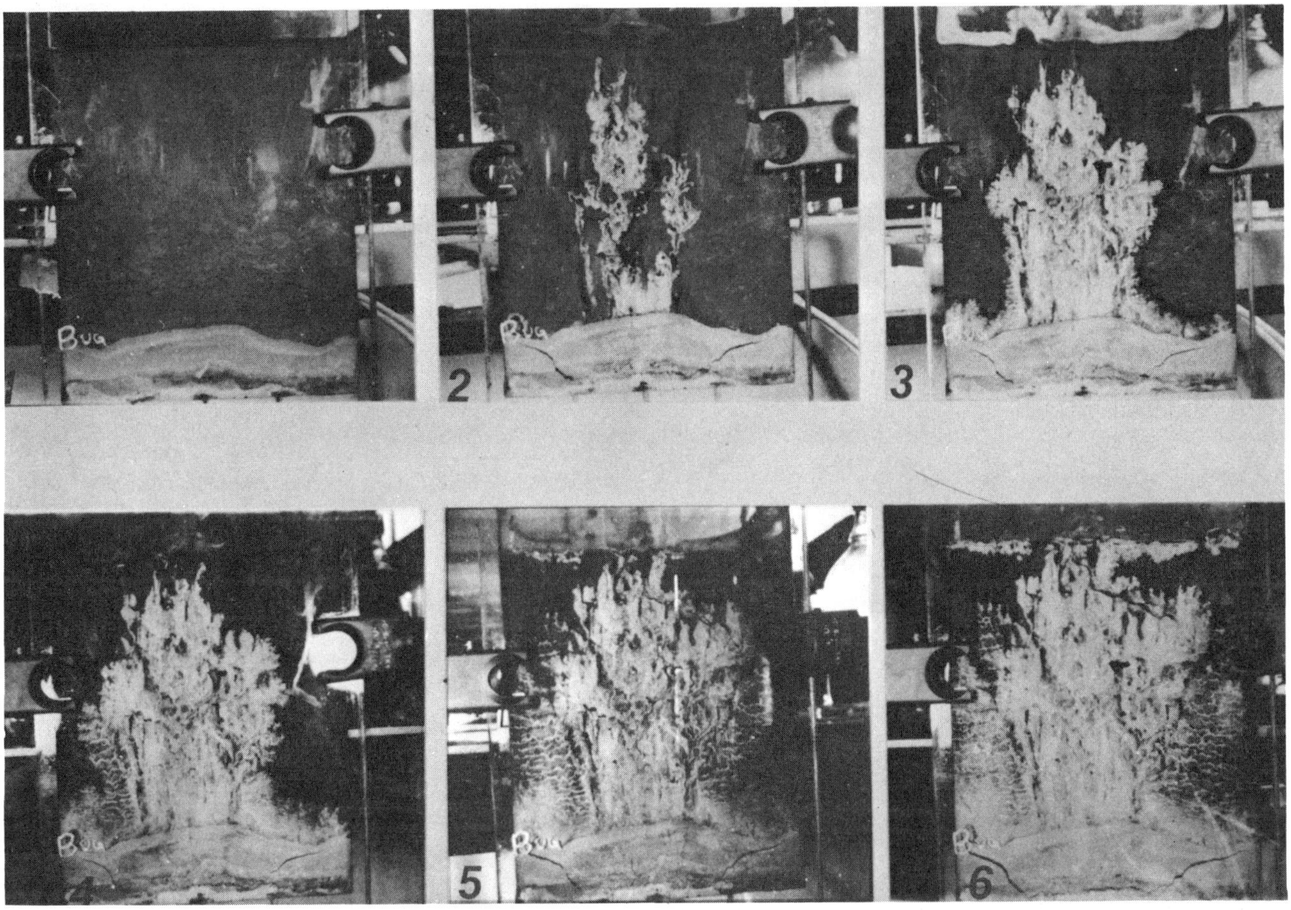

Fig. 1. Microbiologically enhanced oil recovery from a simulated water-wetted reservoir

with various carbonates and silicates as a mineral support, gave comparable results depending on the permeability of the reservoir. Thus, the system appears to be suitable for the investigation of microbiologically enhanced oil recovery and for evaluating the feasibility of microbiological techniques in oil recovery in the field.

This system, consisting of loosely consolidated mineral particles, had the disadvantage that because of its poor consolidation it could not be used to study the effect of flow rates of the reservoir fluids on oil recovery since an increase of the flow resulted in channeling. For the same reason it could not be used to investigate recovery rates at higher hydrostatic pressure heads.

Therefore, a more consolidated system was developed. It consisted of a cement block of a predetermined porosity. Such blocks were obtained by varying the ratio and particle size of aluminum powder, quartz sand, cement, and water in the mixture and by controlling the temperature while curing the block. The permeability was varied from less than 1 md to 5 d with a reproducibility of better than 10%.

The results obtained with such systems have shown that the recovery of oil decreases with an increase in the rate of the flow through the simulated reservoir. The optimum superficial flow velocity was about 15 cm per day. The effect of pressure as applied to the water front did not influence the recovery significantly. The system was further useful in the evaluation of the effect of the size and shape of the microorganism on plugging of the simulated reservoir. Rod-shaped organisms were more effective in decreasing the permeability of the system, and such a decrease was enhanced if a higher hydraulic pressure head was applied to the moving aqueous front. Coccoid organisms were less liable to produce blockages, and the original flow-through could be maintained by an increase of the hydraulic pressure head of the water front. If the microorganisms were labelled with a radioactive material, the actual position and shape of such a biological plug could be determined by autoradiography.

As mentioned before, the effect of the organisms on the reservoir rock components must also be considered. Blocks

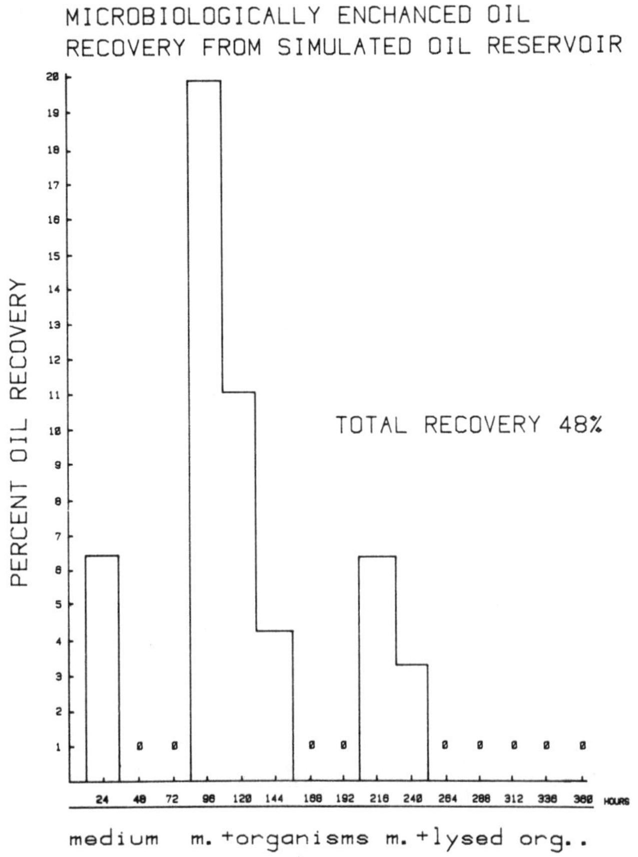

Fig. 2. Quantitative effects of microbiologically enhanced oil recovery from a simulated water-wetted reservoir

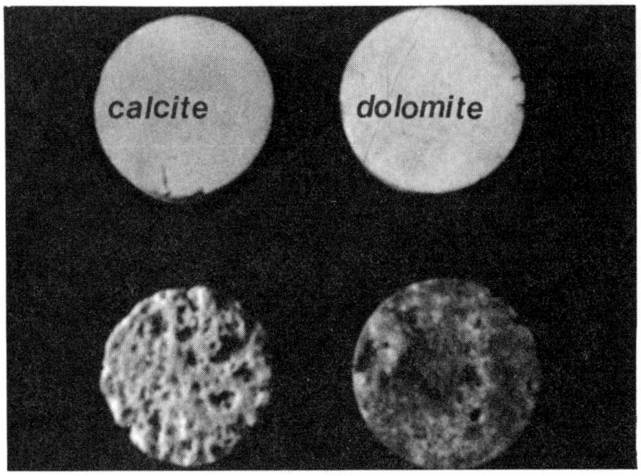

Fig. 3. Effect of biological activity on calcite and dolomite cores

of reservoir rocks obtained from Australian reservoirs were impregnated with a medium capable of supporting bacterial growth under anaerobic conditions in the presence of crude oil. These were inoculated with anaerobic microorganisms and were incubated for a 2-month period at 60°C. Permeabilities and porosities were measured before and after the incubation.

Figure 3 shows discs of calcite and dolomite exposed to such an activity. It is evident that calcite was affected to a greater degree. Its porosity and permeability were increased from 17% to 45% and from 2 d to 47 d, respectively, while those of the dolomite were altered from 24% to 27% and from 2 d to 14 d, respectively. No significant differences in porosities and permeabilities were observed when the experiment was conducted in the absence of microorganisms. These figures have to be taken as an indication only, as the effect of the microorganism under otherwise similar conditions depends on the crystallinity of the mineral components and, in the case of the dolomitic rock, on the degree of dolomitization. The increases in permeability and porosity is due to the biological production of fatty acids and CO_2.

Geobiological processes, such as microbiologically enhanced oil recovery, proceed at slow rates. To accumulate enough data required for the measurements of such rates, a relatively large number of samples must be collected with a minimal disturbance to the rest of the experiment. For instance, data are necessary to establish the rate of changes in porosity and permeability of the sedimentary rock due to the diagenetic processes affecting the mineralogical components, to introduce the microorganism moving with the aqueous phase, or to establish in situ a microbiological population and its own mobility. Furthermore, secondary effects on the processes under investigation, caused by the walls of the containers and other instrumentation, must be minimized. This is possible only by an increase in the size and complexity of the experimental system. A number of such systems have been developed and were used quite successfully in this work.

Figure 4 shows a reaction system where the capacity is 1,000 l. The additional capacity of the fluid reservoirs is in the vicinity of 20,000 l. The system is provided with facilities for temperature variation and three-dimensional fluid flow through the sediments, a coring device for collection of poorly consolidated sediments while preserving their microstrata, illumination when microbial growth depending on photosynthesis is being investigated, and an electronic control for automatic introduction and maintenance of experimental cycles and conditions such as fluid movements, fluid retention, and temperature variation.

This system was used to establish the permeability of microorganisms through specified sediments, their capacity to produce a suitable redox environment for sulfate reduction and the formation of metal sulfides, and consequent effects of plugging on the movement of fluids through the

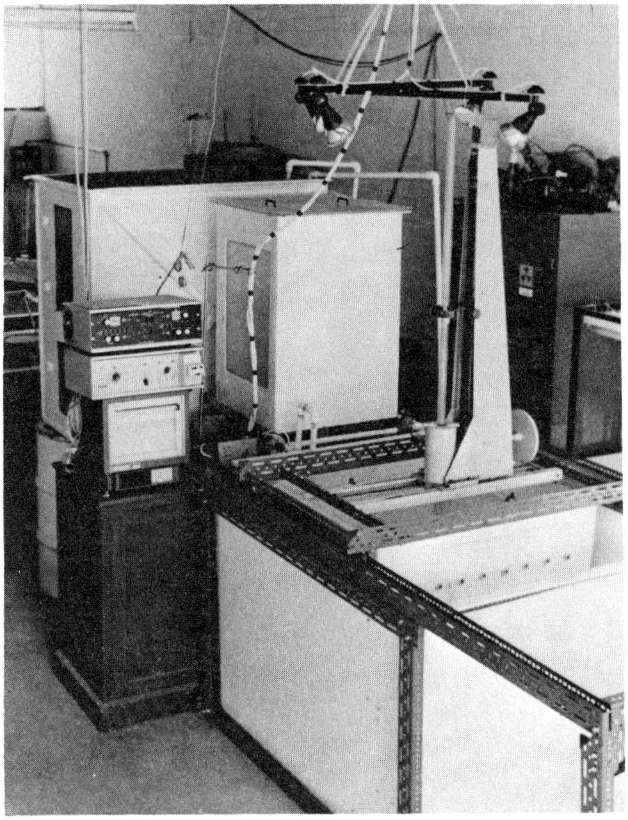

Fig. 4. Simulating system used in the investigation of cementation of interstitial spaces and directional changes of flow of fluids through sediments due to biological activity

sediments. These parameters are of significant importance to enhanced oil recovery. Considerable directional changes in fluid movements were observed when fluids had to pass through new boundaries of differing permeability.[18] Cementation resulting from diagenesis of carbonates, due to alteration of CO_2 level in interstitial fluids, decreased the permeability of the simulated sediments quite considerably in some cases.

All of the experimental systems described so far were able to simulate most of the factors that may influence the activity of microorganisms living under conditions likely to be encountered in an oil reservoir with the exception of pressure.

The hydrostatic pressures encountered in the oil reservoirs may reach several thousand kilopascals. Therefore it is important to establish the effect of pressure on the microorganisms. The metabolic processes of organisms may be changed by increased pressure, as demonstrated by an increase in hydrogen sulfide production in *Desulfatomaculum nigrificans,* by a factor of 10. Pressure can also produce a nutritional requirement for amino acid, or changes in shape and size of the whole organism. Such changes may influence the suitability of an organism for the proposed techniques. Therefore, a chemostat was developed in which microorganisms were grown continuously under anaerobic conditions at high temperatures and pressures. This system is operative up to temperatures of 150°C and pressures up to 20,000 kPa. The composition of the media being used for the bacterial growth is adjustable. Additional components may be added, and the temperatures and pressures may be changed as required. Samples of the organisms can be collected and investigated without interfering with the growth process. Such a system is essential for the investigation of changes induced by the alteration of a single parameter, while other parameters are kept constant. The system developed is shown in Figure 5.

The results obtained with this system are interesting. An organism isolated from Melbourne sewage was originally a slim rod about 8 μm long and 2 μm wide, adapted to a temperature of 65°C and a pressure of 20,000 kPa. Its mean generation time was slightly increased from 2 hr to about 2.5 hr, but there were no additional nutritional requirements. However, changes were observed in the morphology of the organisms. After about 24 hr the shape changed to an elliptic form about 6 μm long and 4 μm wide, approximately doubling the volume. After rapid exposure to atmospheric pressure, the majority of the cells disintegrated within a few minutes. This observation is puzzling, as no vacuoles were detected in the spheroidal forms. If the pressure was adjusted over a period of several hours to the atmospheric value, the microorganisms gradually changed back to their original shapes and were able to propagate and produce cells morphologically indistinguishable from their original forms. The work with this system is in its early stage, but it is hoped that the system will prove to be a useful tool in MEOR investigations.

The purpose of this communication is not to report individual results but to indicate a different approach to the studies of investigating geobiological problems and, in particular, microbiologically enhanced oil recovery.

The results obtained so far indicate that the application of microbiological techniques in the recovery industry could be feasible and could be capable of overcoming some of the difficulties encountered in applying abiological techniques. It is clear at this stage that substantially more work must be done before the full potential of the biological techniques can be established. Furthermore, it must be realized that biological techniques, even if fully developed, will not apply to all reservoirs, as in some cases environment reservoir conditions may limit or completely prevent the desired biological activity.

To evaluate the variety of parameters and to attempt to clarify problems that occur in developing microbiologically enhanced oil recovery techniques, it is necessary to have a

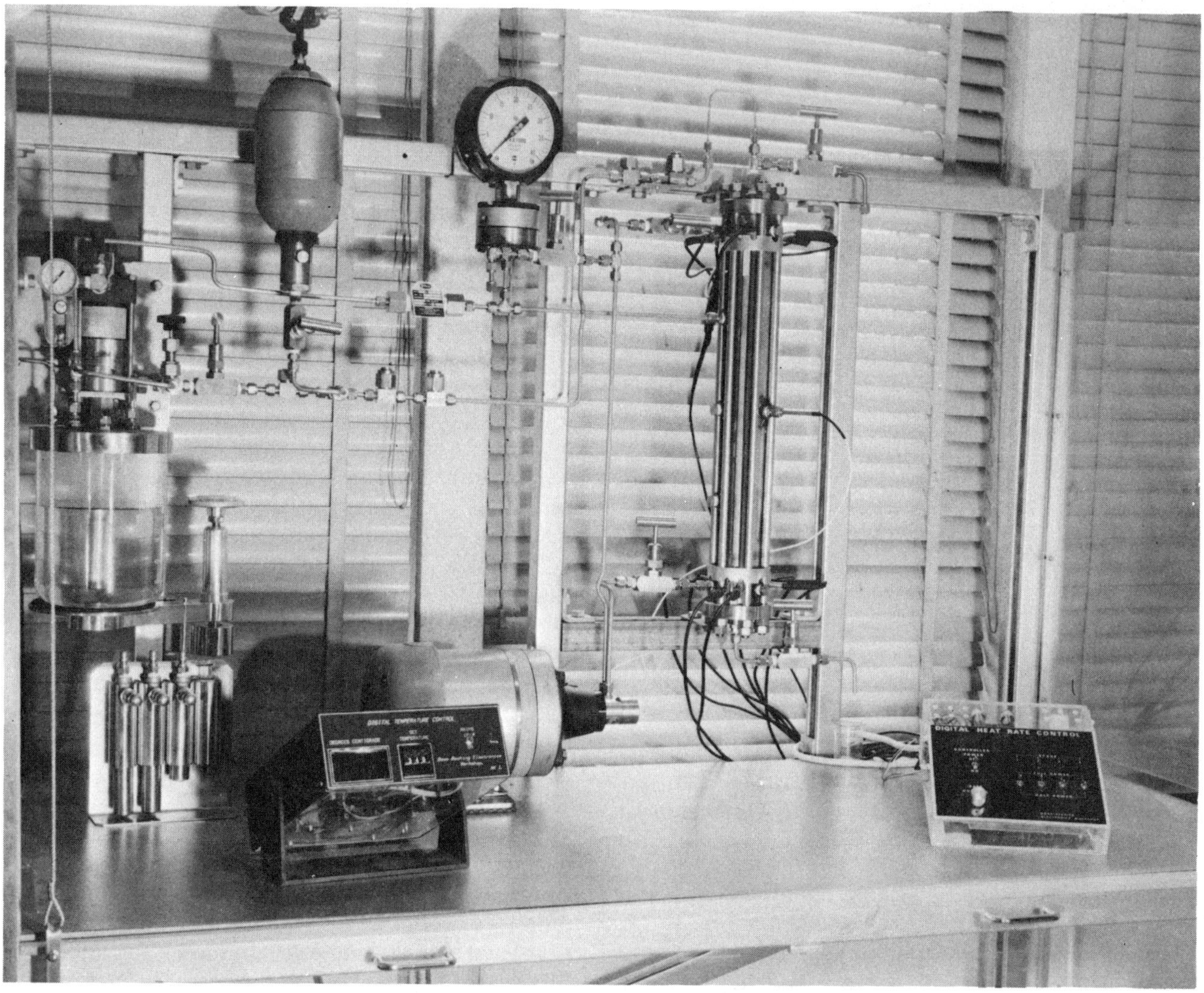

Fig. 5. Chemostat operative under anaerobic conditions at temperatures up to 150°C and pressures up to 20,000 kPa

close cooperation with a number of workers specialized in a variety of skills. Physicochemical, microbiological, mineralogical, geological, and many other skills are required to understand the complexity of a reservoir and its suitability to a particular recovery and to develop suitable techniques leading to a technologically and economically acceptable method.

The problems faced in developing microbiologically enhanced oil recovery techniques are unique. Only a closely cooperating multidisciplinary team, prepared to develop and apply an original approach to the investigation, is likely to bring the technique to economical reality.

ACKNOWLEDGMENTS

The project "In situ Biological Production of Surfactants for Enhanced Oil Recovery" is sponsored by the National Energy Research Development and Demonstration Council of Australia. The Baas Becking Laboratory is supported by the Bureau of Mineral Resources, Commonwealth Scientific Industrial Research Organization, and the Australian Mineral Industries Research Organization. The author wishes to acknowledge the enthusiastic and helpful support of Mrs. V. Partridge.

REFERENCES

1 BUBELA, B. 1970. Chemical and microbiological changes in *Bacillus stearothermophilus* induced by copper. Chemico-biological Interactions *2*:107–116.

2 BUBELA, B. 1973. Effect of copper on the growth characteristics of *Bacillus stearothermophilus*. Zentralblatt für Bakteriologie *128*:274–284.

3 BUBELA, B., and T.C. POWELL. 1973. Effect of copper on the growth characteristics of *Bacillus stearothermophilus*. Zentralblatt für Bakteriologie *128*:457–466.

4 BUBELA, B., I. JOHNS, and J. FERGUSON. 1978. A system for simulation of evaporative sedimentary environments. B.M.R. J. Australian Geology and Geophysics *2*:166–170.

5 JOHNS, I.A., and B. BUBELA. 1979. Multichannel pH-Eh-specific ion monitoring system. B.M.R. J. Australian Geology and Geophysics *4*:398–402.

6 BUBELA, B., and J. FERGUSON. 1973. Apparatus for studies of artificial sediments. J. Sed. Petrology *43*:1167–1170.

7 BUBELA, B. 1981. Banded sulfide ores: The experimental formation of sulfides from flowing liquids. Economic Geology *76*:171–172.

8 BUBELA, B. 1981. A model for sulfide bands formation under epigenetic conditions. B.M.R. J. Australian Geology and Geophysics *6*:117–122.

9 BUBELA and FERGUSON, op cit.

10 BUBELA, B., J. FERGUSON, and P.J. DAVIES. 1975. Biological and abiological processes in a simulated sedimentary system. J. Geol. Soc. Austral. *32*:135–141.

11 DAVIES, P.J., B. BUBELA, and J. FERGUSON. 1977. Simulation of carbonate diagenetic processes: Formation of dolomite, huntite, and monohydrocalcite by the reaction between nesquihonite and brine. Chem. Geology *19*:181–214.

12 FERGUSON, J., B. BUBELA, and P.J. DAVIES. 1975. Simulation of sedimentary ore-forming processes: Concentration of Pb and Zn from brines into organic and Fe-bearing carbonate sediments. Geol. Rundshau *64*:767–782.

13 HALLBERG, R.O., B. BUBELA, and J. FERGUSON. 1980. Simulation of metal chelating in two reducing systems. J. Geomicrobiology *2*:99–113.

14 JOHNS and BUBELA, op cit.

15 MOONEY, J., B. BUBELA, J. FERGUSON, and R.O. HALLBERG. 1978. Simulated system of a sedimentary environment: mathematical treatment. B.M.R. J. Australian Geology and Geophysics *2*:93–100.

16 BUBELA, 1981, Banded sulfide ores: The experimental formation of sulfides from flowing liquids, op cit.

17 BUBELA, 1981, A model for sulfide bands formation under epigenetic conditions, op cit.

18 BUBELA, B. 1980. Some aspects of interstitial water movements in simulated sedimentary systems. B.M.R. J. Australian Geology and Geophysics *5*:257–263.

2 BAROTOLERANCE AND MICROBIAL ENHANCEMENT OF OIL RECOVERY

Robert E. Marquis

University of Rochester

No attempt will be made in this presentation to give a global review of the various environmental extremes in deep oil reservoirs that could affect the activities of microbes used for enhanced recovery. Instead, it seems best to concentrate on one parameter, hydrostatic pressure, and to consider its interactions with the other major environmental factors affecting microbial functions. Most scientists, other than marine biologists, have only very superficial knowledge of the biological effects of hydrostatic pressure. Barobiology is often considered exotic or obscure. However, the biochemical apparatus of all living organisms, including the human species, is adapted to function only within certain limits of pressure, and one's knowledge of biology is lacking if it does not include at least some appreciation for the ways in which organisms have adapted and can adapt to a range of ambient pressures. This knowledge is particularly important for an appreciation of the difficulties faced in microbial enhancement of oil recovery.

It is generally accepted that the environmental conditions in deep oil formations are biologically extreme — one might say hostile to life. Since the focus here will be on hydrostatic pressure, it is wise to start with some review of the pressures that should be considered. For a surface or near-surface formation such as the open tar-sand deposits, the pertinent pressures are only slightly in excess of one atmosphere (atm), and pressure is not a critical environmental parameter for microbial enhancement of recovery. (In this presentation, the standard atmosphere will be used as pressure unit because it is so readily understood. A standard atmosphere is equal to 14.7 pounds per square inch, or 101.3 kilopascals.) However, the richest interval for oil recovery from subterranean formations in 1971 was about 2,130 meters.[1] At this depth, pressures are somewhat in excess of 200 atm, since the pressure gradient is on average about 1 atm per 10 meters depth, although the gradient may be as great at 2.3 atm per 10 meters in overpressured strata. In time, Hunt states the average depth for maximal recovery of oil is expected to increase to 2,400–3,000 meters; that is, to regions with ambient pressures of 240–300 atm. Therefore, it seems that one can reasonably consider pressures of, say, 100–300 atm as most pertinent in terms of enhancement of oil recovery from deep reservoirs.

In microbial barobiology these pressures are deep-sea pressures. The average pressure on the ocean floor is 380 atm.[2] However, the environment in oil formations is very different in other respects from the deep sea. In the ocean the temperature is fairly uniform at about 2°C — on the low side for most microorganisms. In oil formations, in contrast, temperatures are very high for most microbes. For example, in the Texas Gulf Coast formations, temperatures at depths of 2,000–3,000 meters are in the 60–90°C range; in the Louisiana coastal fields, they are slightly higher; and in the Middle Eastern fields, temperatures at these depths are 100–123°C.[3] Certainly, these temperatures can be considered close to or beyond the upper limits for microbial life. In a recent paper, Heinen and Lauwers indicate an upper limit for thermophilic bacteria under pressure of 105°C.[4] This limit possibly can be extended upward by a few degrees through careful choice of growth parameters, but it is unlikely that the limit can be extended beyond 110°C. Of course, the parametric requirements for microbial functions other than growth are often less exacting, and it may be possible to have microbes metabolize in high-temperature formations without growth or cell division.

In the open ocean, pH and salinity are nearly contant, with a pH value of about 8.2 near the surface and about 7.8 at all other depths. Salinity, due to a mixture of salts, ranges from 34–37 parts per thousand (ppm), roughly equivalent to 0.5 M NaCl solution. There is no such uniformity of pH and salinity in marine sediments, especially below the surface layers, and even in marine microbiology one must deal with a wide range of environmental extremes when considering microbial activities in sediments.

In oil formations the pH value generally is in the range from 4 to 9. The ends of this range are extreme for most microbes, and few can grow over this entire range; still, the values do not exclude microbial life. Salinities of reservoir waters tend to be more restrictive biologically. Mean salinities in sandstone reservoirs, due mainly to NaCl, are about 27,000 ppm—somewhat below the sea-water values of 34,000 and 37,000 ppm.[5] In limestone deposits salinities are about 90,000 ppm average, roughly equal to 1.5 M NaCl solution. Therefore, it seems that the most pertinent salinities in regard to microbial enhancement of oil recovery are about 0.5–1.5 M NaCl. Of course, if extremes are considered, the maximum pertinent salinity would approach 6 M.

Information is available on the effects of reservoir waters on the activities of sulfate-reducing bacteria. Kuznetsova and Kuznetsova with Pantskhava report that waters from a Devonian reservoir in the Zhigulevsk field were highly mineralized with a chloride content of 177 g/l or approximately 5.0 M.[6,7] Sulfate-reducing bacteria would not grow in media prepared with this water, nor was sulfide detectable in the water. The E_h value of the water was relatively high, +20 mv. When the water was diluted with fresh water to give a final chloride concentration of 1.69 M, H_2S was produced.

Waters from a Permian reservoir in the Kalinvka field were found to be less highly mineralized, with a chloride content of about 1.3 M. The waters contained high levels of H_2S (1975.6 mg/l), indicative of the activities of sulfate reducers, and the E_h value was low — 205 mv. In the laboratory, sulfate reducers would grow in media prepared with the water from the Permian reservoir, although sulfide production could be enhanced by diluting the waters with fresh water.

The overall conclusion of the work was that the high mineral content of the water from the Devonian formation was inhibitory to sulfate-reducing bacteria. It appeared that the ratio of (Ca^{2+} + Mg^{2+}) to (K^+ + Na^+) also was important, and waters with high ratios of divalent to monovalent cations were particularly inhibitory.

Attention has now been turned to other bacteria for enhanced oil recovery because early field trials, reviewed by Davis, have indicated the sulfate reducers are too destructive to be used in most reservoirs.[8] In fact, the objective is generally to suppress their activities with biocides.

With this background it is worthwhile to consider some specific interactions of environmental parameters affecting microbial growth and function. The major focus will be on growth. Other functions such as respiration, fermentation, and synthesis of biopolymers tend to have wider ranges of tolerance for environmental extremes. In contrast, growth is a very complex cell function with multiple components and many sensitive sites. Some functions, such as the differentiation process of sporogenesis are even more sensitive than is growth to environmental extremes. However, sporogenesis is probably not absolutely required for microbial enhancement of oil recovery, whereas growth is required in most of the proposed operations. Some proposals have involved injecting bacterial spores prepared in the laboratory into formations, e.g., the patented process of Hitzman.[9] It would clearly be necessary for the spores to germinate in the formations. Hydrostatic pressure is itself a germinating agent, so it probably would be unnecessary to add chemical germinants if a deep reservoir were being inoculated. Unfortunately, the germinated cell prior to outgrowth is unusually pressure sensitive, so problems may arise with use of spores for inoculating very deep reservoirs.[10]

A well-known set of interactions between environmental parameters affecting microbial growth is that for temperature and pressure. Before considering this interaction the effects of hydrostatic pressure alone on microbial growth should be reviewed. Figure 1 presents typical growth curves, in this case for the bacterium *Streptococcus faecalis*. For this experiment the bacterium was grown in a tryptone-glucose-yeast-extract medium at 25°C at 1 atm and at 408 atm.[11] It is apparent that the effects of 408 atm pressure included: (*i*) prolonging the lag phase or phase of adjustment; (*ii*) slowing exponential growth; and (*iii*) decreasing the final extent of growth, here assessed in terms

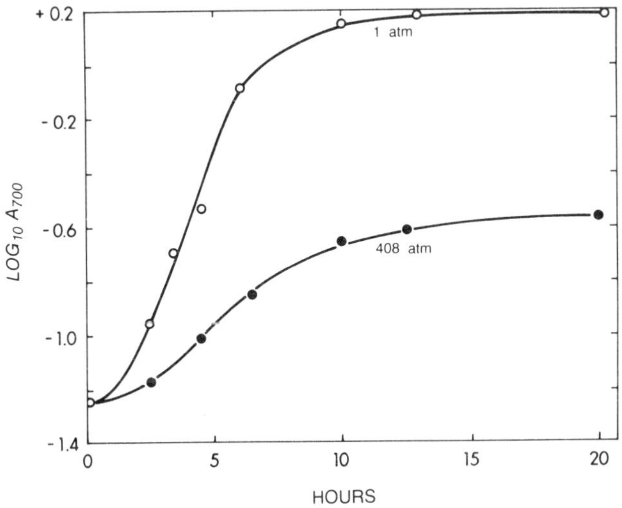

Fig. 1. Growth curves for *Streptococcus faecalis* ATCC 9790 in a tryptone-glucose-marmite medium at 25°C and 1 (○) or 408 (●) atm. A_{700} refers to culture absorbance for light of 700-nm wavelength

of the optical density of the culture for light of 700 nm wavelength. This pattern of response is characteristic of most of the microbes studied. With this particular bacterium the effects of pressure on growth can be related in large part to the effect pressure has on the membrane ATPase, which is involved in movement of protons out of the cell and movement of potassium ions into the cell. Pressure stimulates the hydrolytic activity of the enzyme, so the organism becomes inefficient in its growth under pressure and is unable to produce sufficient ATP for rapid growth—especially in the face of enhanced hydrolysis.

Pressure also acts in a contrasting way to inhibit the nonhydrolytic functions of the enzyme. Potassium pooling within the cells is greatly reduced under pressure. *S. faecalis* produces large amounts of acid, mainly lactic acid, from sugars in the growth medium. In fact glycolysis is the major source of ATP for growth. Because pressure inhibits the proton-extruding functions of the membrane ATPase, the organism becomes hypersensitive to metabolic acids under pressure. For example, growth at 408 atm stops when the culture pH is reduced to a value of approximately 6.1. Neutralization of the acid allows for a new round of growth. However, at 1 atm growth continues until the pH has dropped to a value of about 4.9 This sensitization to metabolic acids is a very common response of bacteria to pressure and is often the major cause of reduced growth under pressure for bacteria that produce metabolic acids. Basically, the bacteria are unable to maintain Δ pH across the cell membrane, with the cytoplasm alkaline compared to the environment, because of the adverse effects of pressure on proton-extruding functions.

A review of the ideal gas equation might lead one to think that pressure and temperature would interfere in some antagonistic manner with microbial growth and metabolism. Further, one might expect to enhance barotolerance simply by increasing pressure. Unfortunately, biological cells are condensed systems and are far from ideal. As Johnson and Lewin found in their study with *Escherichia coli*, the optimal temperature for growth could be increased by a few degrees Celsius by increasing the growth pressure from 1 to 68 atm.[12,13] However, the maximum increase in optimal growth temperature that can be achieved through compression of cultures was only a few degrees. The experimental results of Johnson and Lewin and, subsequently, of others have shown clearly that bacteria become very thermosensitive at higher growth pressures, while they become very barosensitive at higher growth temperatures. In effect, increased temperature and increased pressure do not act antagonistically in their effects on microbial growth but show synergistic inhibitory effects.

Another way to obtain an overview of pressure-temperature interactions affecting microbial growth involves defining some standard growth response and then determining experimentally the various pressure-temperature combinations under which the response occurs. An example is presented in Figure 2. Here the chosen response was the extent of growth at 37°C and 1 atm. One could as easily have chosen the rate of growth under these conditions or rates and extents of growth under other conditions. Actually for the two test bacteria here, *Streptococcus faecalis* and *Bacillus licheniformis*, growing in the complex medium indicated in the figure legend, rate and extent of growth were directly related. Rapid growth was more extensive; slower growth was less extensive. For each point plotted in Figure 2, it was important to determine experimentally just how the extent of growth varied with pressure at a variety of temperatures. Then from the data for any one temperature, an estimate of the pressure at which the extent of growth was equal to that at 1 atm and 37°C could be made. This value was plotted in the temperature-pressure plane of Figure 2. The contours shown in Figure 2 are not complete. Because growth at negative pressures was not examined, the contours do not extend below the zero-pressure line. One might project on purely theoretical grounds that the contours would become closed below zero. Presumably, in considering solely barophilic organisms, one might have obtained closed contours even in the positive pressure plane.

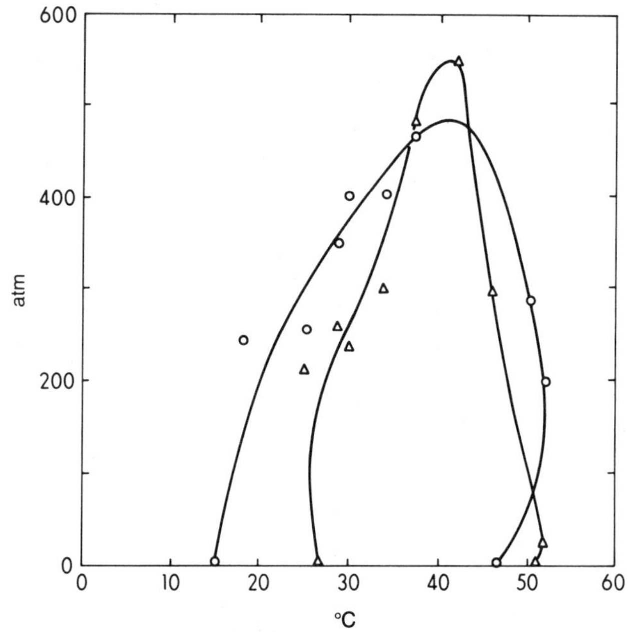

Fig. 2. Interactions of pressure and temperature affecting bacterial growth. Data are shown for *Streptococcus faecalis* (○) growing in tryptone-glucose-marmite broth and for *Bacillus licheniformis* (Δ) growing in the same medium supplemented with 0.1% (w/v) KNO_3. Each point indicates pressure-temperature conditions under which the extent of growth is equal to that at 1 atm and 37°C

The areas within the contours indicate conditions under which growth was equal to or more extensive than that at 37°C and 1 atm. The areas outside of the contours indicate conditions under which growth was less extensive. The usual tendency in these types of contours is for an initial, moderate broadening with increased pressure. In general, the optimal pressure for microbial growth tended to be in the range of 50–200 atm rather than at 1 atm. At higher pressures the contours narrowed as pressure became more and more restrictive for growth. In effect, the temperature range for growth narrowed at higher pressures.

The contours indicate clearly that barotolerance was maximal for both bacteria at a temperature of about 40°C and declined sharply at higher or lower temperatures. Again, barotolerance was reduced at suboptimal or supraoptimal growth temperatures, and the latter reduction is pertinent to the workings of microbes in deep oil formations. The need for testing microbes to be used in enhanced recovery under pressure is apparent here. A high-performance microbe at 1 atm may be nonfunctional at 300 atm. Pressure may be a difficult environmental parameter with which to deal in the laboratory or pilot plant. However, for effective testing prior to field trials, one must use pressurized growth vessels and organisms adapted to grow or function at reservoir pressures.

The inhibitory interactions of pressure and metabolic acids for microbes were mentioned earlier. Pressure sensitizes microbes to alkaline extremes as well. Apparently pressure interferes with the action of the H^+/Na^+ antiporters, which are thought to be important for cells to maintain a relatively acid internal pH when functioning in highly alkaline media.

Recently, a barotolerant variant of *Streptococcus faecalis* has been isolated.[14] This isolation is noteworthy in that previous attempts in a number of laboratories had met with little or no success. The variant isolated was able to grow at pressures as high as 1,000 atm at 37°C, whereas the parent organism had a maximum growth pressure at this temperature of only about 750 atm. Initially it was believed that this variant (called APR-11) was acid tolerant. However, it turned out not to be more acid tolerant but instead had a regulatory defect in its arginine dihydrolase system. When the parent organism carries out glycolysis to produce acids, primarily lactic acid, the arginine dihydrolase system is almost completely repressed in the sense that enzymes for the pathway are not produced. However, in the APR-11 variant the system is not repressed. While the bacterium is producing acids from glycolysis, it is also producing ammonia from arginolysis. The result is partial neutralization of the acids and enhanced barotolerance. The important point here is that barotolerant variants of bacteria can be isolated, but of course one must be sure to isolate variants in which the desirable features of the organism are not altered. Clearly, there is a great need of basic information on the bacterial characteristics that are desirable for enhanced recovery.

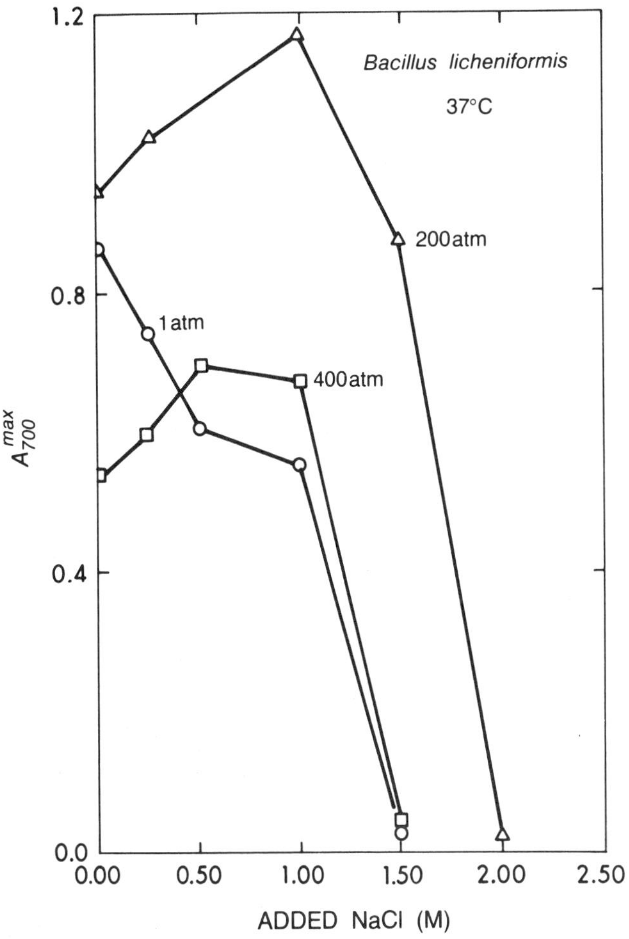

Fig. 3. Enhanced salt tolerance of *Bacillus licheniformis* growing under pressure. The organism was grown in tryptone-glucose-marmite broth with 0.1 percent (w/v) KNO_3 a 37°C and 1 (○), 200 (Δ), and 400 (□) atm

In the interactions of pressure with other environmental factors considered above, the combined parametric influences have been restrictive for growth. The results of recent experiments on the effects of pressure on salt tolerance indicate that growth under pressure is actually less sensitive to the inhibitory effects of NaCl. An example of the data obtained is presented in Figure 3, specifically with *Bacillus licheniformis* growing at 37°C in a complex medium to which the indicated amounts of NaCl had been added. The growth parameter indicated is the maximal absorbance of the cultures for light of 700-nm wavelength. The 1-atm data indicate that *B. licheniformis* is not very halotolerant and that its growth was decreased progressively as more NaCl was added to the medium. The maximum concentration of

NaCl at which any growth occurred was between 1.0 and 1.5 M. In other words the bacteria would not be able to grow at 1 atm in the saline environment, on the average, in a limestone reservoir. However, at 200 atm a very different picture is seen. This pressure enhanced growth slightly in an ordinary medium. Enhancement of growth was greater in a medium with 0.5 M NaCl added and even greater in a medium with 1.0 M NaCl added. The maximum concentration of added NaCl permitting growth at 200 atm was between 1.5 and 2.0 M. Thus, a pressure of 200 atm increased the maximum tolerable NaCl concentration by approximately 0.5 M. While no growth occurred at 1 atm in a medium with 1.5 M NaCl added, nearly full growth compared with that in the base medium occurred at 200 atm.

At 400 atm, growth in the absence of added NaCl was significantly reduced. However, addition of NaCl to the medium up to a concentration of about 1.0 M actually enhanced growth. When the added NaCl concentration was increased to 1.5 M, growth did not occur at either 1 atm or 400 atm.

Similar data were obtained with *S. faecalis*, a moderately salt-tolerant bacterium, and studies are underway with other bacteria — particularly ones such as *B. licheniformis*, which are good candidates for enhanced recovery operations.

Unfortunately, further discussion of the interactions of hydrostatic pressure with other environmental factors affecting microbial functions is not reasonable now simply because of the lack of information. Of particular importance would be information on pressure-induced changes in sensitivity to E_h and to oxygen. Some years ago, ZoBell and Hittle presented data to indicate that bacteria become hypersensitive to oxygen when growing under pressure.[15] On the other hand, extensive studies described in the PhD thesis of Stephen R. Thom indicate that hydrostatic pressure actually has a slight moderating effect on oxygen toxicity.[16] However, if pressure is imposed through use of high-pressure gases such as helium, bacteria do become hypersensitive to oxygen. The hypersensitivity is due specifically to the gas and not to the pressure. Other high-pressure gases such as nitrogen, argon, krypton, and xenon also potentiate the toxic action of oxygen. In closed reservoirs microbes used for enhanced recovery may be exposed to hyperbaric gases. In fact, it seems that gas production may be one of the most desirable features of the organisms chosen for enhanced recovery. Further, these gases may have major effects on the physiology of the organisms, and there is a need for more information on both general and specific biological effects of hyperbaric gases.

Finally, only very sketchy views exist on how microbes might respond to nutrient limitations in deep reservoirs. Some marine microorganisms can adapt to oligotrophic conditions, presumably because they have very low maintenance requirements, possess low saturation constants for the substances needed, and can concentrate nutrients from very dilute solutions. Recently, Novitsky and Morita have isolated from waters of the Antarctic Convergence a psychrophilic vibrio that responds to starvation by decreasing cell size so much that viable cells can pass through a 0.4-μm filter.[17] These tiny cells can survive in starvation conditions for as long as three years. Unfortunately, at this time an understanding of the dynamics of microbes in deep oil reservoirs and whether this sort of adaptation could occur in these subterranean populations falls in the realm of interest and speculation.

CONCLUSION

1) Hydrostatic pressure is a very important environmental parameter to be considered in relation to microbial enhancement of oil recovery in deep formations, especially because of its interactions with other environmental parameters.

2) All strain selection and pilot operations for microbial enhancement of oil recovery should be carried out with cultures at in situ pressures; otherwise, it is possible to be led astray in predicting the behavior of the strains in the chosen formations.

REFERENCES

1 HUNT, J.M. 1979. Petroleum geochemistry and geology. W.H. Freeman and Co., San Francisco.

2 KINNE, O. 1972. Marine ecology. Wiley-Interscience, London.

3 HUNT, op cit.

4 HEINEN, W., and A.M. LAUWERS. 1981. Growth of bacteria at 100°C and beyond. Arch. Microbiol. *129*:127–128.

5 COLLINS, A.G. 1975. Geochemistry of oil field waters. Elsevier Publishing Co., Amsterdam.

6 KUZNETSOVA, V.A. 1960. Occurrence of sulfate-reducing organisms in oil-bearing formations of the Kuibyshev region with reference to salt composition of layer waters. Mikrobiologiya *29*:298–301 (Trans.).

7 KUZNETSOVA, V.A., and E.S. PANTSKHAVA. 1962. Effect of freshening of stratal waters on development of halophilic sulfate-reducing bacteria. Mikrobiologiya *31*:103–106 (Trans.).

8 DAVIS, J.B. 1967. Petroleum microbiology. Elsevier Publishing Co., Amsterdam.

9 HITZMAN, D.O. 1962. Microbiological secondary recovery of petroleum. U.S. Patent 3,032,472.

10 GOULD, G.W., and A.J.H. SALE. 1972. Role of pressure in the stabilization and destabilization of bacterial spores, p. 147–157. In 26th Symposium of the Society for Experimental Biology. Academic Press, New York.

11 MARQUIS, R.E., N. PORTERFIELD, and P. MATSUMURA. 1973. Acid-base titration of streptococci and the physical states of intracellular ions. J. Bacteriol. *114*:491–498.

12 JOHNSON, F.H., and I. LEWIN. 1942. The growth rate of *E. coli* in relation to temperature, quinine and coenzyme. J. Cell Comp. Physiol. *28*:47–75.

13 JOHNSON, F.H., and I. LEWIN. 1942. The influence of pressure, temperature and quinine on the rates of growth and disinfection of *E. coli* in the logarithmic growth phase. J. Cell. Comp. Physiol. *28*:77–97.

14 MARQUIS, R.E, and G.R. BENDER. 1980. Isolation of a variant of *Streptococcus faecalis* with enhanced barotolerance. Can. J. Microbiol. *26*:371–376.

15 ZOBELL, C.E., and L.L. HITTLE. 1976. Some effects of hyperbaric oxygenation on bacteria at increased hydrostatic pressures. Can. J. Microbiol. *13*:1311–1319.

16 THOM, S.R. 1981. Non-narcotic actions of nitrogen, oxygen, nitrous oxide and noble gases on microbial growth. PhD thesis, University of Rochester, Rochester, New York.

17 NOVITSKY, J.A., and R.Y. MORITA. 1976. Morphological characterization of small cells resulting from nutrient starvation of a psychrophilic marine vibrio. Appl. Environ. Microbiol. *32*:617–622.

ADDITIONAL REFERENCES

BUBELA, B. 1978. Role of geomicrobiology in enhanced recovery of oil: status quo. APEA Journal *18(1)*:161–166.

3 PATENTS EMPLOYING MICROORGANISMS IN OIL PRODUCTION

T.R. Jack and B.G. Thompson

Nova Husky Research Corporation, Calgary

Petroleum microbiology, as it pertains to the production of oil, is not a new endeavor but its development has been based on surges of interest. Research activity has been spurred by the recognition of the importance of microorganisms in some specific topical problems. Jet fuel system fouling, effluent treatment, corrosion and hydrogen sulfide (H_2S) production by sulfate-reducing bacteria, and oil-spill cleanup are examples. Currently the need to extend production from oil reservoirs by new enhanced oil recovery (EOR) technologies has renewed research efforts to control and exploit microorganisms in the oil field itself.

With many applied research groups now entering this field, it is appropriate to review the patent literature at this time.

Scope of this review. The principal aim of this review is to provide an outline of the patent literature that illustrates the scope of previous work as it pertains to oil production. It is not the intention here to provide a comprehensive review of all of the patent literature in petroleum microbiology, nor even to provide a comprehensive review of all of the patent literature in petroleum microbiology, nor even to provide a thorough patent search of any specific topic.

It is our hope that this review will give an overview and stimulate the development of new applications for microorganisms and microbial products in the oil field.

Microorganisms in the oil field: patent potential. Much of the effort of petroleum microbiologists has been directed to controlling the detrimental activities of microorganisms in the oil field or to the exploitation of microorganisms in the treatment of various effluent problems. An early text, *Petroleum Microbiology* by J.B. Davis, still provides an excellent introduction to these subjects.[1]

In the future, microorganisms will be of increasing interest as a means to produce chemical agents on a large scale from renewable feedstocks that are not tied to the price of petroleum. A number of products useful in enhanced oil recovery will be targets. The immense size of these markets and the impetus arising from new discoveries in genetic engineering suggest that the use of microbial products in the oil field will be an area of considerable patent activity.[2] Of particular interest will be biosurfactants,[3,4] certain cosurfactants such as butanol,[5] improved mobility control agents and viscosifiers based on polysaccharides and other exopolymers,[6] emulsifying agents for oil cleanup and viscosity control,[7] as well as enzymes capable of upgrading oil quality through processes such as desulfurization and deamination.[8]

THE PATENT LITERATURE

Inhibition of undesirable activity. Clark et al. in a review of the oil reservoirs of nine oil-producing states in the USA concluded that conditions in 4.7–54% of the reservoirs in each state were suitable for supporting microbial growth.[9]

Evidence indicates that indigenous hydrocarbon-utilizing organisms are present in some reservoirs.[10] Certainly after septic water injection during secondary oil recovery has taken place, bacterial populations exist in the reservoir itself. Bacteria growing in a reservoir can lead to formation plugging, biodegradation of injected EOR chemical in situ, and a souring of production wells where sulfate-reducing bacteria are actively generating H_2S in situ.[11] Production

Table 1. Inhibition of undesirable bacterial activity.

Patent No.	*Effective Compound*	*Use In*	*Specific Bacteria (if any) Affected*
U.S. Patent 3,612,178	formaldehyde	displacement/drive fluids, water-oil emulsions	—
U.S. Patent 3,751,364	quaternary, polymerized oxyalkylated amines	submersible oil well pumps	—
U.S. Patent 3,800,872	orthocresol or chromium potassium sulfate	displacement/drive fluids	—
U.S. Patent 3,903,272	2-nitrobutanol phosphate and isopropanol phosphate	displacement/drive fluids	sulfate-reducing bacteria
U.S. Patent 4,049,559	oxygen	waste water	sulfate-reducing bacteria
U.S. Patent 4,077,879	chloride dioxide	waste water containing sulfides	sulfate-reducing bacteria
U.S. Patent 4,231,426	benzene, toluene, xylene	tracer fluids containing nitrate	nitrate-respiring/ assimilating organisms
U.S. Patent 4,249,608	benzene, toluene, xylene or alkyl substituted benzene and xylene	displacement/drive fluids	—
U.S. Patent 4,299,709	benzene, toluene, xylene	tracer fluids	—
CDN Patent 760905	organic substituted quaternary ammonium compounds	underground formation	sulfate-reducing bacteria
U.S. Patent 3,578,499	—	—	—
U.S. Patent 3,759,828	—	—	—

of these organisms from underground or above-ground contamination introduces bacteria to field batteries and piping systems with the subsequent problems of microbially induced corrosion (especially by sulfate-reducing bacteria),[12] pump and filter clogging through slime formation, fouling of interior pipe surfaces, and general cleaning problems in downstream oil and fuel handling facilities (especially where hydrocarbon-utilizing bacteria and fungi are present).

The control of these organisms is marginal in most oil fields and requires constant monitoring and attention.

Patents (Table 1) emphasize the need for including low-cost antibacterial agents in formulations proposed for a variety of purposes. Thus, where nitrate, an ion not normally found in connate waters, is to be employed as a tracer for plotting underground water flow in the reservoir, the formulation is specified to contain up to 2,000 ppm of a substituted benzene bacteriocidal to organisms capable of utilizing nitrate as a nitrogen source and/or electron acceptor (U.S. Patents 4,231,426; 4,299,709). Similarly, where biodegradable polymers are used to control the mobility of aqueous drive fluids intended to enhance oil recovery, antibacterial agents are also proposed (U.S. Patents 4,249,608; 4,299,709; 4,231,426).

Other patents seek to control microbial activity through the manipulation of physical-chemical parameters rather than by the addition of antibacterial agents. Obviously such an approach is more feasible above ground. For example, U.S. Patent 4,053, 396 claims a process for the treatment of refinery waste water wherein organics are absorbed on charcoal and degraded in the presence of a specified amount of oxygen, which suppresses the obligate anaerobes responsible for reducing sulfate to H_2S.

Retention of biocides in the affected area may be a problem in practice. A method for improving the contact between biocidal agents and troublesome bacteria is taught in U.S. Patent 3,578,499, in which a gelling formulation containing a gelling agent, a neutral diluent, a wetting agent, and a dye is used as a carrier for the biocide. The gel in effect increases the residence time of the biocide in the target site.

The cross section of agents and specified target applications presented in Table 1 obviously includes some agents of dubious efficacy (e.g., benzene) or economic feasibility (e.g., formaldehyde). In effect, there are few safe, inexpensive biocides that are environmentally acceptable and effective against established microbial populations — particularly those growing as biofilms on protected surfaces in the reservoir or piping system.

Removal of biomass. The ubiquitous presence of bacteria wherever conditions permit can be undesirable due to their metabolic activity. The preceding section has illustrated various patents dealing with inhibition of bacterial activity. However, in most cases the antibacterial agents employed either stop the growth of the organisms (bacteriostatic agents) or kill the organisms present (bacteriocidal agents). These agents do not remove, disperse, or dissolve the biomass already in place.

In some areas of oil-field operation, such as the well bore region of the reservoir and above ground in production facilities (batteries, piping, etc.), the presence of bacterial biomass (live or dead) can cause severe plugging problems resulting in a loss of permeability for the subterranean formations and reduced flow or even complete clogging of valves, screens, filters, and piping. The biomass responsible for severe plugging problems consists of both live and dead bacterial cells and extracellular products such as polysaccharide slimes that stabilize the plug and trap any further particulate matter present. A number of patents are concerned with methods of clearing this organic debris.

The agents employed are designed to disrupt biomass plugs by both physical and chemical means.

One set of patents (UK, 1,296,031; Canada, 968701) specifies a combination of a chelating agent, polyamine carboxylate, to disrupt bacterial cell wall and slime structures stabilized by interchelating cations cuch as Ca^{2+} as well as an alcohol sulfate surfactant to facilitate dispersion of the biomass and penetration by the chemical agents. In addition the formulation includes 0.5–5.0% alkali metal hydroxide to catalyze the hydrolysis and saponification of biological macromolecules present. Such a formulation is claimed for use above and below ground.

Another basic approach is to employ strong oxidizing agents to disrupt and disperse organic debris. U.S. Patent 3,529,666 claims the sequential use of hydrogen or alkali metal peroxides followed by an aqueous acid injection (5.0% to 3.5% HCl) to remove organic material from geological formations. A similar patent (U.S. 3,528,503) claims potassium permanganate (2–20%) as the oxidizing agent in the same regime.

To control the corrosion induced by such potent agents, U.S. Patent 3,482,636 claims the use of hypohalites (hypochlorite and hypobromite) as the oxidizing agent but includes water-dispersible soluble silicates or hydroxides (pH 13) as corrosion inhibitors. Again a strong acid solution is sequentially injected after the oxidizing agent.

Use of these strongly oxidizing and extreme pH solutions can disrupt organic matter but can also lead to corrosion problems, oxygen formation underground, and disturbance of the inherent oxidation-reduction potential in the well bore region.

Dispersion of the disrupted organic matter is presumably effected by a return to production from the treated well or by subsequent water flooding (U.S. Patent 3,529,666).

Production of oil: microbially enhanced oil recovery (MEOR). Very little is known about the exact mechanisms of oil release in microbially enhanced oil recovery. However, growth of microbes in situ may have a number of potentially important interactions with the inorganic matrix and the oil present in a formation. Growth on organic substrates (injected or natural) and/or inorganic substrates can create metabolic products that fall into five general classes: organic acids, inorganic acids, gases, surfactants, and biopolymers. Organic acids can dissolve carbonates, which can increase the permeability of a reservoir and release CO_2. Inorganic acids, such as sulfuric acid produced by sulfur-oxidizing bacteria, can dissolve carbonate formation with the same effect. Gases produced as metabolic byproducts from organic substrate utilization include H_2, CH_4, and CO_2. Gases produced as metabolic byproducts from inorganic anions include N_2 from NO_3^- reducing bacteria and H_2S from $SO_4^=$ reducing bacteria. Produced gases could repressurize a reservoir and when dissolved in oil reduce its viscosity. The production of H_2S from $SO_4^=$ could also dissolve sulfate-containing minerals to increase the permeability of a reservoir or remove inorganic matrices. Surfactants and biopolymers formed in situ may cause the release of hydrocarbons. Direct microbial modification of hydrocarbons could remove or alter fractions of the hydrocarbon and may reduce viscosity. Many patents employ facets of the above theoretical framework as potential production tools.

The use of actively growing bacteria in situ in the oil reservoir as agents of oil release was first proposed in the 1940s. MEOR patents issued up to the early 1960s (U.S. Patents 2,413,278; 2,641,566; 2,660,550) have been critically and comprehensively reviewed by Davis.[13]

Most of the early patents were initially based on the use of anaerobic sulfate-reducing bacteria from the genus *Desulfovibrio* (U.S. Patents 2,413,278; 2,641,566; 2,660,550). While these organisms are capable of multiplying and dispersing in susceptible oil reservoirs, the noxious H_2S necessarily produced in their metabolism and the inconsistencies in their ability to release oil in laboratory experiments resulted in the inclusion of other genera in later patent claims.[13]

The most extensively tested direct MEOR approach is based on a patent by Updegraff (U.S. Patent 2,807,570), which claims the use of injected obligate and facultative anaerobes capable of fermenting or metabolizing injected water-soluble carbohydrate in situ to generate gases. Members of the genera of *Aerobacter, Escherichia, Clostridium,* and *Bacillus* are specifically cited as suitable organisms.

Analyses of extensive production records from field tests utilizing this technique in Poland indicate that the enhanced oil production obtained arises from (*i*) the short-term effects of the metabolically active cultures and (*ii*) a long-term effect (years in duration) probably arising from a well bore cleaning action by the organic acids, solvents, and surfactants produced during the fermentation.[14] The short-term effect seemed to be due mainly to the repressurization of the reservoir by fermentation gases and an increase in permeability achieved through carbonate dissolution by the organic acids.[15,16]

A major concern with the injection of live bacteria into any underground formation is that the biomass buildup in the well bore region will reduce formation permeability with a consequent loss in fluid flow (injectivity). Damage to the organisms themselves during injection may also be significant. Hitzman in U.S. Patent 3,032,472 has circumvented these concerns by describing the inoculation of a reservoir by bacterial spores that are smaller, more robust, and more stable to storage and handling than the corresponding vegatative cells. *Clostridium* and *Desulfovibrio* are genera specifically cited as suitable spore-forming bacteria.

The potential plugging of the production zone also prompted the patenting of an alternate method of introducing viable organisms from the families *Bacteriaciae* and *Pseudomonadaceae* into reservoirs having a water-oil contact interface. In U.S. Patent 3,185,216, bacteria are injected into the water-bearing zone below the oil production zone through a well extended into the water zone for this purpose. The organisms are selected from groups capable of utilizing hydrocarbons and producing gases such as CO_2 or CH_4 in situ. These organisms so introduced will then migrate to the water-oil interface and proliferate with concomitant gas formation, which in turn enhances production from the overlying oil-bearing zone.

A number of patents base their claims on the direct alteration of the oil itself through bacterial action. The validity of patent claims based on the anaerobic alteration of pristine hydrocarbons in situ is questionable, since significant, fast, anaerobic utilization of hydrocarbons has not been demonstrated.[17] Jones (U.S. Patent 3,332,487) claimed the injection of oxygen or air with aerobic bacteria capable of oxidizing low molecular weight hydrocarbons. Heat from the ensuing biodegradation can be removed by the use of cooling water or the heated metabolic products can be pushed as a slug through the reservoir to a producer well. Oxygen-tolerant anaerobes are cited as biological agents for this same purpose; the presumption is that appropriate anaerobes could attack and degrade the cells and extracellular products left behind by the anerobes after oxygen injection was stopped.

A similar process is claimed in U.S. Patent 3,340,930 wherein the aqueous phase of an aerobic culture of hydrocarbon-utilizing bacteria or yeasts is injected into the oil reservoir. Consequent further recovery of oil arises through the action of the organisms and the metabolic products that might include bioemulsifiers and surfactants.

Exploitation of microbial plugging to alter the permeability characteristics of a reservoir under water flood is claimed in Netherlands Patent 89580. In this case, injection and growth of bacteria in situ are claimed to result in the selective reduction of permeability in high-permeability channels developed during the water-flooding process. Injected bacteria would preferentially enter these channels and in due course plug them. In subsequent water flooding the flow of injected water through these zones would be thus reduced, and improved oil production would arise through both an increase in the sweep efficiency for surrounding oil-rich zones and a reduction in the fraction of water coproduced with the oil at the producer well.

Use of a bacterial cellular solution as a driving fluid to displace oil from the reservoir is claimed in U.S. Patent 3,650,326. In this process the increasing viscosity of the bacterial suspension reduced the mobility of the water phase to match the oil mobility more closely. This reduces fingering by the aqueous drive fluid, which normally leads to premature water breakthrough at the producer well.[18]

Patents claiming the direct use of microorganisms as agents of production from oil sands, oil shale, and other fossil fuels include both the liberation of hydrocarbon from its host matrix and its conversion into an upgraded fuel. The breakdown of oil shale particle size by the action of unspecified anaerobes (U.S. Patent 4,085,972), treatment of kerogen-type oil shale (U.S. Patent 2,641,565), and the actual fermentation of ring compounds in fossil fuels including oil shale under anaerobic conditions (U.S. Patent 3,826,308) have been described. The efficacy of such processes is not clear.

A major obstacle to the recovery of organic bitumen and kerogen from oil shale is that the organics are locked up into an impermeable inorganic matrix composed of carbonates. U.S. Patent 3,982,995 claims the use of bacteria capable of strong acid production as agents for carbonate dissolution. Anaerobic rumen organisms and aerobic chemoautotrophic bacteria (thiobacilli or *Ferrobacillus sulfooxidans*) that live by oxidizing inorganic materials such as reduced sulfur compounds or ferrous salts are specified in the claims. The action of the rumen bacteria results from the formation of

Table 2. Representative polysaccharides from the patent literature.

Organisms	*Polysaccharide*	*Reference*
Arthrobacter stabilis, NRRL B-3225	—	U.S. Patent 3,632,570
Azotobacter indicus var. *myxogenes* ATCC 21423	heteropolysaccharide-7	Canadian Patent 945091 U.S. Patent 3,915,800 U.S. Patent 3,960,832
Erwinia tahitica ATCC 21711 reclassified as *Klebsiella pneumoniae* ATCC 21711	heteropolysaccharide-10 19% uronic acid 39% glucose 29% galactose 13% fructose	U.S. Patent 3,933,788 U.S. Patent 4,186,025 U.K. Patent 1411014
Klebsiella pneumoniae ATCC 31314	heteropolysaccharide-21 33% mannose 29% glucose 21% galactose 17% glucuronic acid {5.7% acetyl and 4.9% pyruvate	U.S. Patent 4,211,774 U.S. Patent 4,247,639
Various fungi including *Sclerotium* sp. and *Corticium* sp.	"Scleroglucan"	U.S. Patent 3,301,848
Xanthomonas campestris *Xanthomonas begoniae*	Xanthan	U.S. Patent 3,020,207
Methylomonas	38–49% glucose 11–20% fructose 7–21% mannose 11–18% galactose 10–20% uronic acid	U.S. Patent 4,233,438
Hansenula holstii	phosphorylated mannan	U.S. Patent 3,312,279

organic acids (formic, acetic, or propionic) while sulfuric acid is the agent in the case of the *Thiobacilli.*

Separation of bitumen from oil sands such as the Athabasca oil sands in Alberta involves the release of a viscous hydrocarbon from an intimate association of sand, clay, and water. The value of biosurfactants in the separation process has been assessed in recent studies, but the general potential of bacteria in this regard was recognized even in the earliest patents (U.S. Patents 2,413,278; 2,641,566; 2,907,389).

Biopolymers in enhanced oil recovery. Most enhanced oil recovery techniques are based on the displacement of oil from the reservoir by the injection of fluids into the production zone through an injection well with oil being produced at adjacent production wells. Since large volumes of fluid are required, these are usually aqueous in nature.

Unfortunately, water tends to move more readily than oil underground. By fingering through the oil phase, it will break through to the production well, bypassing much of the oil in the reservoir and resulting in poor sweep efficiency. This problem can be minimized by reducing the mobility of the water phase through the inclusion of chemical agents such as water-soluble polymers.[18]

Polysaccharides produced by microorganisms can control mobility by increasing the water phase viscosity. These products have a promising future in the EOR market due to the wide variety of complex biopolymers available. Their unique physical characteristics are not available in simple chemical polymers such as polyacrylamide.

The potential market for polysaccharides used in EOR is enormous and the patent literature reflects its importance.[19] Table 2 summarizes some of the polysaccharides described in recent patents.

In some cases, commercialization of specific polysaccharides such as xanthan or scleroglucan has led to patented improvements in the production procedures (U.S. Patent 3,436,311) and in physical properties and stability through formulation or modification (U.S. Patents 4,104,193; 3,729,460; 4,218,327; 3,256,271; 4,078,607; 4,256,590; Canadian Patent 675,416).

Patents based specifically on the use of xanthan gum, or *Xanthomonas*-produced biopolymers (collectively called xanthan in this review), are the most abundant. U.S. Patent 3,305,016 describes the basic process. The polymer is rarely used as a sole viscosifier but instead is often combined with other products or modified chemically prior to use.

U.S. Patent 3,801,502 describes the treatment of *Xanthomonas* fermentation products with alcohols, ketones, phenols, and nonionic surfactants to produce adjuncts for water flooding. U.S. Patent 4,008,766 claims the formulation of xanthan with anionic (amide-linked sulfonates) and nonionic (polyethylene oxide) surfactants. Improved thermal stability of the polymer by formulation with aliphatic alcohols and alkali earth carbonates is claimed in U.S. Patent 4,141,842.

Polyvalent metal ions in combination with xanthan generate gels through interchelation of the polymer (U.S. Patent 3,908,760). U.S. Patent 4,048,079 also claims gel formation through interchelation of xanthan by multivalent chromium ions, but a novelty here is the introduction of chromium in a hexavalent noninterchelating state followed by its reduction to Cr^{3+}, which will form gels.

Various combinations of chemical polymer (polyacrylamide or its derivatives) and xanthan have been proposed (U.S. Patents 4,039,028 and 4,128,482). In part the improvement embodied here arises from the difference in behavior of these polymers. Polyacrylamide, cheaper but salt sensitive (U.S. Patent 3,707,187), can be protected by a buffer slug of biopolymer that is salt tolerant. Polyacrylamides control mobility of the water phase by reducing formation permeability through absorption, while xanthan acts as a viscosifier. Thus these polymers are in some senses complementary.

Colloids have also been combined with xanthan to yield injection fluids. Use of the whole culture broth from a *Xanthomonas* fermentation for EOR is claimed in U.S. Patent 4,119,546, which describes the resulting solution as a hydrophilic colloid suspension of particle size less than 3 μ. It should be noted that *Xanthomonads* are plant pathogens, which will surely restrict the use of suspensions of live cells in the field.[20] Colloidal silica has also been claimed as an additive to xanthan (U.S. Patent 3,956,145).

U.S. Patent 4,296,203 employs a mutant of *Xanthomonas* capable of producing a pyruvate-free version of the biopolymer.

In addition to being recommended as a viscosifier to improve oil recovery by water flooding, xanthan and its derivatives have also been claimed as an ingredient to improve microemulsion flooding wherein complex mixtures of chemicals are used to generate a stable microemulsion slug in which oil is solubilized rather than being simply pushed by the water phase as in water flooding (U.S. Patent 3,719,606).

Instructions for the preparation of xanthan formulations suitable for EOR injection are detailed in German Patent 2,725,069 and French Patent 2,330,697.

Injectivity problems encountered with the injection of xanthan into oil wells have been countered by claims of pretreatment with proteolytic enzymes to remove protein contaminants (U.S. Patent 3,966,618) and homogenization (U.S. Patent 3,436,346).

More general patents based mainly on polysaccharides make analogous claims to those described for xanthan above.

Phosphorylated mannan produced by the yeast *Hansenula holstii* has been claimed as a water-flooding viscosifier similar to xanthan (U.S. Patent 3,312,279).

In two cases, polysaccharide production by hydrocarbon-utilizing microorganisms is claimed (U.S. Patents 3,655,512 and 3,965,985). In the latter case, use of the polymeric gum for the selective plugging of an oil-bearing formation is specified in the patent claims (U.S. Patent 3,965,985).

Colloid formulations that incorporate carbon black or other particles including dead cells with a polysaccharide and a polyethylene oxide surfactant for drive fluids are claimed in U.S. Patent 3,618,664.

Combinations of biopolymers with polyacrylamides are also claimed (U.S. Patents 3,707,187; 3,724,545).

Polysaccharides to be used for permeability control agents are claimed to be modified by free radical additions of substituted mono-olefins (U.S. Patent 3,844,348).

The above review of polysaccharides in the patent literature has focused on the use of such agents as viscosifiers to control water phase mobility in water flooding or more sophisticated (microemulsion) EOR techniques and to act as plugging agents to reduce high-permeability zones or channels underground, possibly through gel formation. Two further miscellaneous applications of polysaccharides designed to enhance oil production have been patented.

Incompetent underground formations can result in the undesirable movement of unconsolidated sand. Sand control in the well bore region is described in U.S. Patent 3,861,467. A permeable cement barrier is put in place between the formation and the well bore by the injection of a slurry of sand, cement, water, and xanthan gum in a specified composition.

Use of xanthan and galactomannan gel as a hydraulic fracturing agent is taught in U.S. Patent 4,067,389. Once the fracture has formed, injection of halide salt reduced the viscosity of the salt-sensitive biopolymer formulation, allowing its subsequent removal from the fracture by backflowing the injection well (U.S. Patent 4,067,389).

Miscellaneous uses. A number of miscellaneous uses of bacteria in oil production have been patented.

A very early patent (U.S. Patent 1,998,756) suggested the use of various organic jellifying agents in well treatment. While these were not specified to be of microbial origin, bacteria were included in the formulation to degrade the jelly after its use as a carrier for acid or paraffin solvent in the well treatment process.

One problem associated with EOR techniques that are based on the use of surfactants in situ is the loss of the surface active agents through absorption onto mineral surfaces. Hitzman (U.S. Patent 3,326,286) claims the use of dead microbial cells as carriers for nonionic surfactants (polyethylene oxides or their thioanalogs) with the aim of achieving a better dispersion of the surfactants in the reservoir.

Other patented uses for bacteria exist. U.S. Patent 3,650,326 describes the use of bacterial cell suspensions treated with formaldehyde and anionic surfactants to provide a water-flooding agent in which the viscosity increases with time. U.S. Patent 3,598,181 teaches a similar process wherein bacterial cultures including pseudomonads are treated with an anionic surfactant to create a high-viscosity fluid for water flooding. A similar use of whole cell suspensions is claimed in U.S. Patent 4,096,073 wherein a suspension of gram-negative cells is treated with base (sodium hydroxide) at moderate temperatures to generate a microbial viscosifier.

A problem arising in the use of in situ combustion EOR techniques is to determine the movement of the underground fire front. Where such a technique is employed in very shallow reservoirs, U.S. Patent 3,489,217 claims a method of following the areal expansion of the hot zone by sampling surface soils for microbial populations capable of utilizing the low molecular weight, volatile hydrocarbons that presumably are driven to the surface by heat from the in situ combustion process. The areal distribution of such microbes is construed to be identical with the areal expansion of the hot zone underground.

Single-cell protein production involves the aerobic growth of dense microbial cultures on a cheap substrate. The gases generated from this process are rich in CO_2, and Hitzman has patented the idea of using the CO_2 produced for EOR in an appropriate CO_2 flooding scheme (U.S. Patent 4,261,420).

Other applications: drilling, workover, and completion. Bacterial exopolymers have been suggested as viscosifiers for drilling muds, well bore fluids, oil-well fracturing fluids, and well completion fluids, either alone or in combination with other agents. The use of xanthan gum is the most common for these purposes. U.S. Patents 3,882,029 and 3,989,632 claim that the combination of water-insoluble/slowly oil-soluble particles with an aqueous solution of xanthan gum, hydroxyethyl cellulose, and chrome lignite make efficient well completion and workover fluids. Effective nonclay well bore fluids containing xanthan gum and heat-stabilized with calcined dolomite are claimed in U.S. Patent 4,290,899. U.S. Patent 3,625,889 also claims xanthan as a useful component of well completion fluids. Xanthan, in combination with guar gums and starch, is claimed as being a viscosifier of drilling muds.

The exopolymers of many other bacteria, including species in the genera *Pseudomonas, Methanomonas, Arthrobacter, Corynebacterium, Bacillus, Mycobacterium, Actinomyces, Nocardia, Micrococcus, Rhodobacillus, Chromatium, Serratia, Rhizobium, Aerobacter, Escherichia,* and *Streptococcus,* have also been suggested for use as viscosifiers in drilling muds (U.S. Patent 3,965,985).

Effluent control. The production and transportation of oil and oil products brings oil and oil effluents into contact with many aspects of the environment, such as ground waters, and surface fresh- and salt-water bodies. The disposal of these effluents and wastes as taught by patents occurs either by direct microbial degradation, by the physical dispersion of the effluent for removal by microbial emulsifying agents, or by a combination of these two actions.

Many microbially produced emulsification agents useful for removing oil have been described. U.S. Patent 3,997,398 describes a polysaccharide emulsifying agent produced by *Corynebacterium hydrocarboclastus,* capable of emulsifying mixtures of kerosene and tanker oil. U.S. Patent 4,230,801 describes the production of an anionic emulsifying agent, α-emulsions, produced by *Arthrobacter* sp. ATCC 31012 (later reclassified as an *Acinetobacter* sp.) that can emulsify hydrocarbon substrates that contain aliphatic and cyclic components particularly well. U.S. Patent 4,234,689 teaches the use of these emulsifying agents for cleaning oil-contaminated vessels, oil spill management, and use in EOR.

The direct microbial attack on oil effluents is carried out in a number of contexts. Enclosed containers in contact with hydrocarbon products can be cleaned in this fashion. The degradation of residual hydrocarbons in the tanks of an oil tanker either by native or added oil-degrading microorganisms of the *Arthrobacter* sp. (*Acinetobacter*) under aerobic conditions has been taught by U.S. Patent 3,941,692 and Canadian Patent 1,023,288. Both patents claim also that the microbial products could be recovered for uses such as single-cell protein. U.S. Patent 3,769,164 and Canadian Patent 978,114 claim the removal of oil contamination by oil-degrading microorganisms from environments such as the open sea, inland fresh waters, and beaches, as well as tanker bottoms and storage tanks. The organisms employed were claimed to be edible and nonpathogenic and were found to be members of the genera of *Candida, Aureobasidium, Myrothecium, Cladosporium, Saccharomyces, Aspergillus,* and *Rhodotorula* (Canadian Patent 978,114). Microorganisms have also been claimed for the degradation of oil effluents on open waters (U.S. Patent 3,105,014 and Canadian Patents 997,290 and 1,017,693), in drilling effluents (U.S. Patent 4,035,289), and in ground waters (U.S. Patent 3,846,290). The protection of exposed rock surfaces such as beaches and shorelines before the arrival of an oil spill is taught in Canadian Patent 1,021,708, wherein bacteria selected from the genera of *Micrococcus, Arthrobacter,* and *Achromobacter* are mixed with an inert absorbing surface and are freeze-dried. For application the formulation is mixed with an aqueous phase and is applied to rock surfaces prior to exposure to oil. Presumably the bacteria applied may decompose oil effluents on the treated rock surface, and residual surfactant or emulsifying capacity will prevent adherence of the oil to the rock surface.

Emulsifying agents, whether of microbial or nonmicrobial origin, may aid in the attack on hydrocarbon effluents by microorganisms by acting to increase considerably the surface area (and hence the attackable area) of the oil. Two examples of bioemulsifying agents that could perform this function have already been mentioned (U.S. Patents 3,997,398 and 4,230,801). Canadian Patent 992,894 teaches the use of alkaline or alkaline earth nontoxic silicates, carbonates, phosphates, phosphites, or hypophosphites as emulsifying agents that enhance microbial degradation of oil products. Canadian Patent 998,952 describes the use of mineral and organic acids and their alkali or alkaline earth salts as emulsifying agents for petroleum products for the same function.

Pipelining. Water and particulate matter deposition in pipelines transporting crude oils and refined petroleum products is a major problem. U.S. Patent 4,216,026 claims that the passage of a Bingham plastic fluid plug, consisting of water and xanthan gum that may be cross-linked with a multivalent metal, through the pipeline will remove this water and debris. The water and debris are collected by the plug.

It has also been suggested that pipelining viscous heavy oils could be facilitated by the emulsification of water into the oil to cut its viscosity using biodegradable biological emulsifying agents. On delivery these emulsions would be broken to reclaim the oil.

CONCLUSION

Patents already exist for the use of microorganisms or their products in most aspects of oil production. Yet these patents, covering forty years of activity in the field of petroleum microbiology, only provide a background for new research efforts based on the concepts of using bacteria as direct agents of oil recovery in situ and of EOR chemical production in large-scale, biomass-based facilities on the surface. The further development of biocatalysts for oil upgrading and petrochemical conversions also lies ahead. These applications may be incredibly enhanced by the rapidly developing field of genetic engineering, since the stimulation of oil production is of vital economic concern and can offer the financial promise necessary for sophisticated research and development.

PATENT REFERENCES

United States

US 1,998,756. J.J. Grebe and S.M. Stoesser. Assignee: The Dow Chemical Co. Treatment of deep wells. Issued Apr. 23, 1935.

US 2,413,278. C.E. Zobell. Assignee: American Petroleum Institute. Bacteriological processes for treatment of fluid-bearing earth formations. Issued Dec. 24, 1946.

US 2,641,565. R.T. Sanderson. Assignee: The Texas Co. Treatment of shale. Issued June 9, 1953.

US 2,641,566. C.E. Zobell. Assignee: Texas Development Corp. Recovery of hydrocarbons. Issued June 9, 1953.

US 2,660,550. D.M. Updegraff and G.B. Wren. Assignee: Socony-Vacuum Oil Co. Inc. Secondary recovery of oil by *Desulfovibso.* Issued Nov. 24, 1953.

US 2,807,570. D.M. Updegraff. Assignee: Socony Mobil Oil Co. Inc. Recovery of petroleum oil. Issued Sept. 24, 1957.

US 2,907,389. D.O. Hitzman. Assignee: Phillips Petroleum Co. Recovery of oil from oil sands and the like. Issued Oct. 6, 1959.

US 3,020,207. J.T. Patton. Assignee: Jersey Production Research Co. Thickening agent and process for producing same. Issued Feb. 6, 1962.

US 3,032,472. D.O. Hitzman. Assignee: Phillips Petroleum Co. Microbial secondary recovery. Issued May 1, 1962.

US 3,105,014. W.H. Harrison. Assignee: Tidewater Oil Co. Bacterial treatment of media containing hydrocarbons and sulfides. Issued Sept. 24, 1963.

US 3,185,216. D.O. Hitzman. Assignee: Phillips Petroleum Co. Use of bacteria in the recovery of petroleum from underground deposits. Issued May 25, 1965.

US 3,256,271. R.G. Schweiger. Assignee: Kelco Co. Method of improving viscosity characteristics of *Xanthomonas* hydrophilic colloids and esters produced thereby. Issued June 14, 1966.

US 3,301,848. F.E. Halleck. Assignee: The Pillsbury Co. Polysaccharides and methods for production thereof. Issued Jan. 31, 1967.

US 3,305,016. G.P. Lindblom, G.D. Ortloff, and J.T. Patton. Assignee: Esso Production Research Co. Displacement of oil from partially depleted reservoirs. Issued Feb. 21, 1967.

US 3,312,279. D.O. Hitzman. Assignee: Phillips Petroleum Co. Method of waterflooding by using phosphorylated mannan. Issued Apr. 4, 1967.

US 3,326,286. R.R. Harvey. Assignee: Phillips Petroleum Co. Oil recovery in water flooding. Issued June 10, 1967.

US 3,332,487. L.W. Jones. Assignee: Pan American Petroleum Corp. Aerobic bacteria in oil recovery. Issued July 25, 1967.

US 3,340,930. D.O. Hitzman. Assignee: Phillips Petroleum Co. Oil recovery process using aqueous microbial drive fluids. Issued Sept. 12, 1967.

US 3,436,311. R.B. Ferguson and J.D. Westover. Assignee: The Pillsbury Co. Fungal polysaccharide composition and method for making same. Issued Apr. 1, 1969.

US 3,436,346. J.D. Westover and R.B. Ferguson. Assignee: The Pillsbury Co. Process for preparing filterable aqueous polysaccharide solutions. Issued Apr. 1, 1969.

US 3,482,636. C.W. Crowe. Assignee: Dow Chemical Co. Method of lessening the inhibitory effects of fluid flow due to the presence of solid organic substances in a subterranean formation. Issued Dec. 9, 1969.

US 3,489,217. R.E. Gilchrist. Assignee: Tenneco Oil Co. Method of determining the movement of heat in a subterranean formation using bacterial analysis. Issued Jan. 13, 1970.

US 3,528,503. C.W. Crowe. Assignee: Dow Chemical Co. Method of improving permeability of geological formations by removal of organic material therefrom. Issued Sept. 15, 1970.

US 3,529,666. C.W. Crowe. Assignee: Dow Chemical Co. Method of improving permeability of geological formations by removal of organic material therefrom. Issued Sept. 22, 1970.

US 3,578,499. H.R. Crotty, C.R. Coffey, and T.C. Tesdahl. Assignee: W.R. Grace and Co. Gelling composition for general purpose cleaning and sanitizing. Issued May 11, 1971.

US 3,598,181. E.H. Wegner. Assignee: Phillips Petroleum Co. Oil recovery employing viscosifiers produced by the action of anionic surfactants on bacterial cultures. Issued Aug. 10, 1971.

US 3,612,178. W.F. Germer and C.W. Stringer. Method of recovering oil using flow stimulating solution. Issued Oct. 12, 1971.

US 3,618,664. R.R. Harvey. Assignee: Phillips Petroleum Co. Oil recovery method using an aqueous fluid drive. Issued Nov. 9, 1971.

US 3,625,889. C.D. Branscum. Assignee: Phillips Petroleum Co. Well completion fluids. Issued Dec. 7, 1971.

US 3,632,570. J.W. Gill. Assignee: Hercules Inc. Polysaccharide process. Issued Jan. 4, 1972.

US 3,650,326. D.O. Hitzman. Assignee: Phillips Petroleum Co. Hydrocarbon recovery employing aqueous medium driving fluid having increased viscosity. Issued Mar. 21, 1972.

US 3,655,512. T. Suzuki and K. Tanaka. Process for producing polysaccharides by fermentation. Issued Apr. 11, 1972.

US 3,707,187. B.L. Knight. Assignee: Marathon Oil Co. Flooding method using salt-insensitive polymers for better mobility control. Issued Dec. 26, 1972.

US 3,719,606. W. Askew and H. Froning. Assignee: Pan American Petroleum Corp. Microemulsion of increased viscosity for improved oil recovery. Issued Mar. 6, 1973.

US 3,724,545. B. Knight. Assignee: Marathon Oil Co. Economical mobility control in oil recovery processes. Issued Apr. 3, 1973.

US 3,729,460. J. Patton. Modified heteropolysaccharides. Issued Apr. 24, 1973.

US 3,751,364. T. Bellos. Assignee: Petrolite Corp. Method of protecting metal surfaces against abrasive wear in pumps with polyquaternaries. Issued Aug. 7, 1973.

US 3,759,828. H. Harrison. Assignee: Texaco Inc. Soluble oil compositions. Issued Sept. 18, 1973.

US 3,769,164. E. Azarowicz. Assignee: Bioteknika International, Inc. Microbial degradation of petroleum. Issued Oct. 30, 1972.

US 3,800,872. R. Friedman. Assignee: Getty Oil Co. Methods and composition for recovery of oil. Issued Apr. 2, 1974.

US 3,801,502. D.O. Hitzman. Assignee: Phillips Petroleum Co. Water-flood bacterial viscosifier. Issued Apr. 2, 1974.

US 3,826,308. A.L. Compere-Whitney. Assignee: Imperatix. Process for producing product from fossil fuel. Issued July 30, 1974.

US 3,844,348. C. Stratton. Assignee: Phillips Petroleum Co. Method of controlling subterranean formation permeability employing graft-polymerized bacterial gum polysaccharide compositions. Issued Oct. 29, 1974.

US 3,844,350. B. Knight and J. Rhudy. Assignee: Marathon Oil Co. Oil recovery using polymer preslug. Issued Oct. 29, 1974.

US 3,846,290. R. Raymond. Assignee: Sun Research and Development Co. Reclamation of hydrocarbon contaminated ground waters. Issued Nov. 5, 1974.

US 3,861,467. B.G. Harnsberger. Assignee: Mobil Inc. Permeable cementing method. Issued Jan. 21, 1975.

US 3,882,029. P.W. Fisher, J.P. Gallus, and D.S. Pye. Assignee: Union Oil Company of California. Well completion and workover fluid. Issued May 6, 1975.

US 3,903,272. E.S. Littman and H.C. Noe Jr. Assignee: Nalco Chemical Co. Microbiocide process and composition effective against sulfate reducing bacteria waterflooding operations. Issued Sept. 2, 1975.

US 3,908,760. R.L. Clampitt and J.E. Hessert. Assignee: Phillips Petroleum Co. Method of treating subterranean formations with aqueous gels. Issued Sept. 30, 1975.

US 3,915,800. K.S. Kang and W.H. McNeely. Assignee: Kelco Co. Polysaccharide and bacterial fermentation for its preparation. Issued Oct. 28, 1975.

US 3,933,788. K.S. Kang, G.T. Veeder, and D.D. Richey. Assignee: Kelco Co. Polysaccharide and bacterial fermentation process for its preparation. Issued Jan. 20, 1976.

US 3,941,692. D. Gutnick and E. Rosenberg. Cleaning of cargo compartments. Issued Mar. 2, 1976.

US 3,956,145. C.A. Christopher Jr. and A. Satter. Assignee: Texaco Inc. Fluid for injection into a subterranean reservoir to displace hydrocarbons in the reservoir. Issued May 11, 1976.

US 3,960,832. K.S. Kang and W.H. McNeely. Polysaccharide and bacterial fermentation process for its preparation. Issued June 1, 1976.

US 3,965,985. D.O. Hitzman. Assignee: Phillips Petroleum Co. Microbial synthesis from aldehyde containing hydrocarbon derived products. Issued June 29, 1976.

US 3,966,618. G.T. Colegrove. Assignee: Merck & Co. Inc. Clarification of xanthan gum. Issued June 29, 1976.

US 3,982,995. T.F. Yen, M.O. Appleman, and J.E. Findley. Assignee: The University of Southern California. Method of converting oil shale into a fuel. Issued Sept. 28, 1976.

US 3,989,632. P.W. Fischer, J.P. Gallus, and D.S. Pye. Assignee: Union Oil Co. of California. Reducing the loss of fluid through the walls of a well penetrating a permeable subterranean formation during completion, workover, and other operations. Issued Nov. 2, 1976.

US 3,997,398. J.E. Zajic and E. Knettig. Assignee: Canadian Patents and Development Ltd., Ottawa, Canada. Emulsifying agents of microbial origin. Issued Dec. 14, 1976.

US 4,008,766. J.G. Savins. Assignee: Mobil Corp. Oil recovery by waterflooding employing a biopolymer-surfactant system. Issued Febr. 22, 1977.

US 4,035,289. J. Gratacos, M. Guillerme, A. Sirvins, and B. Tramier. Assignee: Nationale Elf Aquitaine (Production), France. Purifying treatment for effluents of mineral oil drillings. Issued July 12, 1977.

US 4,038,206. C.L. Karl. Assignee: General Mills Inc. Hydroxyalkyl bean/*Xanthomonas* hydrophilic colloid blends. Issued July 26, 1977.

US 4,039,028. R.K. Knight. Assignee: Union Oil Co. of California. Mobility control of aqueous fluids in porous media. Issued Aug. 2, 1977.

US 4,048,079. R.L. Clampitt and J.E. Hessert. Assignee: Phillips Petroleum Co. Aqueous gels and uses thereof. Issued Sept. 13, 1977.

US 4,049,559. N. Grier, R.A. Dybas, and R.A. Strelitz. Assignee: Merck & Co. Inc. Composition for secondary and tertiary oil recovery. Issued Sept. 20, 1977.

US 4,053,396. A. Clamen, J.M. Fernbacher, and R.V. Trense. Assignee: Exxon Research and Development Co. Control of aerobic biological growth in activated carbon waste water treatment. Issued Oct. 11, 1977.

US 4,067,389. J.G. Savins. Assignee: Mobil Corp. Hydraulic fracturing technique. Issued Jan. 10, 1978.

US 4,077,879. C.I. Smeck. Assignee: Olin Corp. Treatment of sulfide-bearing waters with chlorine dioxide. Issued Mar. 7, 1978.

US 4,078,607. H.W. Carter, C.A. Christopher, and T. Jefferson. Assignee: Texaco Inc. Oil recovery process using improved thickened aqueous flooding liquids. Issued Mar. 14, 1978.

US 4,085,972. S. Ghosh and D.L. Klass. Assignee: Institute of Gas Technology. In situ mining of fossil fuel containing inorganic matrixes. Issued Apr. 25, 1978.

US 4,096,073. D.O. Hitzman. Assignee: Phillips Petroleum Co. Microbial viscosifiers. Issued June 20, 1978.

US 4,104,193. W.H. Carter, C.A. Christopher, and T. Jefferson. Assignee: Texaco Inc. Importing superior viscosity to aqueous polysaccharide solutions. Issued Aug. 1, 1978.

US 4,105,461. J.S. Racciato. Assignee: Merck and Co. Inc. Thickening compositions containing xanthan gum, guar gum, and starch. Issued Aug. 8, 1978.

US 4,119,546. W.W. Wernau. Assignee: Pfizer Inc. Process for producing *Xanthomonas* hydrophilic colloid, product resulting therefrom, and use thereof in displacement of oil from partially depleted reservoirs. Issued Oct. 10, 1978.

US 4,128,482. R.K. Knight. Assignee: Union Oil Co. of California. Mobility control of aqueous films in porous media. Issued Dec. 5, 1978.

US 4,141,842. M.K. Abdo. Assignee: Mobil Corp. Waterflooding oil recovery process employing stabilized biopolymers. Issued Febr. 27, 1979.

US 4,186,025. K.S. Kang, G.T. Veeder, and D.D. Richey. Assignee: Merck & Co. Inc. Aqueous polysaccharide composition. Issued Jan. 29, 1980.

US 4,211,774. K.S. Kang and G.T. Veeder. Assignee: Merck & Co. Inc. Bacterial polysaccharide S-21 and complex thereof. Issued July 8, 1980.

US 4,216,026. P.R. Scott. Assignee: Shell Oil Co. System for removing fluid and debris from pipelines. Issued Aug. 5, 1980.

US 4,218,327. S.L. Wellington. Assignee: Shell Oil Co. Stabilizing the viscosity of an aqueous solution of polysaccharide polymer. Issued Aug. 19, 1980.

US 4,230,801. D.L. Gutnick and E. Rosenberg. Assignee: Biotechnologie Aktiengesellschaft für Emulsar, Basil, Switzerland. Production of α-emulsans. Issued Oct. 28, 1980.

US 4,231,426. W.H. Carter and J.L. Sides. Assignee: Texaco Inc. Method of using tracer fluids for enhanced oil recovery. Issued Nov. 4, 1980.

US 4,233,438. P.A. Myers and D.J. Westlake. Assignee: British Petroleum Co. Ltd. Heteropolysaccharide biopolymer. Issued Nov. 11, 1980.

US 4,234,689. D.L. Gutnick, E. Rosenburg, and Y. Shabtai. Assignee: Biotechnologie Aktiengesellschaft für Emulsan, Basil, Switzerland. Production of α emulsans. Issued Nov. 18, 1980.

US 4,247,639. K.S. Kang and G.T. Veeder. Assignee: Merck & Co. Inc. Bacterial polysaccharide. Issued Jan. 27, 1981.

US 4,249,608. W.H. Carter. Assignee: Texaco Inc. Polymer-containing fluid and an oil recovery method using the fluid. Issued Febr. 10, 1981.

US 4,256,590. A.I. Laskin and L.A. Haslund. Assignee: Exxon Research and Engineering Co. Modified heteropolysaccharides and their preparation. Issued Mar. 17, 1981.

US 4,261,420. D.O. Hitzman. Assignee: Provesta Corp. Enriched oil recovery using carbon dioxide. Issued Apr. 14, 1981.

US 4,290,899. S.T. Foster, T.D. Foster Jr., and T.R. Malone. Assignee: Union Carbide Co. Thermal stabilizer for non-clay wellbore fluids. Issued Sept. 22, 1981.

US 4,296,203. W.C. Wernau. Assignee: Pfizer Inc. *Xanthomonas* biopolymer for use in displacement of oil from partially depleted reservoirs. Issued Oct. 20, 1981.

US 4,299,709. W.H. Carter and J.L. Sides. Assignee: Texaco Inc. Tracer fluids for enhanced oil recovery. Issued Nov. 10, 1981.

Canada

CDN 675416. G.P. Lindblom and J.T. Patton. Assignee: Jersey Production Research Co. Substituted heteropolysaccharides. Issued Dec. 3, 1963.

CDN 760905. J.A. Kepley. Assignee: Continental Oil Co. Subterranean formation treatment. Issued June 13, 1967.

CDN 945091. K.S. Kang and W.H. McNeely. Assignee: Kelco Co. Polysaccharide and bacterial fermentation process for its preparation. Issued Apr. 9, 1974.

CDN 968701. T.S. Felmann, H.B. Carlile Jr., and C.F. Blankenhorn. Assignee: Shell Canada Ltd. Dispersing cellular micro-organisms. Issued June 3, 1975.

CDN 978114. E.N. Azarowica. Assignee: Bioteknika International Inc. Microbial degradation of petroleum. Issued Nov. 18, 1975.

CDN 992894. G.H. Salomone. Method and composition for the emulsification and the degradation of petroleum products and fertilizers thus obtained. Issued July 13, 1976.

CDN 997290. P.M. Townsley. Assignee: John Dunn Agencies Ltd., Canada. Material for biological degradation of petroleum. Issued Sept. 21, 1976.

CDN 998952. P. Fusey. Assignee: Societe Anonyme: Banque pour l'Expansion Industrielle "Banexi," France. Composition for bringing mineral oil, expecially petroleum products, into the form of a biodegradable emulsion. Issued Oct. 26, 1976.

CDN 1017693. W. Marconi. Assignee: Sran Progetta SpA, Italy. Degradation of petroleum products on water surfaces. Issued Sept. 20, 1977.

CDN 1021708. R.R. Mohan, E.R. Bucker, J. Nixon, and G.H. Byrd Jr. Assignee: Exxon Research and Engineering Co. Method of protecting beaches against oil contamination by means of microorganisms. Issued Nov. 29, 1977.

CDN 1023288. D. Gutnick and E. Rosenburg. Assignee: Petroleum Fermentations Inc. Microbial cleaning of cargo compartments. Issued Dec. 27, 1977.

United Kingdom

UK 1396031. T.S. Felmann, H.B. Carlile Jr., and C.F. Blankenhorn. Dispersing and/or dissolving cellular micro-organisms. Issued May 29, 1975.

UK 1411014. K.S. Kang, G.T. Veeder, and D.D. Richey. Assignee: Kelco Co. Microbiological production of a heteropolysaccharide. Issued Oct. 22, 1975.

The Netherlands

NETH 89580. J. Van Heiningen, H. Jan De Haan, and J.D. Jansen. Assignee: N.V. De Bataafche Petroleum Maatschappij. Werkwijze voor het winnen van aardolie uit gesteente. Issued Nov. 15, 1958.

France

FR 2330697. L.A. Naslund and A.I. Laskin. Assignee: Exxon Research and Engineering Co. Treatment of a heteropolysaccharide. Issued June 3, 1977.

Germany

GER 2725069. L.A. Naslund and A.I. Laskin. Assignee: Exxon Research and Engineering Co. Modified polysaccharides useful in petroleum recovery. Issued Nov. 23, 1978.

REFERENCES

1 DAVIS, J.B. 1964. Petroleum microbiology. Elsevier, N.Y.

2 LANPHIER, E.O. 1981. The impact of genetic engineering on energy. International Resource and Development Inc., Norwalk, Connecticut.

3 GERSON, D.F., and J.E. ZAJIC. 1979. Microbial biosurfactants. Process Biochemistry *14*:20–29.

4 COOPER, D.G., and J.E. ZAJIC. 1980. Surface-active compounds from microorganisms. Advances in Applied Microbiology *26*:229–253.

5 WALTON, M.T., and J.L. MARTIN. 1979. Production of butanol-acetone by fermentation. *In* H.J. Peppler and D. Perlman (ed.), Microbial technology, 2nd ed., vol I. Academic Press, N.Y.

6 SANDVIK, E.I., and J.M. MAERKER. 1977. Application of xanthan gum for enhanced oil recovery, p. 242–264. *In* P.A. Sandford and A. Laskin (ed.), Extracellular microbial polysaccharides. ACS symposium series, vol. 45. ACS, Washington.

7 ZAJIC, J.E., and C.J. PANCHAL. 1976. Bioemulsifiers. CRC Critical Reviews in Microbiology *5*:39–66.

8 FINNERTY, W.R. 1980. Microbial desulfurization and denitrogenation. A paper presented at the Second International Chemistry Conference of the North American Continent. Las Vegas, Nevada, August 28, 1980.

9 CLARK, R.B., D.M. MUNNECKE, and G.E. JENNERMAN. 1980. In situ microbial enhancement of oil production. A paper presented at the SIM annual meeting. Flagstaff, Arizona, August 10–15, 1980.

10 ANONYMOUS. 1972. Bacteria have destroyed 10% of the world's crude. World Oil, February:28–29.

11 GRULA, M., and E. GRULA. 1982. Study of the biodegradation of compounds used in enhanced oil recovery. U.S. Department of Energy Contract No. DE-AT19-78BC 30201. Enhanced oil recovery and improved drilling technology progress review. U.S. Department of Energy, Bartlesville Energy Technology Center, Oklahoma.

12 TATNALL, R.E. 1981. Fundamentals of bacteria induced corrosion. Materials Performance *70*:32–38.
13 DAVIS, op cit.
14 KARASKIEWICZ, J. 1974. Zastosowanie Metod Mikrobiologicznych W Intensyfikacji Eksploatacji Karpackich Zloz Ropy Naftawej. PRACE, Instytutu Naftowego, Wydawnictwo. ''Slask'' Katowice, Cracow, Poland.
15 DAVIS, op cit.
16 JACK, T.R. 1980. Enhanced oil recovery by microbial action. A paper presented at the Second International Chemistry Conference of the North American Continent. Las Vegas, Nevada, August 28, 1980.
17 WARD, D.M., and T.D. BROCK. 1978. Anaerobic metabolism of hexadecane in sediments. Geomicrobiology Journal *1*:1–9.
18 SANDVIK and MAERKER, op cit.
19 LANPHIER, op cit.
20 BUCHANAN, R.E., and N.E. GIBBONS (ed.) 1975. Bergey's manual of determinative bacteriology, 8th ed. Williams and Wilkins, Baltimore, Maryland.

ADDITIONAL REFERENCE

MALIK, K.A. 1978. Microbial removal of organic sulphur from crude oil and the environment: some new perspectives. Process Biochemistry *13*:10–35.

4 IMPORTANT ANALYTICAL METHODS ON INTERFACIAL PHENOMENA RELATING TO MICROBIAL ENHANCED OIL RECOVERY

J.E. Zajic, L. Jones, and S.B. Wilson

University of Texas at El Paso

W. Seffens

Petroleum Bioresources, El Paso

In studying the effectiveness of microbially enhanced oil recovery (MEOR), one must focus upon surface phenomena within the system. Whether this focus is upon the surface of a solid, microbe, oil or water, the boundary or interface between these individual phases (Figure 1) can be described in terms of interfacial free energy (γ). Solvent-air and liquid-liquid interfacial energies are determinable by direct means while γ liquid-solid and solid-air must be evaluated by indirect means.

This paper reviews three simple techniques utilized in the study of interfacial energies and will include discussions of the following methods: (*i*) the ring method, (*ii*) the spinning drop method, and (*iii*) the contact angle method.

THE RING METHOD

The ring method has been used for many years as a method of directly measuring the surface tension of a liquid

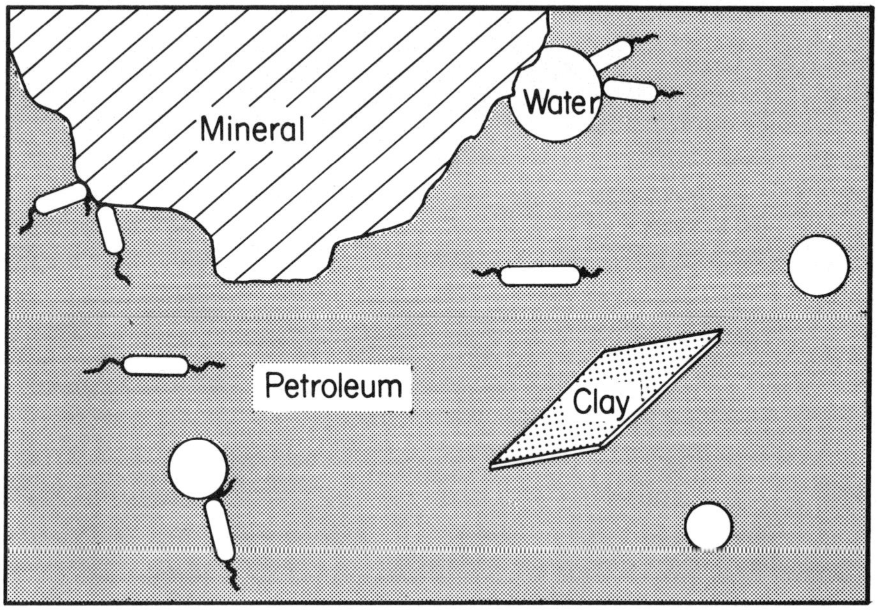

Interfaces:	γ:
Water/Petroleum	$\gamma_{W/P}$
Mineral/Petroleum	$\gamma_{M/P}$
Mineral/Water	$\gamma_{M/W}$
Mineral/Bacteria	$\gamma_{M/B}$
Bacteria/Water	$\gamma_{B/W}$
Bacteria/Petroleum	$\gamma_{B/P}$

Fig. 1. Phases present in typical petroleum reservoirs

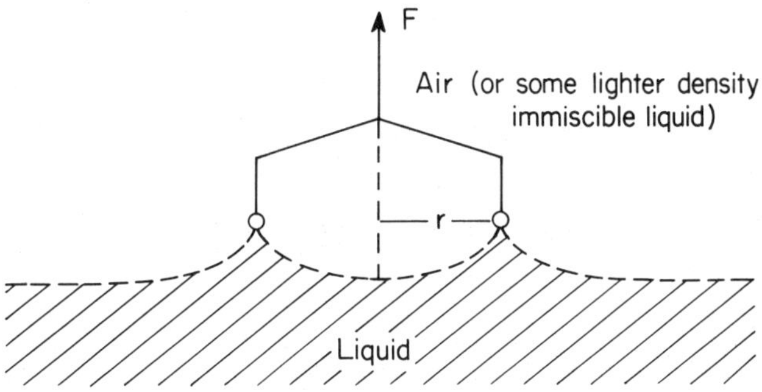

Fig. 2. General force diagram for a du Noüy tensiometer

or, more precisely, the liquid-vapor interfacial free energy (γ_{in}). Attributed to du Noüy, this method is performed by pulling a ring through a wetting liquid and measuring the force (strain gauge) required to pull the ring through the interface and to detach the ring from the surface of the liquid.[1] This implies that the ring method could also be used to determine the interfacial free energy between two immiscible liquids, provided that the lower phase preferentially wets the ring (Figure 2).

As the ring ascends through the interface, the weight (W_ℓ) of the liquid being pulled upward is equated to the liquid-vapor interfacial free energy ($\gamma_{\ell v}$) multiplied by the periphery of the ring.

$$W_\ell = \gamma_{\ell v} 4\pi r \tag{1}$$

The force (F_ℓ) required for detachment from the surface is actually a measurement of the strain upon the wire supporting the ring. Upon detachment,

$$F_\ell = W_\ell = \gamma_{\ell v} 4\pi r \tag{1a}$$

so that the interfacial free energy can be found by rearranging to:

$$\gamma_{\ell v} = \frac{F_\ell}{4\pi r} \tag{2}$$

where F_ℓ = strain upon the ring supporting wire. This assumes that the vector of interfacial free energy ($\gamma_{\ell v}$) acts vertically. However, in many situations this assumption has been shown to be in error.[2] Generally, the interfacial free energy does not act exactly vertically due to the meniscus volume held within smaller rings. A correction factor (β) is necessary that depends upon ring size.[2,3]

$$\gamma_{\ell v} = \frac{F\beta}{4\pi r} \tag{3}$$

However, this error is minimized through the use of large rings of fine wire such as 3.0-cm diameter rings made of wire with a 0.02-cm radius.[4]

The ring method is generally used for measuring interfacial free energies greater than 1.0 mN/m, although its use can be extended downward to 0.5 mN/m with a fair amount of accuracy.

Typical experiments using a bacterial culture of *Corynebacterium fascians* in a mineral salts medium with 2% sucrose and 1% yeast extract gives surface and interfacial readings as shown in Figure 3. Curve A is for water, 73 mN/m at 20°C, while curve B is the surface tension of a whole broth of *C. fascians.* Curve C is the interfacial tension between a whole broth of *C. fascians* and hexadecane. The vertical axis measures the strain upon the ring-supporting wire, while the horizontal axis is a measure of the vertical distance the ring travels before it breaks away from the surface. The maximal force (the peak of each strain vs. travel curve) is that which satisfies equation (2) or (3) and is the surface or interfacial tension ($\gamma_{\ell v}$).

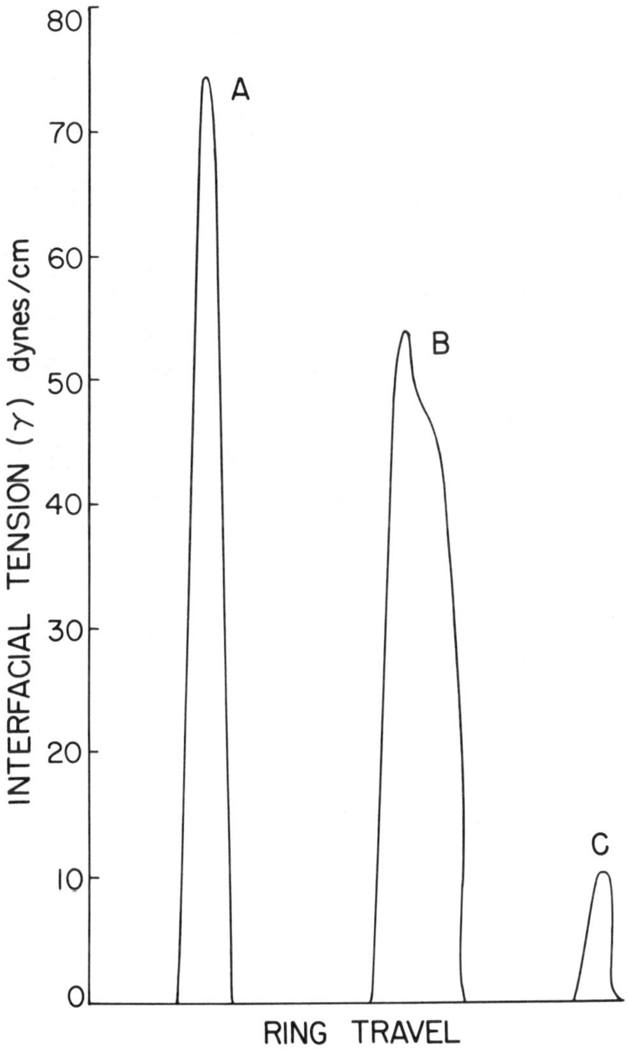

Fig. 3. Measurement of interfacial tension using the ring method. Measurement of γ_{1v} for H_2O (A). Measurement of γ_{1v} for *C. fascians* culture in mineral salts medium, 2% sucrose and 1% yeast extract (B). The value $\gamma\ell/\ell$ for the above *C. fascians* culture was measured against an upper hexadecane phase (C)

THE SPINNING DROP METHOD

The spinning drop technique enables one to determine extremely low interfacial free energy values based upon the shape of a drop being rotated in a selected medium under controlled conditions.

A drop of liquid is suspended within an immiscible liquid of greater density inside a glass tube (approx. 8 cm × 0.02 cm). When this tube is rotated, the drop aligns itself along the axis of rotation. As the rotational velocity is increased, a centrifugal force develops that opposes the interfacial free energy effect of minimizing the surface area. As a result the spherical drop deforms into a prolate ellipsoid.

At a sufficiently high rotational speed, the ratio of drop length to drop width becomes greater than 4.0. At this point it is assumed and has been shown that the drop can be treated as an elongated cylinder and the interfacial free energy (γ) can be calculated without having to solve for the exact drop shape.[5]

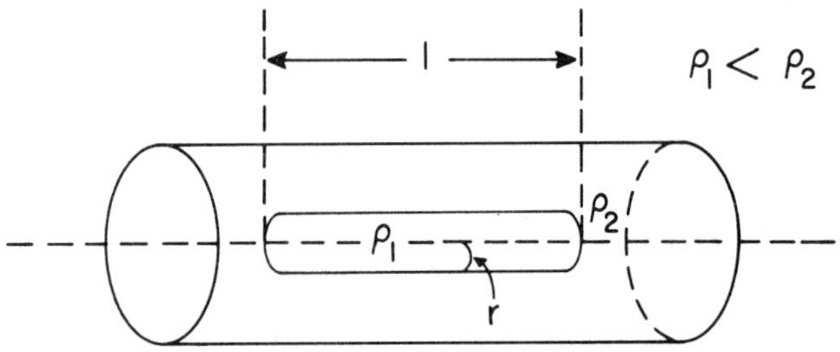

Fig. 4. Pictorial diagram of liquid cylinder typically observed in the spinning drop apparatus. Density of liquid I (ρ_1) is less than liquid 2 (ρ_2); r = radius of ρ_2, which is lighter density

The resulting centrifugal force affecting the two liquids can be described as:

$$F_1 = \omega^2 r \rho_1$$

$$F_2 = \omega^2 r \rho_2$$

The net centrifugal force effecting a volume element within the cylinder is then:

$$F_2 - F_1 = \omega^2 r \Delta\rho \tag{4}$$

where $\Delta\rho$ is the difference in the liquid densities and ω is the rotational velocity.

The potential energy per unit volume due to the centrifugal force distortion at a radius r from the axis of revolution can be described as:

$$E = \frac{\omega^2 r^2 \Delta\rho}{2} \tag{5}$$

Calculating this energy over the entire cylindrical drop of length ℓ, one obtains the energy of the drop due to its rotation:

$$E_R = \ell \int_0^r \frac{(\omega^2 r^2 \Delta\rho)}{2} 2\pi r dr = \frac{\pi\omega^2 r^4 \Delta\rho\ell}{4} \tag{6}$$

The interfacial free energy of this cylinder is $E_i = 2\pi r\ell\gamma$, where γ is the interfacial free energy per unit area between the two liquids. Thus, the total energy of the drop can be described as:[6]

$$E_T = E_R + E_i = \frac{\pi\omega^2 \Delta\rho r^4 \ell}{4} + 2\pi r\ell\gamma \tag{7}$$

Once the spinning drop has achieved an equilibrium value, $dE_T/dr = 0$, equation (8) can be solved for γ:

$$\gamma = \frac{\omega^2 \rho \Delta P r^3}{4} \tag{8}$$

This result is generally known as Vonnegut's equation for the interfacial energy of a spinning drop.[7]

The spinning drop method can also be performed using a bubble of gas suspended in a liquid where the γ calculated is the liquid-vapor interfacial free energy ($\gamma_{\ell v}$).

Applications of the spinning drop method have shown accurate measurements of interfacial free energies down to 0.001 mN/m.[8-11] This method was used to investigate interfacial free energies of a system composed of a *Corynebacterium fascians* bacterial culture at various NaCl concentrations measured against a drop of hexadecane (see Figure 5).

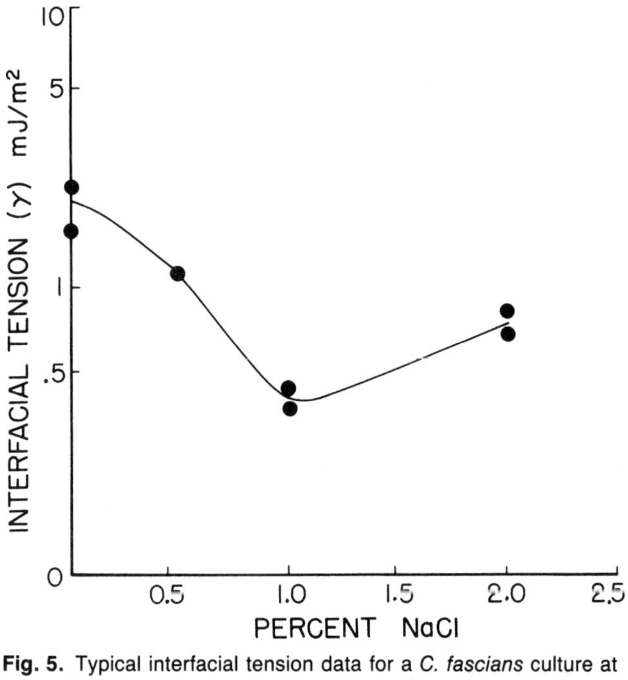

Fig. 5. Typical interfacial tension data for a *C. fascians* culture at increasing salt concentrations utilizing a spinning drop hexadecane

THE CONTACT ANGLE METHOD

The contact angle method allows one to study the interfacial phenomena of a solid-liquid-vapor system. Once the surface tension ($\gamma_{\ell v}$) of the liquid used is known, the method enables the calculation of both the solid-vapor interfacial free energy (γ_{sv}) and the solid-liquid interfacial free energy ($\gamma_{s\ell}$).

A drop of liquid placed on a uniform flat surface will spread over the surface, or the drop will form an angle (θ) with the solid surface and not spread.

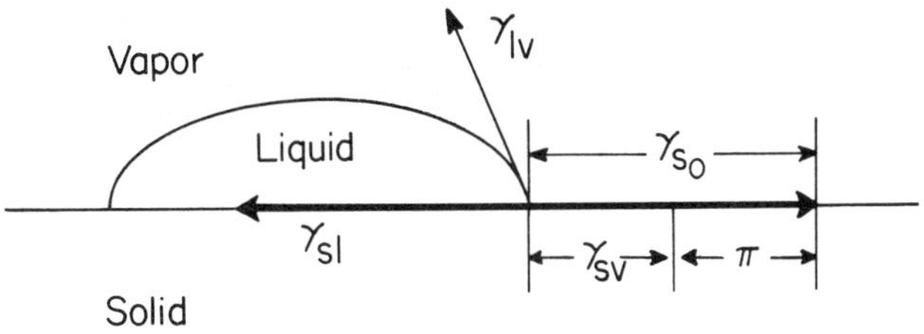

Fig. 6. Diagram showing the classic tension or energy relationships between a drop, surface of a smooth solid, and a vapor (gaseous phase). γ_{1v} = liquid-vapor surface tension; γ_{s1} = solid-liquid interfacial tension; γ_{s_0} = solid-surface interfacial energy; γ_{sv} = solid-vapor interfacial energy; π = equilibrium film pressure

Young's equation describes the balance of forces for this latter phenomenon:[12]

$$\gamma_{sv} - \gamma_{s\ell} = \gamma_{lv} \cos\theta \tag{9}$$

where:

γ_{sv} = solid-vapor interfacial free energy (mN/m)
$\gamma_{s\ell}$ = solid-liquid interfacial free energy (mN/m)

The contact angle (θ) can be obtained directly by measuring the angle from the solid-liquid interface through the liquid to the line tangent to the liquid-vapor interface. This measurement can be made through the use of a goniometer telescope.[13] Changes in drop volume can produce hysteresis between an advancing maximum and a receding minimum contact angle. In practice, however, one studies the more relevant measurement to the specific application.[14]

Once the contact angle (θ) is obtained and the surface tension of the liquid ($\gamma_{\ell v}$) is known, both the solid-vapor interfacial free energy (γ_{sv}) and the solid-liquid interfacial free energy (γ_{sl}) can be calculated. This is accomplished by solving, simultaneously, an equation developed by Neumann et al.:[15]

$$\gamma_{s\ell} = \frac{\sqrt{\gamma_{sv}} - \sqrt{\gamma_{\ell v}}}{1 - 0.015\sqrt{\gamma_{sv}\gamma_{lv}}} \tag{10}$$

and Young's equation (11):

$$\gamma_{sv} - \gamma_{s\ell} = \gamma_{\ell v} \cos\theta \tag{11}$$

The results of contact angle measurements using saline and α-bromonaphthalene upon monolayers of bacterial cells is shown in Table 1.

The solid-vapor interfacial free energy (γ_{sv}) is not a measure of the solid surface free energy. The solid-vapor interfacial free energy (γ_{sv}) is the net value of the solid surface free energy (γ_{s_0}) after subtracting for the effect of the equilibrium film pressure. This equilibrium film pressure component (π) represents the energy of the vapor phase in contact with the surface of the solid.[16]

The solid surface free energy (γ_{s_0}) of materials can be calculated by using selected liquids with known surface tensions (γ_{lv}):[17]

$$\gamma_{s_0} = \frac{\gamma_{lv}(\cos\theta + 1)^2}{4} \tag{12}$$

This application has lead to the determination of the solid surface free energies for a number of minerals.[18]

SUMMARY

In the effort to increase technology in the field of enhanced oil recovery, the collection of information at the level of interfacial phenomena is of vital importance. Through the use of the methods described, one is able to study the ability of microbial systems to lower interfacial energies and effectively aid in the release of oil from mineral surfaces.

Table 1. Contact angle measurements of liquids upon bacterial cell monolayers.

Cell Monolayer Culture Type	Liquid Contact Angle° *Saline*	*α-Bromonaphthalene*	Calculated Interfacial Free Energy $\gamma B/S$ (mN/m)	$\gamma B/\alpha B$
Zymomonas mobilis 2% agar 1% glycerol	25	56	0.62	3.55
Streptococcus pyogenes 2% agar 1% glycerol	27	33	0.81	0.55

REFERENCES

1 du NOÜY, P.L. 1919. A new apparatus for measuring surface tension. J. Gen. Physiol. *1*:521.

2 HARKINS, W.D., and H.F. JORDAN. 1930. A method for the determination of surface and interfacial tension from the maximum pull on a ring. J. Am. Chem. Soc. *52*:1751.

3 FREUD, B.B., and H.Z. FREUD. 1930. Theory of the ring method for surface tension. J. Am. Chem. Soc. *52*:1772.

4 DAVIES, J.T., and E.K. RIDEAL. 1961. Interfacial phenomena. Academic Press, New York.

5 CAYIAS, J.L., R.S. SCHECHTER, and W.H. WADE. 1975. The measurement of low interfacial tension via the spinning drop technique. ACS Symposium Series No. 8:234.

6 ADAMSON, A.W. 1976. Physical chemistry of surfaces. John Wiley and Sons Inc., New York.

7 VONNEGUT, B. 1942. Rotating bubble method for determination of surface and interfacial tensions. Rev. Sci. Instrum. *13*:6.

8 DOE, P.H., M.M. EL-EMARY, W.H. WADE, and R.S. SCHECHTER. 1979. The influence of surfactant structure on low interfacial tensions. ACS Symposium Series *No. 91*:17.

9 MORGAN, J.C., R.S. SCHECHTER, and W.H. WADE. 1979. Ultra-low interfacial tension and its implication on tertiary oil recovery. Solution chemistry of surfactants Vol. 2. Plenum Publishing Corporation, New York.

10 CHAN, K.S., and D.O. SHAN. 1980. The molecular mechanism for achieving ultra low interfacial tension minimum in a petroleum surfactant/oil/brine system. J. Dispersion Sci. and Tech. *1*:55.

11 CAYIAS, SCHECHTER, and WAVE, op cit.

12 YOUNG, T. 1855. Miscellaneous works, Vol. 1. G. Peacock (ed.), J. Murray, London.

13 FOX, H., and W. ZISMAN. 1950. The spreading of liquids on low energy surfaces. I. Polytetrafluoroethylene. J. Colloid Sci. *5*:514.

14 GERSON, D.F., and J.E. ZAJIC. 1979. Immobilized microbial cells. ACS Symposium Series *No. 106*:29.

15 NEUMANN, A.W., C.F. GILLMAN, and C.J. van OSS. 1974, Phagocytosis and surface free energies. J. Electroanal. Chem. *49*:393.

16 ROSS, R., and J. OLIVER. 1959. On physical adsorption. Wiley Interscience, New York.

17 BECHER, P. 1977. Interaction parameter calculations from contact angle data. J. Colloid Interfacial Sci. *59*:429.

18 ZAJIC, J.E., W. SEFFENS, and N. ROGEN. 1982. Surface energy study of Utah tar sand and spreading tension, a new technique. Second international conference on heavy crude and tar sands Vol. 1.

5

SURFACE ENERGY DETERMINATION OF MINERALS USING CONTACT ANGLE MEASUREMENTS

J.E. Zajic and W. Seffens

Petroleum Bioresources, El Paso

The surface energy of solids is a characteristic interfacial factor that determines adsorption, wetting, and adhesion properties of mineral surfaces. A direct method for its measurement has not been established because of the ability of solids to support shear stress and not to display obvious surface tension phenomena. Several techniques exist that give estimates of the surface free energy of certain classes of solids, although none are broadly applicable.

This study examines the procedures applicable to surfaces of clays and various minerals using the contact angle of α bromonaphthalene (αBN). From these results a study has been developed that has application to petroleum recovery and ore flotation. Correlations that are based on mineral surface free energy are more general than wettability, since the latter is dependent on fluid considerations. There is considerable technical and theoretical interest in cataloging petroleum reservoir rock in terms of water or oil wettability. This discussion seeks to expand this interest to consider surface free energies of rock samples for use in correlating performances of enhanced oil recovery methods.

CONTACT ANGLES ON MINERALS AS RELATED TO OIL RECOVERY

Of the energy options being developed, tertiary oil recovery based on some type of fluid injection process shows great promise. Typically, a new oil well first produces spontaneously and is pumped until oil flow becomes uneconomical. Reservoir pressure and oil viscosity are important factors in the recovery process. Additional oil may be recovered by pumping water into the formation for secondary recovery (water flooding) and pumping out at other points a mixture of oil and water. Eventually this also becomes uneconomical and 30–70% of the original oil may remain. Tertiary oil recovery refers to processes designed to extract this residual oil, which is typically dispersed through the capillary system of the porous oil-bearing reservoir. Processes using surfactants are becoming important with recoveries of additional oil effective under diverse reservoir conditions.[1]

Displacement efficiency is determined in part by the geometry of the void space and by the interfacial properties characterizing the fluid and solid phases involved. If the formation rock is completely water wet ($\theta = 0$), water will try to envelop all reservoir grains and force oil out into a pore channel. Even though some oil might still be trapped, the recovery would be high. On the other hand, if all solid surfaces of the formation rock were completely oil wet ($\theta = 180°$), oil would envelop all grains and force water into the middle of the pore channel. In this extreme case, recovery would be very low by secondary water flooding. Thus, water flooding operations depend upon the contact angle of the system and may be changed from $>90°$ to $<90°$ through the use of caustic or surface-active agents.[2]

In practice when wettability is considered to be important, laboratory data are obtained for preserved cores and fluids at conditions considered to simulate satisfactorily those in the given reservoir. Such data cannot be readily reproduced from one laboratory to another and are difficult to interrelate or quantify. Systematic study under controlled wettability conditions is needed to develop an understanding of the role of wettability in oil recovery and other displacement or transport phenomena.

A reservoir is basically a complex system of interconnected capillaries of various size and shapes. Consider as a model a horizontal two-branch capillary with a capillary flow branching in front of a solid particle with a channel above and below it, meeting behind the particle. If a pressure gradient ΔP_a is applied, water will flow more readily through the capillary with the larger diameter. Thus, a certain volume of oil may be trapped in the smaller capillary when the water reaches the upstream fork. The sum of forces acting on the trapped oil may be used to find the total pressure drop ΔP_t across the ganglia, including capillary forces, according to the formula:

$$\Delta P_t = \Delta P_a + \frac{4\gamma_{wo}\cos\theta_A}{D_A} - \frac{4\gamma_{ow}\cos\theta_R}{D_R} \quad (1)$$

with θ_A and θ_R the advancing and receding contact angles, respectively, and D_A and γ_{wo} the pore diameter and interfacial tension at the advancing interface. The terms D_R and γ_{ow} are the pore diameter and interfacial tension at the receding oil-water interface. As noted by Langnes et al., the receding capillary term is almost always greater than the advancing (middle) term because $\theta_R < \theta_A$.[3] Thus, capillary pressure hinders flow, since the volumetric flow rate is proportional to the total pressure drop ΔP_t and the applied pressure ΔP_a is reduced by the sum of the capillary terms. The velocity and direction (forward being positive) of flow can be predicted from Darcy's law:

$$\nu = D\Delta P_t/32\mu L \quad (2)$$

with velocity ν in cm/sec, in a channel of length L cm, and mean diameter D with a fluid viscosity μ centipoises. Owing to the random nature of the porous reservoir material, characterization of the geometric parameters L, D_A, and D_R is possible only for model systems.

If a surfactant were added to reduce γ_{wo} to some minimal value γ_{min}, the oil globule would move as ΔP_t becomes negative. The quantity of trapped oil is therefore dependent upon the value of $\gamma\cos\theta$ at each end of the globule as well as upon the imposed pressure ΔP_a.

This shows the importance of further understanding the advancing and receding contact angles in porous media and their relationship to minimal interfacial tensions (γ_{min}).

In the case of petroleum and the formation water encountered in oil reservoirs, the contact angle measured through the water phase is often far from zero.[4] This situation can arise from the adsorption of polar constituents, which are present in petroleum in considerable quantities and are found upon the high-energy surface of the reservoir rock. When such adsorption occurs on a high-energy surface, it becomes comparable to a typical low-energy surface in terms of hydrophobicity. A major objective in oil recovery process, therefore, is to adjust these interfacial parameters to permit maximal fluid displacement at minimal cost.

Treiber et al. somewhat arbitrarily chose water advancing contact angles in the ranges 0° to 75°, 75° to 105°, and 105° to 180° to classify reservoirs as respectively water wet, intermediately wet or oil wet.[5] The distribution of wettabilities for 30 sandstone reservoirs for which advancing contact angles were reported had the following classification: for sandstone reservoirs, 7% were intermediate wet, 43% were water wet, and 50% were oil wet. A reclassification by Morrow according to spontaneous inhibition behavior designates the water-wet group to possess advancing contact angles of 0° to 62°. The intermediate-wet group ranges from the water-advancing angle of 62° up to a water-receding angle of about 133°, at which condition oil should spontaneously displace water. For the same set of sandstone reservoirs, 33% were intermediate wet, 40% were water wet, and 27% were oil wet.

CONTACT ANGLES AND SURFACE FREE ENERGY

Contact angles in air and in immiscible liquids have been determined for a large number of liquids on solid surfaces. Contact angle measurements are useful because they correlate with properties such as adhesion, surface energy, and wetting, all of which are important in separation processes. Contact angle values depend upon temperature, the surface material, surface roughness, surface contamination, liquid properties, and aging of the physical system.[7,8] The latter arises from the slow adsorption of liquid into the outer layers of the solid at the interface following the first occurrence of wetting. Contact angles are easily measured and good reproducibility is possible. For an excellent review of techniques, see Newmann and Good.[9]

The magnitude of the contact angle is determined by the balance of forces at the boundary line between liquid, solid, and second fluid, which is normally air. This balance of forces is described by Young's equation:

$$\gamma_{lv}\cos\theta = \gamma_{sv} - \gamma_{sl} \quad (3)$$

where γ_{lv}, γ_{sv}, and γ_{sl} are specific interfacial free energies for the liquid-vapor, solid-vapor, and solid-liquid interfaces, respectively, and θ is the angle measured of the liquid drop profile with the solid surface at the line of contact. $\cos\theta$ and γ_{lv} are the only two experimentally measurable quantities. The two terms on the right side of equation 3 are not independent of the fluid.

Becher has noted that for a selected class of fluids, it is possible to determine the surface free energy of a solid γ_s without considering the unobtainable γ_{sv} and γ_{sl} terms.[10] A contact angle measurement with a single fluid in which dispersion forces predominate and which has a sufficiently high liquid surface free energy to ensure a non-zero contact angle is enough to determine γ_s from the following expression:

$$\gamma_{lv}(\cos\theta + 1)^2/4 = \gamma_s \quad (4)$$

Many liquids are acceptable for this determination, including tetrabromoethane (γ_{lv} = 49.7 mJ/m^2) or α-bromonaphthalene (αBN, γ_{lv} = 44.6), and tricresyl phosphate (TCP, γ_{lv} = 40.9). This approach contrasts the critical surface tension method of Fox and Zisman, which requires a series of measurements to be made on homologous liquids.[11] The surface free energy of that liquid which just spreads (that is, $\theta = 0$) is equated with the surface free energy of the solid.

The surface free energy of a solid γ_s has been suggested to be a separable component of the interfacial terms γ_{sv} and γ_{sl}.[12,13] It is assumed that both adjacent solid and fluid phases have nonbulk properties within the interfacial region due to perturbed molecular interactions that can be described by free energy terms. Let γ_A^B represent the specific interfacial free energy of phase *A* in contact with phase *B*, so that for solid-liquid (S/L) and solid-vapor (S/V) interfaces:

$$\gamma_{SL} = \gamma_S^L + \gamma_L^S \tag{5}$$

and

$$\gamma_{SV} = \gamma_S^L + \gamma_V^S \tag{6}$$

For a solid that is unperturbed (i.e., it is essentially rigid) by the presence of a fluid phase in contact with it:

$$\gamma_S^L = \gamma_S^V = \gamma_S \tag{7}$$

The term γ_V^S can be found experimentally from vapor adsorption isotherms, and γ_L^S can be derived theoretically from a model treatment.[13] Therefore, the surface free energy of a solid γ_s can be related to the interfacial free energy terms of Young's equation.[14]

Since γ_s is not directly measured, its precision is dependent upon the uncertainty in θ. Analysis of the propogation of error in equation 4 for a standard error in θ, denoted ${}^e\theta$, follows from:

$${}^e\gamma_s = {}^d\gamma_s/d\theta \cdot {}^e\theta = \frac{\gamma_{lv}}{2}(\cos\theta + 1)\sin\theta \cdot {}^e\theta \tag{8}$$

For the system αBN at $\theta = 30°$, the error in θ is multiplied by a factor of 21; while for mercury (Hg, γ_{lv} = 476 mJ/m^2) at $\theta = 135°$, the error factor is about 49. This highlights the necessity for obtaining reliable contact angle data upon the surfaces to be studied.

EXPERIMENTAL RESULTS

The surface free energy γ_s of a series of minerals and clays has been evaluated using equation 4 from Becher and using the direct visual goniometer method by Fox and Zisman.[15,16] The apparatus used is a Gaertner goniometer, having rotatable 90° crosshairs with vernier to 3 minutes. Surfaces of samples (except for clays) were washed with detergent solution, then ethanol, then distilled water prior to examination. The fluids utilized for contact angles were used as delivered without additional purification, and all possessed low enough vapor pressures to avoid evaporation during the experiments. Mercury contact angles exhibited no change overnight, while αBN and TCP showed a decrease of 1°–5° in three hours. Typically, a sample analysis took no longer than 5 minutes to complete. Five or six liquid drops were examined in an advancing mode by adding a small amount of liquid to a drop already on the sample surface. Drop volumes were approximately 10 μl. Greatly irregular drops were ignored if the ratio of the maximal diameter to the minimal diameter exceeded 1.2, approximately.

CLAY MINERAL STUDY

Clays are composed of a number of different minerals in varying proportions of small particle size. Clay minerals are essentially crystalline hydrous aluminum silicates. In some, Mg^{+2} or Fe^{+3} substitute in part for Al^{+3}, and alkalis or alkaline earths may be present. Five clay mineral types were examined as listed in Table 1. Liquid mercury metal on a series of API-49 clay mineral standards (Ward's Natural Science Establishment) was examined only since αBN was absorbed. All surfaces were shaved by a knife edge to produce a smooth surface prior to use. Nineteen samples of various clays were examined in total.

Table 1. Clay mineral types.

1. Kaolinite—$Al_4(Si_4O_{10})(OH_8)$—derived by alteration of aluminum silicates, particularly feldspar.
2. Halloysite—two forms, one with kaolinite composition, the other with $Al_4(Si_4O_{10})(OH)_8$–$4H_2O$.
3. Montmorillonite—pyrophyllite structure with sandwiched sheets of molecular water containing free cations.
4. Bentonite—altered volcanic ash with montmorillonite the dominant clay mineral.
5. Illite—a mica-like clay mineral, differing from the micas in having less substitution of Al for Si, in containing more water, and in having K partly replaced by Ca and Mg. Chief constituent in many shales. Intermediate between montmorillonite clays and true micas.

Table 2 summarizes the clay mineral data. No significant difference can be observed between the various members of a clay, but there are differences between the clay minerals. Kaolinite has the highest average contact angle with mercury, followed by illite then bentonite and montmorillonite.

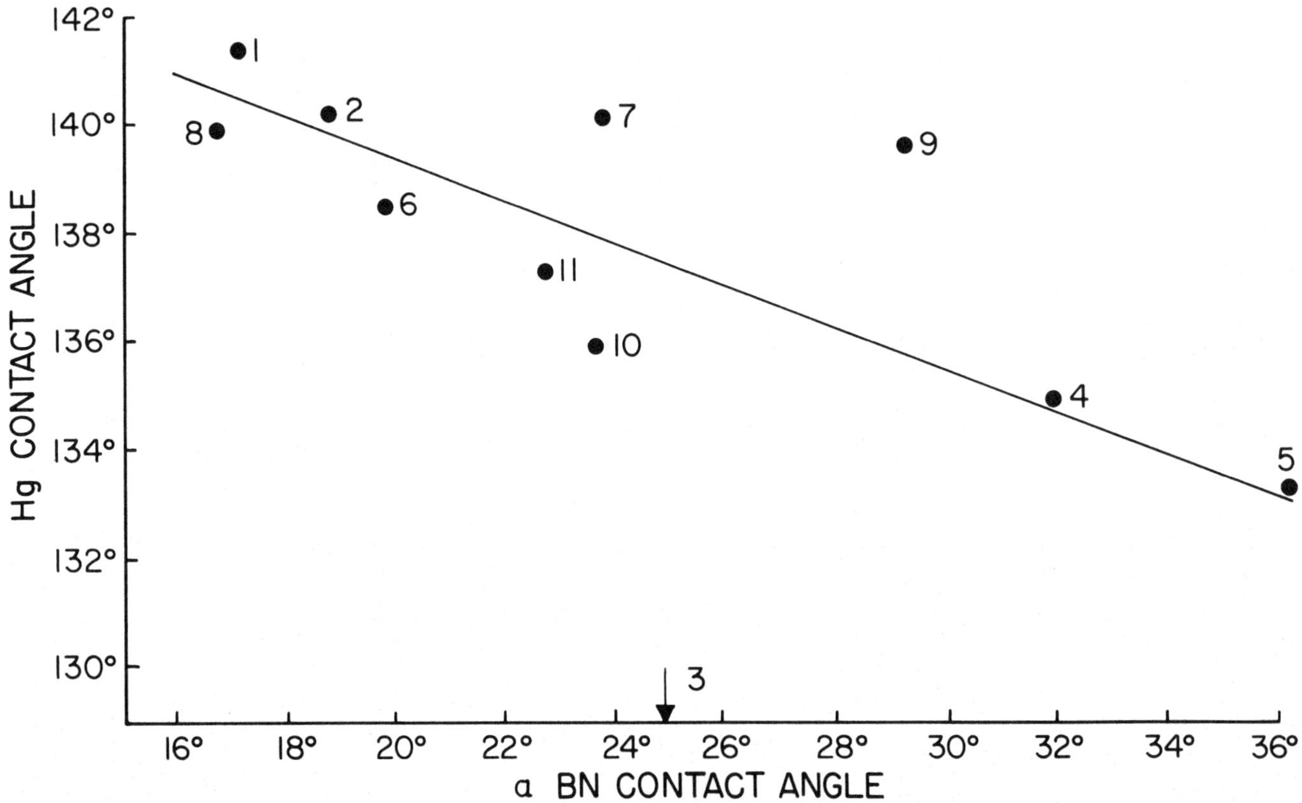

Fig. 1. Relationship between contact angles of mercury and bromonaphthalene on various mineral samples. Data from Table 3. Mineral numbers refer to labels in Table 3. Error bars are not included.

Table 2. Mercury contact angle results on clay minerals.

		Contact Angle, °	*Standard Deviation*	*% Error*
Kaolinite	2	147.5	5.78	3.9
	4	138.0	3.99	2.9
	5	144.0	4.76	3.3
	7	154.8	3.08	2.0
	9	149.8	6.03	4.0
	17	151.5	3.56	2.3
	H-9	142.9	4.38	3.1
Bentonite	20	137.0	6.91	5.0
	23	143.1	3.84	2.7
	24	137.8	6.07	4.4
	25	137.8	5.93	4.3
	26	147.6	4.65	3.2
	27	142.5	4.24	3.0
	31	140.7	2.80	2.0
Montmorillonite	22	133.8	5.08	3.8
	22a	135.8	4.62	3.4
	00a	140.0	3.40	2.4
Illite	35	130.2	6.08	4.0
	36	144.3	4.78	3.3

The bentonite and montmorillonite possess similar mercury contact angles, and indeed they are chemically very similar.

MINERAL STUDY

A series of minerals were examined and contact angles measurements were made using mercury and αBN, as shown in Table 3. The two carbonates tested (rhodachrosite and calcite) have similar low contact angles, while two metal oxides (columbite and cassiterite) have high contact angles. To test the applicability of mercury for use in finding surface free energies utilizing equation 4, θ(Hg) vs. $\theta(\alpha$BN) for each mineral are plotted in Figure 1. A lincar corrclation is possible, but several points are far removed from the linear portion of the curve. More data are needed to establish this correlation and the cause of deviations.

PHYSICAL PROPERTY CORRELATION WITH MINERAL CONTACT ANGLE DATA

Correlations between contact angle and/or mineral surface free energy and mineral physical properties described as hardness, cleavage, and specific gravity have been

Table 3. Mineral contact angle data and computed surface free energies.

Mineral	Hg Contact Angle Number of Measurement	Hg Contact Angle $\bar{x}$, °	αBN Contact Angle Number of Measurement	αBN Contact Angle $\bar{x}$, °	γ_s (mJ/m^2) (From αBN Data)
1. Talc	10	141.4	8	17.6	42.5
2. Stibnite	9	140.3	10	18.7	42.3
3. Muscovite	10	129.8	10	25.2	40.5
4. Calcite	12	134.9	10	31.9	38.1
5. Rhodachrosite	10	133.4	12	36.2	36.4
6. Apatite	10	138.5	12	19.7	42.0
7. Columbite	10	140.2	12	23.8	40.9
8. Microcline	9	139.0	12	16.9	42.7
9. Cassiterite	8	139.6	10	29.3	39.1
10. Grossularite garnet	10	135.9	6	23.6	40.9
11. Samarskite	12	137.2	4	22.7	41.2

examined. Although the results are preliminary, an interesting correlation has been observed between hardness and surface energy as measured by αBN contact angles.

There are several measures of crystal hardness. The scratch hardness scale, established by Mohs in 1812, is constructed so that a mineral with a certain hardness will scratch one of lesser hardness. Scratching is accomplished by penetrating the point of a crystal or crystal aggregate into the face of another crystal. Like other physical properties of minerals, scratch hardness is dependent on crystal structure. The stronger the binding force between component atoms, the harder the mineral.

Abrasion hardness is a quantitative measure of complex mechanical processes. If different minerals are abraded under completely identical conditions, different weight losses are obtained. Emery or silicon carbide grains are customarily used for abrasion to cause plastic deformation, but in addition they penetrate the surface of the crystal and tear out small fragments. Abrasion hardness is dependent upon abrasion fluid since work is done in opposition to surface tension (γ_{sl}). The abrasion hardness of quartz in octyl alcohol is only half as great as in water. Thus, abrasion hardness can be used to measure the relative surface energies of a crystal.[17] Various hardness scales from different sources have been collected and plotted in Figure 2 as a function of scratch hardness. It is evident that scratch hardness is logarithmic with respect to other, perhaps more physical, measures of hardness.

Mineral data are plotted in Figure 3 against scratch hardness. Two groups of data appear by noting that points 1–5 possess perfect or good cleavage. By using a least-squares method, the data were fitted to two linear lines. The upper line has slope −0.268 with intercept 17.04 and correlation coefficient 0.60, while the lower line has slope −0.402 with intercept 18.50 and correlation coefficient 0.95 (significant). A similar plot with specific gravity instead yields scattered data points.

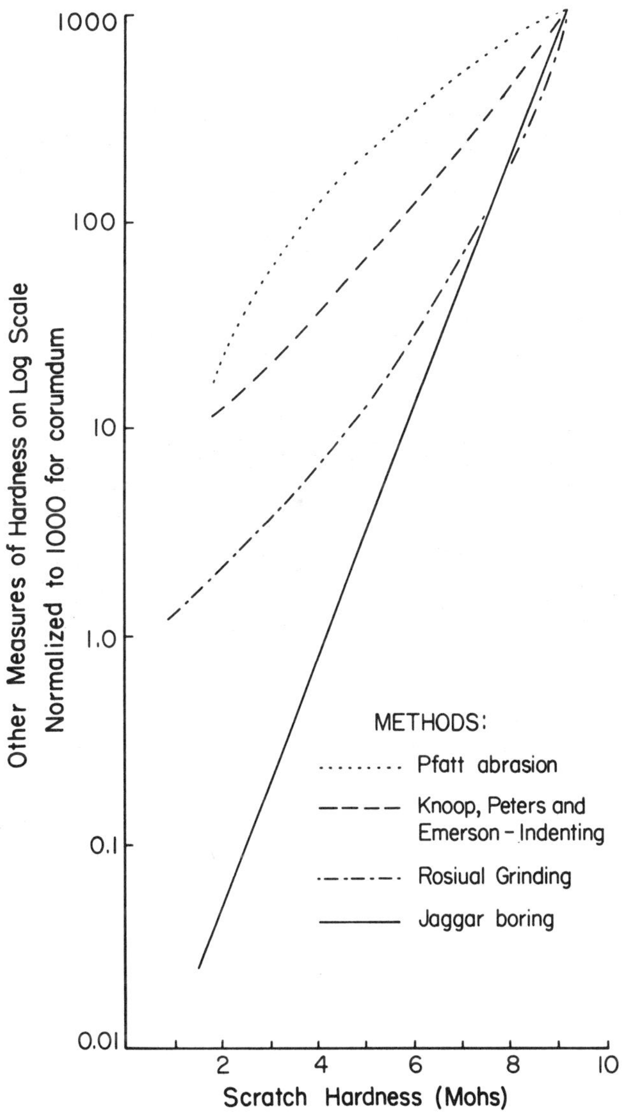

Fig. 2. Comparison of various hardness scales to the scratch hardness of Mohs.

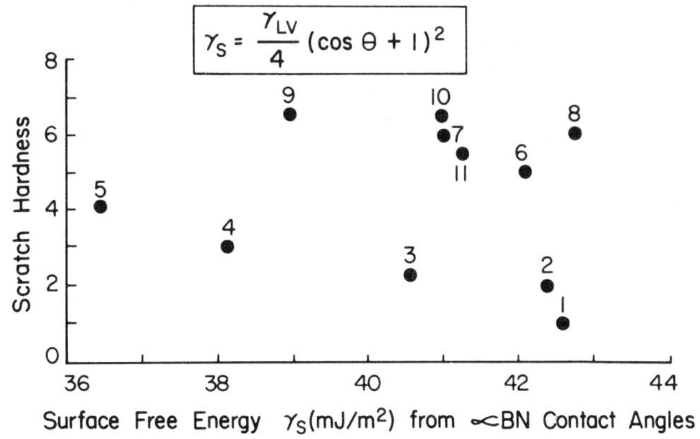

Fig. 3. Correlation between hardness, cleavage, and surface free energy. Labels on data points are from Table 3.

SUMMARY

Evaluation of mineral surfaces in terms of interfacial free energies is important when considering displacement efficiency for enhanced oil recovery. The measurement of contact angles using α-bromonaphthalene is rapid and reproducible on mineral specimens. From such data the surface free energy can be determined. Values of the surface free energy of solid materials have been examined for any correlations for a series of clays and minerals. The surfaces proved to be low in surface free energy and correlate with surface hardness.

REFERENCES

1 GOGARTY, W. and H. SURKALO. 1972. Field test of micellar solution flooding. J. Pet. Tech. *24*:1161.

2 EHRLICH, R., H. HASIBA, and P. RAIMONDI. 1974. Alkaline water flooding from wettability alteration–evaluating a potential field application. J. Pet. Tech. *26*:1335–1343.

3 LANGNES, G., J. ROBERTSON, and G. CHILINGER. 1972. Secondary recovery and carbonate reservoirs. Elsevier, New York.

4 MELROSE, J. 1970. Interfacial phenomena as related to oil recovery mechanisms. Can. J. Chem. Eng. *48*:638.

5 TREIBER, L., D. ARCHER, and W. OWENS. 1972. Laboratory evaluation of the wettability of fifty-five oil producing reservoirs. Soc. Petr. Eng. J. *12*:531.

6 MORROW, N. 1976. Capillary pressure correlations for uniformly wetted porous media. J. Can. Petr. Tech. *15*:49–69.

7 ELLIOT, G., and A. RIDDIFORD. 1962. Dynamic contact angles and rates of adsorption. Nature. *195*:795.

8 PONTER, A., and P. THORNLEY. 1965. Measurements of contact angles in the presence of mass transfer. Nature. *207*:1288.

9 NEUMANN, A., and R. GOOD. 1979. Techniques of measuring contact angles. Surface and Colloid Science. *11*:31–91.

10 BECHER, P. 1977. Interaction parameter calculations from contact angle data. J. Colloid Interface Sci. *59*:429.

11 FOX, H., and W. ZISMAN. 1950. Spreading of liquids on low-energy surfaces. I. Polytetrafluoroethylene. J. Colloid Sci. *5*:514–531.

12 GOOD, R. 1979. Solids: contact angles and surface free energy. Surface and Colloid Sciences. *11*:1–29.

13 SEFFENS, W., 1981. A model treatment of solid/liquid interfacial energies for non-zero contact angle systems. Thesis for Univ. of Texas at El Paso.

14 EHRLICH, HASIBA, and RAIMONDI, op cit.

15 BECHER, op cit.

16 FOX and ZISMAN, op cit.

17 CORRENS, C. 1969. Introduction to mineralogy, 2nd ed., p. 113. Springer, New York.

6 THE EFFECT OF MICROORGANISMS ON THE PERMEABILITY AND POROSITY OF PETROLEUM RESERVOIR ROCK

David M. Updegraff

Colorado School of Mines

Anyone who has used an old-fashioned porous clay or unglazed porcelain filter to remove bacteria from liquids knows that microbial cells plug porous media. According to a 1938 review by Morton, the use of such filters dates back to work by Tiegel in 1871 who employed porous clay.[1] This was refined in 1884 to unglazed porcelain filters by Chamberland and the Berkefeld and Mandler filters made from kieselguhr (infusorial earth). The earlier workers believed that the filters removed bacteria by simple mechanical sieving; therefore, the pores of the filter were simply too small to allow the bacteria to pass through.

Later workers discovered that several other physicochemical factors were also involved, so filters with mean pore diameters from 8–15 times the length of rod-shaped bacterial cells would still filter such cells from a suspension. Some of the factors involved influence surface adsorption, particularly the surface charges on particles in the filter and the surface charges on the bacterial cells. These charges are influenced strongly by pH, ionic strength, and the presence of cations and anions in solution, particularly polyvalent ions. Also, surface-active materials in solution may adsorb onto the filter medium and change the ability of the filters to remove particulate matter. For example, it is known that viruses pass through filter candles more readily when suspended in media containing surface-active organic compounds than in inorganic media.

The early workers clearly showed that when bacterial suspensions were injected into porous clay, kieselguhr, or porcelain bacteriological filters, the flow rate under conditions of constant pressure differential across the filter declined with the increased volume of fluid injected. In fact, the volume was not the significant factor; rather, the number of bacterial cells injected were significant. Observations indicated that plugging effects were due to bacterial cells occupying a larger and larger portion of the pore volume of the porous medium as the bacterial suspension was injected.

Thus, it was clear that both permeability, which is proportional to flow rate under these conditions, and porosity, which is the percentage of the bulk volume of the solid medium that is attributable to void space, are reduced by bacteria trapped in the pores of the material. Peragallo used microscopic methods to evaluate the pore diameters of bacteriological filters.[2] A piece of the ceramic filter material was ground very thin and cemented to a glass microscope slide. Examination under ordinary bright field illumination clearly revealed roughly spherical pores from 15–20 μm in diameter in effective bacteriological filters. This technique did not permit any estimate of the degree of tortuosity of the pores or the minimum pore entry diameters, which are much more important in determining filtration ability and plugging than are the maximum pore diameters. It was not until Ritter and Drake developed the mercury injection method of determining pore size distribution that this problem was resolved.[3] Indeed, this method opened up a new era in the understanding of the structure of petroleum reservoir rocks.

PORE SIZE DISTRIBUTION OF PETROLEUM RESERVOIR ROCKS

Ritter and Drake's method involves plotting the volume of mercury injected against the absolute pressure, as seen in Figure 1. An equation is then developed, relating the pore entry radius, r, to the volume of mercury injected, V_o–V, and the injection pressure, p:

$$d(r) = \frac{P}{r}\,\frac{d(V_o - V)}{dp}$$

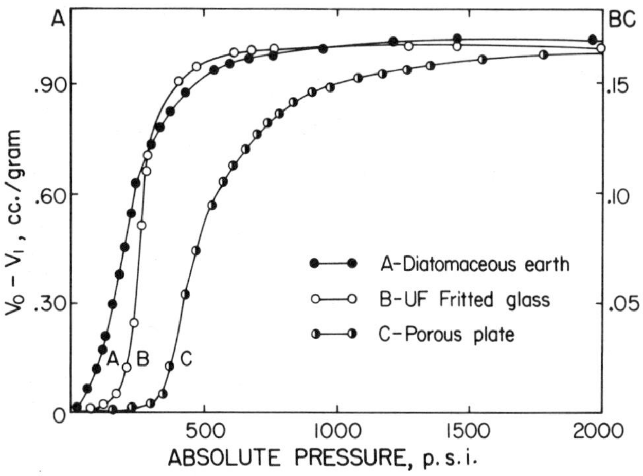

Fig. 1. Ritter and Drake's method of plotting the volume of mercury injected against the absolute pressure. *Adapted from a figure in "Pore Size Distribution in Porous Material,"* Ind. Eng. Chem., *17*:782–786

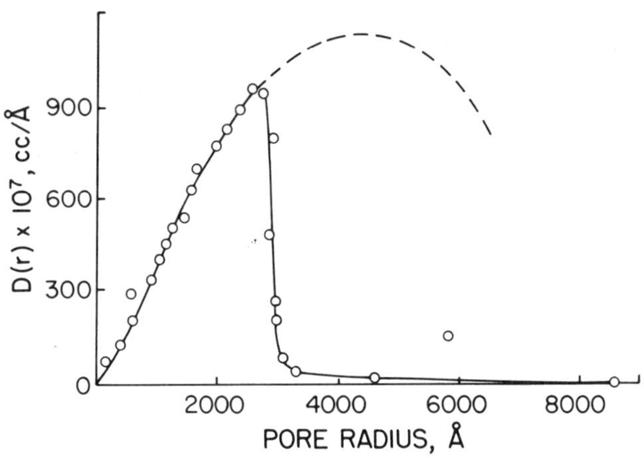

Fig. 3. Curve for fine porous ceramic material

Values of the derivative in the above equation were obtained by graphical differentiation and were employed to construct a pore size distribution function curve. The curves for diatomaceous earth, ultrafine fritted glass, and fine porous ceramic material (all employed as bacteriological filters) are given in Figures 2 and 3. Note that the fritted glass and the porous plate have a negligible amount of their pore volume with a radius of greater than 5,000 Å (0.5 μm). This explains why such materials are good bacteriological filters, since the minimum pore entry diameter of less than 1 μm is of the same order of size as most common bacteria. Thus, their efficacy as bacteriological filters can be ascribed to simple sieving action.

The determination of such pore size distribution curves for samples of petroleum reservoir rock is now routinely determined in oil company laboratories. Theoretically it should be possible to predict the degree of bacterial plugging of a given reservoir rock by bacterial cells of a given size from the pore size distribution curve. In actual practice, however, this is difficult, since the pore size distribution of reservoir rock is extremely irregular. Updegraff and Davis (1954) observed that pores must be at least twice the diameter of cocci or short bacilli in order for the cells to pass through the rock without serious plugging. Thus, an inspection of the pore size distribution curve will lead to the conclusion that injecting bacteria will cause serious plugging if more than 90 percent of the pore volume is attributable to pores less than 1 μm in diameter. Many reservoir rocks fall within this range; therefore, this is a major limitation on the use of in situ microbial enhanced oil recovery. Davis and Updegraff observed empirically that reservoir rocks containing an appreciable fraction of pores larger than 3 μm will pass large numbers of sulfate-reducing bacteria up to 0.6 μm in diameter and 3μm long without serious plugging.[4] Such reservoir rocks generally have a high permeability, greater than 100 md, provided that they also have a high porosity, greater than 15 percent.

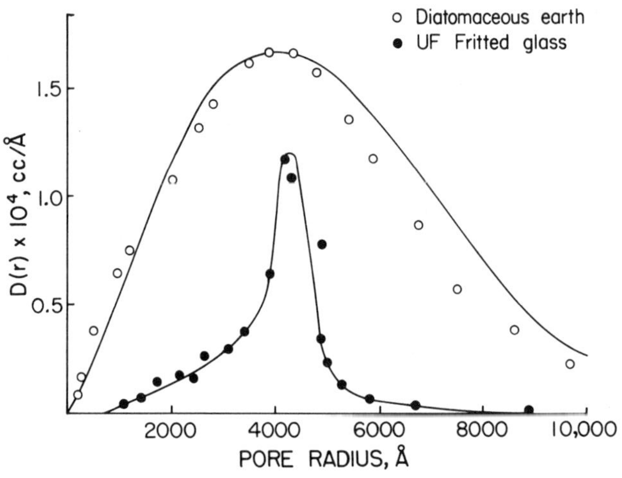

Fig. 2. Curves for diatomaceous earth and ultrafine fritted glass

MICROBIAL PLUGGING OF PETROLEUM RESERVOIR ROCK

As Torrey has observed, early in the history of the petroleum industry it was recognized that primary recovery methods alone — meaning naturally flowing or pumped oil wells — resulted in the recovery of only a minor part of the oil present in a petroleum reservoir.[5] The percentage of the oil originally present in the reservoir that is recovered by primary production varies widely, but 30 percent is an average value.

Beginning about 1888, air and gas injection were initiated to augment the reservoir pressure; later, about 1900, water injection was initiated in the Bradford field, Pennsylvania. The early history of water injection remains obscure, since the practice was illegal under state law until after 1907. In any event the practice of water flooding, in which water is injected into a series of wells in order to force oil up out of nearby recovery wells, resulted in recovering as much or more oil from the Bradford field and nearby oil fields in New York as had been recovered in the primary phase. Since these early efforts, the use of secondary recovery by injection of gases and water has increased tremendously in the United States.

Early workers were also probably aware of the plugging effect of biological slimes. The first experimental study of this biological clogging or plugging effect on petroleum reservoir rock appears to have been conducted by a group of petroleum engineers at the University of Texas at Austin. Merkt describes in-detail studies on microorganisms found in water and oil from the Luling and East Texas oil fields.[6] The effect of injecting such bacteria into cores of porous limestone and sandstone was examined. These and other findings were published by Plummer, Merkt, Power, Savin, and Tapp.[7] An elaborate radial permeability apparatus was used in these experiments. The cylindrical cores, 21½ inches in diameter and 3 inches long, were drilled down the center with a 0.5-inch hole and were mounted in brass core holders between rubber gaskets. The use of a cellulose glue and brass bolts holding the apparatus together prevented leakage. Water flowed either radially through the core from the center hole outward or alternatively from the outside in to the center hole. The pressure differential between the inside and the outside of the core was measured, using mercury manometers, as well as the flow rate. This permitted the permeabilities of the cores to be calculated from the following formula:

$$K = \frac{Q\,\mu\,\log_e R_e/R_w}{2\,\pi\,h\,\Delta P}$$

where:

K = permeability, darcies
Q = flow, cc/sec
R_e = radius of drainage area, cm
R_w = radius of the well, cm
h = thickness of the core, cm
ΔP = pressure change across core, atm
μ = viscosity, cp

Cores of oolitic limestone from the Walnut formation were employed in the first series of tests. The permeabilities were first measured using distilled water. Then a mixed culture of iron bacteria grown in Lieske's mineral salts medium inoculated with water from a local peat bog was injected. Microscopic examination revealed the presence of *Gallionella, Spirophyllum,* and *Leptothrix,* as well as considerable precipitated ferric hydroxide. The injection of 9–13 liters of this culture reduced the permeabilities of three test cores by 70–81 percent but did not plug any of them completely. The initial permeabilities of the cores ranged from 0.4–0.7 md. The injection of 10% hydrochloric acid into these cores, a cleanup procedure frequently used in water injection wells, restored the permeability to the original value or slightly greater.

Experiments were also carried out on waste water from the Luling oil field, which was shown by microscopic and cultural studies to contain algae, algal precipitates of calcium carbonate, diatoms, sulfur bacteria (*Thiobacillus thioparus, Thiobacillus thiooxidans,* and *Thiospirillum*), sulfate-reducing bacteria, hydrogen sulfide, and sulfur precipitates. Flowing 6–18 liters of this water through three Walnut limestone cores resulted in a decrease of 6 to 33 percent from the original permeabilities of 0.7–2.1 md. Upon sectioning the cores, the inside was found to be coated with a thick, black coating containing iron sulfide.

Cores of unconsolidated silica sand were also subjected to the same treatment. The injection of 5 gallons of distilled water decreased the permeability from 6,960 to 76.8 md. This was attributed to the silica effect, caused by the swelling of certain clays when contacted by distilled water. Salt water did not produce this effect. Injection of 5 gallons of the Luling water decreased permeability from 76.8 to 3.4 md.

A similar study was made with East Texas oil-field water before and after the cleanup procedure used in the East Texas water-flooding project. The water was a fairly strong brine, 65,000 mg/l total solids. The untreated water was a black, soup-like fluid containing much FeS and iron and sulfur bacteria, including *Leptothrix*, and sulfate-reducing bacteria. The treated water was perfectly clear. Both waters showed some plugging effect on Walnut limestone cores, but the untreated water had a much more severe effect, decreasing the permeability from 1.1 to 0.31 md. In a second experiment consolidated sandstone cores from the Rodessa field of East Texas and Louisiana were employed. Again the silica effect was evident, and the distilled water permeability dropped from 241 to 50 md. The untreated water caused a permeability drop from 50 to 1.9 md.

Beck studied the bacteria that were believed to be responsible for major plugging problems in the Bradford Allegheny field and in East Texas and Kansas fields as well.[8] These water-flooding projects showed the presence of *Desulfovibrio*, iron bacteria, and *Pseudomonas*. Beck attributed the plugging to both bacterial cells and products formed by bacterial action, particularly iron sulfide and iron hydroxide. Several germicides were evaluated for preventing the problem, and formaldehyde was recommended for inhibiting and controlling sulfate-reducing bacteria.

O'Bryan and Ling carried out another laboratory study on cores of Edwards limestone using the general methods described in the earlier work by Merkt.[9,10] A synthetic brine was prepared of a composition similar to the produced water from the Luling oil field. The cores were sterilized by heating to 200°F for 18 hours and then were dried by passing dry air through them for 12 hours. Permeability to dry air was then measured. *Desulfovibrio desulfuricans* was cultured in the synthetic brine to which 1 g/gal of sodium lactate was added as a nutrient. The population of bacteria increased to 4.5×10^6 cells/ml over a period of 60 days. The cores were saturated first with synthetic Luling brine, then with oil. Finally 100 ml of the *Desulfovibrio* culture were flowed through one set of four cores, while 100 ml of the synthetic brine were flowed through four control cores. The cores were stored for approximately four months under these conditions and were then cleaned by extraction with solvent to remove oil and with boiling distilled water to remove precipitated salts. The cores were dried, and permeability to dry air was measured again. All of the cores that were treated with brine alone showed no decrease or a slight increase in permeability, while those that contained the bacterial culture showed slight decreases in permeability, from 1.5 to 17 percent. Initial permeabilities of the cores ranged from 1.6-6.3 md. The authors attributed the decline in permeability to the precipitation of iron sulfide within the pores. Another possibility was that the permeability decrease may have been caused by plugging of the pores by cells of the bacteria. This was not considered.

Fekete carried out a study on the plugging of Berea sandstone cores by bacteria, later published by Hart, Fekete, and Flock.[11,12] The cores, 2.5 cm in diameter and 7–9 cm long, were mounted in plastic and three taps were drilled through the plastic wall to the surface of the core at equal spacings in order to monitor pressure. *Bacillus subtilis* (1×4 μm) was grown in a nutrient medium, killed by autoclaving and suspended in distilled water. The clean, dry core was evacuated and saturated with distilled water, and permeability was measured. Then the bacterial suspension was injected under a constant pressure of 40 psi, and permeability was monitored by flow-rate measurements. The ratio of permeability to initial permeability was plotted against the volume of bacterial suspension injected. Some bacteria penetrated entirely through each core, but most of the plugging was near the injection end. Permeability was reduced to 80 percent of the initial value by the injection of 45 pore volumes of a suspension containing 0.1×10^6 cells/ml, while the same volume of suspension containing 1.0×10^6 or 10×10^6 cells/ml reduced the permeability to less than 10 percent of the original value.

Crawford pointed out that the pore size distribution curves of petroleum reservoir rocks often show that a major fraction of the pore volume may be made up of pores of the same size range as bacterial cells.[13] Under such conditions the injection of water containing bacteria is bound to result in a serious degree of plugging. Tests showed that the more permeable zones had a much greater percentage reduction in permeability than the less permeable zones. Thus, partial bacterial plugging may result in a beneficial effect, namely a reduction of permeability stratification. This type of plugging, referred to as selective plugging, reduces fingering in a water flood and results in a higher percentage of the petroleum-bearing formation being swept by the water, hence higher oil recovery. It is possible to plug most sandstone cores to a substantial extent by bacteria, but Updegraff and Wren pointed out that very high permeability rocks with a high percentage of large pores will pass enormous numbers of sulfate-occurring bacteria without serious plugging.[14]

Kalish, Stewart, Rogers, and Bennett carried out a careful study of the effect of bacteria on Berea sandstone permeability.[15] The study included cores of high (278–400 md), medium (130–162 md), and low (17.7–48.3 md) permeability, and bacteria of different sizes and shapes (*Pseudomonas aeruginosa*, 0.5×1.5 μm; *Proteus vulgaris*, $0.5 \times 1 \times 1$ to 3 μm; *Micrococcus roseus*, spheres 0.8 μm; and *Bacillus cereus*, 1×6 μm). Cores were 2.54 cm in diameter and 4.08 or 8.16 cm long. They were mounted in epoxy resin; then three pressure-transmitting channels were drilled through the plastic into the core to permit monitoring permeability at different distances from the injection end of the core by measuring pressure at constant injection rates. This practice permitted a continuous monitoring of permeability vs the volume of bacterial suspension injected. Bacteria were grown on agar media, rinsed with calcium chloride brine containing 25,000 mg/l of calcium chloride, centrifuged down, and washed and suspended in fresh calcium chloride brine. The bacteria were killed with merthiolate, and suspensions of known bacterial concentration were injected while permeability was monitored. Some runs were also carried out with viable bacteria.

In every case bacteria decreased the permeability of the cores . Larger bacteria and those forming aggregates of cells brought about much more severe plugging. The adverse plugging effect was due to a skin of plugged rock not more than a few inches from the injection face. Many of the bacteria filtered out on the inlet face of the core, but in all cases a few bacteria penetrated all of the way through the core samples. Bacteria penetrated high-permeability cores more readily than low-permeability cores. Microscopic examination of sections of cores containing bacteria stained with crystal violet showed that penetration was often very irregular, with movement from two or three times as far in one part of the core as in other sections. The plugging effect could be alleviated partially by increasing the injection pressure, supposedly as a result of moving the bacteria lodged in fine capillaries or shearing the cells into smaller pieces. The injection of hydrochloric acid was highly effective in removing bacterial plugging and restoring the original high permeability. The pore size distribution curves of these sandstone cores showed that the greatest percentage of the pore volume included the very small pores up to 0.5 μm in diameter in every case. The median pore radii were 5.5–6.0 μm in the high-permeability cores, 4.5–5.0 μm for the medium-permeability cores, and 3.4–4.0 μm for the low-permeability cores. Thus, in all cases there was an ample number of pores present that were large enough to permit the passage of even the largest bacteria used.

Raleigh and Flock carried out a study on formation plugging by killed cells of *Bacillus subtilis* (1 $\times$ 4 μm) using four distinct types of rock.[16] The methods employed were similar to those described earlier by Kalish et al.[17] Bacteria were suspended in 25 percent sodium chloride brine (to eliminate clay swelling caused by fresh water and to retard leaching in the limestone cores) at 3 $\times$ 10^7 cells/ml. The rock types employed were Berea sandstone (initial permeability, 139 md), Cardium sandstone (initial permeability, 44.1 md), Indiana limestone (initial permeability, 5.7 md), and Devonian limestone (initial permeability, 21.4 md). In every case permeability was reduced drastically by bacterial injection, just as in the earlier study by Kalish et al., although in no case was the plugging complete. Neither permeability nor irreducible water saturation correlated with the rate of plugging. The authors attempted to correlate permeability reduction with the pore geometry factor deduced from the shape of the capillary pressure curve using the equation:

$$\frac{(V_b)\,P_c}{(V_b)\,P_\infty} = e^{G/\log P_c/P_d}$$

where:

$(V_b)\,P_c$ = fractional bulk volume occupied by mercury at pressure P_c

$(V_b)\,P_\infty$ = fractional bulk volume occupied at infinite pressure or total interconnected pore volume

P_c = mercury/air capillary pressure, psia

P_d = mercury/air extrapolated displacement pressure, psia

G = pore geometry factor

The pore geometry factor did not correlate with the rate of plugging, nor was it possible to correlate it with any combination of fundamental rock properties. This is probably attributable to the extremely nonhomogeneous nature of petroleum reservoir rock.

Allred reviewed work on the control of microbial corrosion and plugging of oil reservoirs during water flooding.[18] A water-flooding project in California, where the flooding medium was sea water containing 20–150 million bacteria per milliliter, reduced the permeability of the cores by about 80 percent. The same water treated with mercuric chloride showed no reduction in permeability. This simple observation indicated that the injected bacteria did not cause the plugging, since the mercuric chloride did not remove the bacteria, but simply prevented them from multiplying within the pores of the rock after they were injected. The plugging was attributed to multiplication of the bacteria within the porous reservoir. This particular reservoir had a pore size distribution of 1–12 μm of minimum pore entry diameter. It was emphasized that severe plugging could be brought about by iron bacteria, slime-forming bacteria, algae, and sulfate-reducing bacteria.

Sulfate reducers are particularly troublesome because they corrode the inside surfaces of iron and steel pipes, tanks, pumps, and other oil-field equipment. Anaerobic corrosion causes great direct economic damage in the petroleum industry.[19] It also causes indirect problems by contributing to the plugging of petroleum reservoir rock during water flooding. Iron goes into solution at the anodic areas of the metal surface; electrons then flow to cathodic areas where H^+ is reduced to H_2. The bacteria, *Desulfovibrio* species, can use H_2 as an energy source and sulfate ion as an electron acceptor. They therefore grow in water in the pipes, reducing sulfate to H_2S, which then reacts with ferrous ions to give a black precipitate to FeS. Thus, the plugging effect is due to both bacterial cells and precipitated iron sulfide. Germicides and antiseptics are frequently added to the water in water-flooding projects to inhibit this process.

Sharpley reviewed the relationship between the numbers and kinds of bacteria in injection water used for water flooding and the degree of plugging of injection wells.[20] Sulfate-reducing bacteria, iron bacteria, and slime-forming bacteria were described as the worst offenders. A count of greater than 10,000 per milliliter of any kind of bacteria is regarded as indicating a potential for plugging by the American Petroleum Institute. Sharpley cited the finding that 30,000 bacteria/ml gave serious plugging of a reservoir rock with a permeability of 20 md.

PENETRATION OF BACTERIA THROUGH PETROLEUM RESERVOIR ROCK

Myers and McCready studied the rate of penetration of living *Serratia marcescens* into cores of two types of sandstone and two types of limestone.[21] The bacteria were grown in Bacto nutrient broth containing 1 percent glucose and 10 microcuries of ^{32}p. The cells were centrifuged down, washed, and resuspended in sterile distilled water. The core samples were cylindrical in shape, 4.08 cm in diameter, 17.8–35.6 cm long, and were cut parallel to the bedding planes of the sandstone. They were mounted in epoxy resin and dried in vacuo, and the lower end of the core was immersed in the suspension of radioactive living bacteria containing 3.7×10^9 cells/ml. No pressure was applied to the liquid, except for a few inches of water, but the cores rapidly imbibed the liquid by capillary action. The rate of motion slowed down markedly with time, and 61–104 ml of suspension was taken into the core. The upper surface of each core was swabbed every 4 hours, and the recovered liquid was tested for radioactivity with a Geiger counter and for living bacteria by nutrient agar plate counts. Results of both methods agreed and demonstrated that the bacteria passed through the cores in 24–48 hours. The cores were then split longitudinally and were evaluated by radioautography and by coating the split half core with nutrient agar. In all cases the water penetrated well ahead of the bacteria and some plugging undoubtedly occurred, although no permeability measurements were made to confirm this. Bacteria passed completely through all of the cores — even Late Mesozoic sandstone with a permeability of less than 0.1 md and Mississippian limestone with a permeability of less than 0.1 md.

Myers and Samiroden employed the same method to study the penetration of *Serratia marcescens* into 35 cores, some partially saturated with oil.[22] Limestone from the Pekisko formation and the Beaver Lake formation as well as sandstone from the Cardium, Viking, and Belly River formations were employed. Permeabilities varied widely from less than 0.1 to 306 md. In all except the Pekisko and Cardium cores, bacteria penetrated 6–76 mm of core material in 48 hours or less. Results with Pekisko and Cardium cores were unusual in that bacteria penetrated well into some of the cores but not at all into others. The rate and extent of penetration showed no correlation with permeability or porosity of the cores or whether the cores contained crude oil.

Clark has reinvestigated the question of transport of living bacterial cells through cores of Berea sandstone.[23] Cells were suspended in 2 percent sodium chloride plus 0.01 percent calcium chloride and were injected into cylindrical core samples 5 cm in diameter and 19 cm long with permeabilities of 200–400 md. Surprisingly, some bacteria penetrated the cores while others did not. The size of the bacterial cells did not appear to be a factor, but ionic concentrations did influence the results.

SUMMARY

Since the earliest application of water flooding for the secondary recovery of oil, scientists have appreciated that microorganisms can plug water injection wells. The degree of plugging varies widely, depending upon the numbers and kinds of bacteria present in the water and upon the permeability, porosity, and pore size distribution of the reservoir rock.

In general, large numbers of bacteria create more serious plugging than smaller numbers, and large-sized microbial cells plug worse than small ones. Also, products produced by bacterial action, particularly ferric hydroxide and ferrous sulfide, create plugging under certain conditions. Bacteria penetrate much more readily through cores of high permeability than those of low permeability, yet the percentage decrease in permeability caused by the bacteria is greater in cores of high permeability. This could have a beneficial effect on oil recovery by water flooding because high-permeability zones could be plugged by bacteria, resulting in less permeability stratification (and hence less fingering of water through the formation) and higher sweep efficiency.

Permeability and porosity are related to pore size distribution, and this parameter can be determined by injecting mercury into a clean core sample by high pressure. Pore size distribution curves of petroleum reservoir rock, in which the minimum pore entry diameter is plotted against the volume of pores of that size, show that most petroleum reservoir rocks contain a large fraction of their pore volume in the same size range as most common bacteria, approximately 1 μm in diameter. Thus plugging can be explained as a phenomenon of simple mechanical sieving in which the bacteria become lodged in the narrowest portion of the pore simply because they are too large to pass through. Injection wells rarely plug completely, and it is possible to overcome the problem to some extent by raising the injection pressure or by treating the well with hydrochloric acid.

Petroleum reservoir rocks vary so widely in composition, porosity, permeability, and pore size distribution that it is impossible to make quantitative statements about the degree to which bacteria penetrate the rock or the degree of bacterial plugging produced. Some important qualitative observations can be made, however. Reservoir rocks with an appreciable fraction of their pore volume composed of pores with a minimum pore entry diameter of 3 μm or more will permit the passage of large numbers of sulfate-reducing bacteria 0.6 by 3 μm in size without a serious degree of plugging. Such rocks are usually of high permeability, greater than 100 md.

Large microbes such as algae, yeasts, and fungi can plug formations of very high permeability, but such organisms are relatively easy to eliminate from water used for flooding. Bacteria are the worst offenders, and iron bacteria that accumulate ferric hydroxide, slime-forming bacteria, and sulfate-reducing bacteria are particularly objectionable. Sulfate-reducing bacteria cause anaerobic corrosion of iron and steel, utilize the hydrogen produced by the corrosion reaction as a source of energy, and reduce sulfate to hydrogen sulfide. The sulfide then reacts with ferrous iron produced in the corrosion and forms a precipitate of iron sulfide, which may be a worse plugging agent than the bacterial cells. Motile anaerobic bacteria such as sulfate-reducing bacteria, when supplied with suitable nutrients, can penetrate through a core of high-permeability rock in which there is no fluid flow at rates of up to one inch per day. When water containing bacteria is forced through a core of reservoir rock, the linear rate of water movement is ordinarily greater than the rate of movement of the bacteria; accordingly, the concentration of bacteria is much greater at the injection face than further into the core. Nevertheless, bacteria have been shown to penetrate through cores of both sandstone and limestone with permeabilities ranging from 0.1 md to thousands of millidarcies. Curiously, some bacteria penetrate through cores, while others of the same size do not. Inorganic ions and other solutes affect this process, probably by influencing adsorption processes. The failure of bacteria to penetrate readily through tight formations of low permeability and the consequent plugging represent important limiting factors in the application of microbes in enhanced oil recovery.

Davis concluded, on the basis of a literature review, that bacterially enhanced oil recovery should only be considered for reservoirs with a permeability of greater than 100 md, while Bubela puts this figure at 150 md.[24,25] Estimates of this type are very difficult to make because of the wide variation in size and surface properties of different kinds of bacteria, as well as the wide variation in properties of different petroleum reservoirs. The safest way to proceed would be to conduct adequate tests with cores from the reservoir in question and with the particular bacterial species of choice to ensure that the organisms penetrated the formation without serious plugging.

Another important mechanism of plugging is the multiplication of bacteria within the petroleum reservoir. Unfortunately, very little is known of this process. The process requires the presence of an energy source for the bacteria plus all of the essential nutrient elements. The extent of multiplication generally depends on the concentration of the energy source; and in the presence of a high concentration of energy source and nutrients, bacterial multiplication can lead to severe plugging of low-permeability formations.

REFERENCES

1 MORTON, H.E. 1938. Bacterial filters and filtration techniques. Am. J. Clin. Path., Tech. Suppl. *2*:185–205.

2 PERAGALLO, I. 1937. Recherches experimentales sur les bougies filtrantes. Ann. Inst. Pasteur *58*:48–57.

3 RITTER, H.L., and L.C. DRAKE. 1945. Pore size distribution in porous material. Ind. Eng. Chem., Anal. Ed. *17*:782–786.

4 DAVIS, J.B., and D.M. UPDEGRAFF. 1954. Microbiology in the petroleum industry. Bacteriol. Rev. *18*:215–238.

5 TORREY, P.D. 1950. A review of secondary recovery in the United States, p. 3–29. *In* American Petroleum Institute, Secondary Recovery of Oil in the United States, 2nd ed. American Petroleum Institute, N.Y.

6 MERKT, JR., E.E. 1943. The effect of bacteria on the permeability of oil reservoir rocks. M.S. Thesis, University of Texas, Austin, Texas.

7 PLUMMER, F.B., E.E. MERKT, JR., H.H. POWER, H.J. SAVIN, and P. TAPP. 1944. Effect of certain microorganisms on the injection of water into sand. Petroleum Technol. Pubs., AIME No. *1678*:1–13.

8 BECK, J.V. 1947. Prevention of microbiological growths in water flood operations. Producers Monthly Oct. 1947: 21–25.

9 O'BRYAN, O.D., and T.D. LING. 1949. The effect of the bacteria *Vibrio desulfuricans* on the permeability of limestone cores. Texas Journal of Science *1(3)*:117–128.

10 MERKT, op cit.

11 FEKETE, T. 1959. The plugging of bacteria in sandstone systems. M.S. Thesis, University of Alberta, Edmonton, Alberta, Canada.

12 HART, R.T., T. FEKETE, and D.L. FLOCK. 1960. The plugging effect of bacteria in sandstone systems. Can. Inst. Min. Trans. *63*:318–324.

13 CRAWFORD, P.B. 1962. Continual changes observed in bacterial stratification rectification. Producers Monthly. Feb. 1962:12.

14 UPDEGRAFF, D.M., and G.B. WREN. 1954. The release of oil from petroleum-bearing materials by sulfate-reducing bacteria. Appl. Microbiol. *2*:309–322.

15 KALISH, P.J., J.A. STEWART, W.F. ROGERS, and E.O. BENNETT. 1964. The effect of bacteria on sandstone permeability. J. Pet. Technol. *16(7)*:805–814.

16 RALEIGH, J.T., and P.L. FLOCK. 1965. A study of formation plugging with bacteria. J. Pet. Technol. *17(2)*:201–206.

17 KALISH et al., op cit.

18 ALLRED, R.C. 1976. Microbial problems in secondary oil recovery. *In* The role of microorganisms in the recovery of oil. NSF/RA-770201. 1976 Engineering Foundation Conference.

19 UPDEGRAFF, D.M. 1955. Microbial corrosion of iron and steel. Corrosion *11(10)*:44–48.

20 SHARPLEY, J.M. 1961. Bacteria in flood water: what are they–what they mean. Petroleum Engr. *33(2)*:B-55–B-67.

21 MYERS, G.E., and R.G.L. McCREADY. 1965. Bacteria can penetrate rock. Can. J. Microbiol. *12*:477–484.

22 MYERS, G.E., and W.D. SAMIRODEN. 1967. Bacterial penetration in petroliferous rocks. Producers Monthly *31(4)*:22–25.

23 CLARK, J.B. 1981. The use of microorganisms in enhanced oil recovery processes. DOE Contract No. DE-AC-19-80BC10300. Quarterly Report. U.S. Department of Energy, Washington, D.C.

24 DAVIS, J.B. 1967. Petroleum microbiology. Elsevier, New York.

25 BUBELA, B. 1978. Role of geomicrobiology in enhanced recovery of oil: status quo. APEA Journal *1978*:161–165.

ADDITIONAL REFERENCES

MELROSE, J.C. 1965. Wettability as related to capillary action in porous media. Soc. Petrol. Eng. J. *5*:259–271.

RALEIGH, J.T. 1962. The effect of rock properties on bacteria plugging in reservoir rocks. M.S. Thesis, University of Alberta, Edmonton, Alberta, Canada.

7 PRELIMINARY INVESTIGATION ON THE TRANSPORT PROBLEMS OF MICROORGANISMS IN POROUS MEDIA

Long-Kuan Jang, John E. Findley, and Teh Fu Yen

University of Southern California

The transport problems of microorganisms in porous media are encountered in several engineering practices. Of current interest is an MEOR (microbial enhanced oil recovery) process to recover heavy oil. This method was successfully demonstrated in Eastern Europe by purposely injecting candidate microbial species and nutrient medium into some selected shallow reservoirs.[1,2] It is possible through microbial action that candidate microbes might be used on existing fields to recover crude oil that is too costly to recover by conventional means.

In MEOR processes, the active species, i.e., the candidate microbes, are introduced into the reservoir with the flooding medium. In such a process fundamental factors must be examined: the level of microbial cell penetration in the reservoir, the rate of migration, cell aggregation and accumulation, and cell multiplication.

The field work completed in Eastern Europe indicated that positive results are obtained only with deposits with high porosity and high permeability, i.e., over 100 md.[2,3] Although not explained by the original authors, it might be imagined that the spreading of cells and the subsequent effects of microbial action are retarded to various extents by the bacterial retention when the fluid containing them is injected into the formation rock. It is also generally true that cells and/or cellular aggregates, if any, are more easily retained by formation rocks with lower permeability and lower porosity.

Only a few qualitative or semiquantitative studies on problems regarding the migration of bacteria in rock have been published.[4,5] This paper concentrates on a quantitative study of MEOR applications using formation rocks with higher permeability and porosity. In this case, the bacterial cells may be retained mainly by the rock grain surface. Other retention mechanisms, such as straining and plugging in smaller pores, might occur to a lesser extent than in situations described as a tight formation. In the mechanism of surface retention, the bacterial cells approach the rock surface in the fluid stream by laminar flow, by their own mobility, or through Brownian motion. The subsequent adsorption on the rock surface then occurs.

Generally speaking, the rate of bacterial spreading in a highly permeable formation is mainly affected by the following factors:

1. Porous media, including the porous structure, mineral composition, and the wettability of rock minerals
2. Hydrodynamics, including the injection rate and the fluid flow regime in the porous passages
3. Surface physicochemical and microbiological properties, including the number of available sites for retention on the rock surface, the interactive forces between cells and rock grain surface, the effect of medium composition, and the specific attachment due to cellular wall structures such as extracellular slimes, flagella, and pili
4. Other minor or secondary effects such as the straining and plugging of the smaller pores, the aggregation of bacterial cells, and the formation of cellular dendrites in the pores of the porous media

In the preliminary study, negatively charged polymeric microspheres were used to simulate bacterial cells under nongrowth conditions. The microsphere-containing fluid was injected into a Berea sandstone core. Analysis of the effluent samples indicated a strong interaction between microspheres and Berea sandstone. Realizing the influence of retention on microsphere transport in the porous media, experiments were carried out to characterize some of the surface properties of microspheres and sandstone rock. The electrophoretic mobility of microspheres was measured. The mineral type and overall elemental composition were analyzed using X-ray diffraction and EDAX (energy dispersive analysis of X-ray) techniques. The influence of various additives to the aqueous solution on the adsorption of microspheres onto silica sand is also presented.

NEGATIVELY CHARGED MICROSPHERES

Bacterial cells are known to adsorb onto anion exchange resins. They can also be eluted by varying the ionic strength and/or pH values of the influent aqueous solution. In this sense bacterial cells can be viewed as macroanions bearing a net negative charge on the surface under neutral pH values. A hypothetical cell is represented in Figure 1.

From the viewpoint of surface charge characteristics, negatively charged microspheres were adopted to simulate

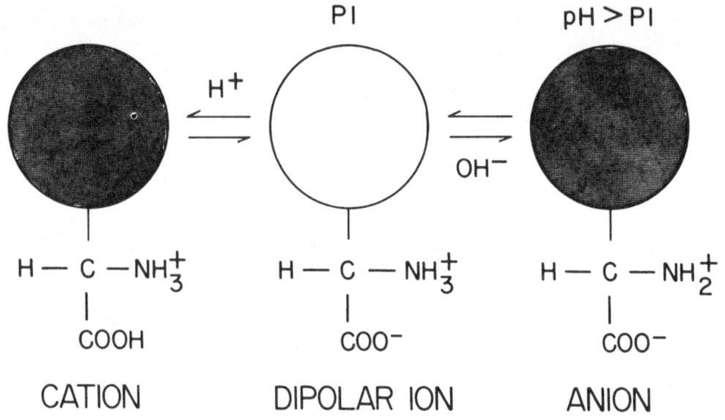

Fig. 1. Dipolar character of hypothetical microbial cell

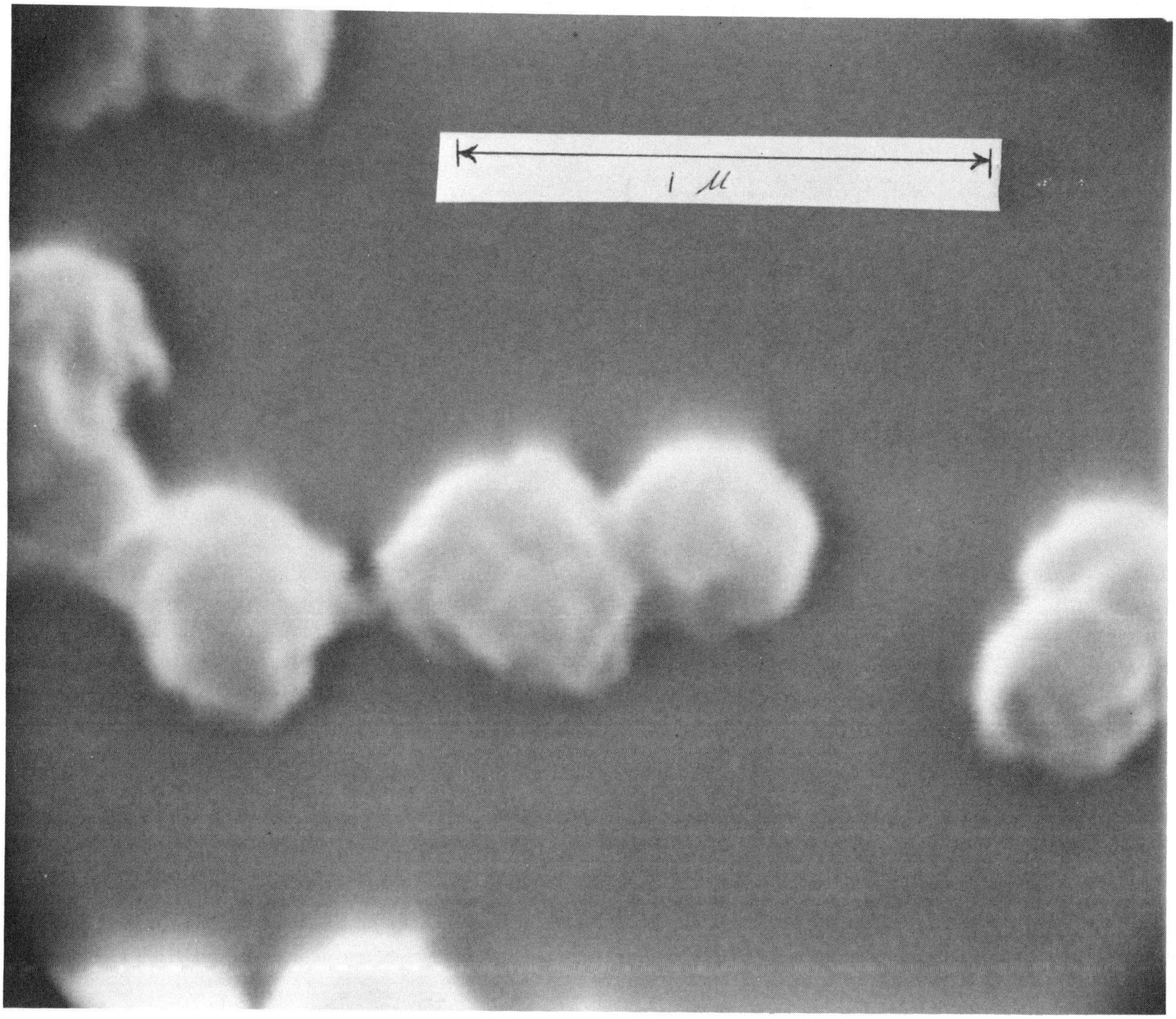

Fig. 2. SEM picture for charged microsphere particles (100K)

bacterial cells in the preliminary study. Preparation was by emulsion copolymerization of 2-hydroxyethyl methacrylate, methacrylic acid, bis-acrylamide, allylamine, and polyethylene oxide according to the procedure of S.P.S. Yen et al.[6] Since polymerization takes place in aqueous solution, a high concentration of hydrophilic hydroxyl and carboxyl groups is present on the surface of the microspheres. At neutral and alkaline pH, the carboxyl groups ionize, imparting a net negative charge to the spheres. The SEM (scanning electron micrograph) of microspheres is shown in Figure 2. Other reasons for use of microspheres are as follows:

1. The sphere can be made with variable density, due to the aqueous solubility nature of the monomer. In this case, the specific gravity of the spheres is equivalent to cells.
2. The controlling factor of polymerization is from the coalescence of self-nucleated particles with water solubility. Size can be controlled.
3. Fluorescent dyes and other labeling agents can be easily bonded to the microsphere.

A core flooding experiment was carried out by injecting a microsphere suspension at 2×10^7 microspheres/ml distilled water into a Berea sandstone core (6 in. long × 1 in. diameter) with porosity 21% and permeability 400 md. The apparatus is shown in Figure 3. The pressure drop was kept at 5 psia throughout the experiment. The accumulated volume versus time is plotted in Figure 4. The visible light absorbance of effluent samples indicated that within the sensitivity limit of the spectrophotometer, negligible amounts of microspheres were detected. Meanwhile, the curve in Figure 4 shows that the retained microspheres accumulated in the pores to such an extent that the intrinsic properties of the porous media were changed. Therefore, the porosity, the permeability, and the flow rate decreased gradually.

It was felt that for highly permeable rock such as Berea sandstone the strong retention of microspheres might arise from strong interactions between microspheres and porous media. Some of the important surface properties were investigated and the results are presented in subsequent sections. The role of other factors, such as porous structure, hydrodynamics, and various retention mechanisms, remains to be assessed.

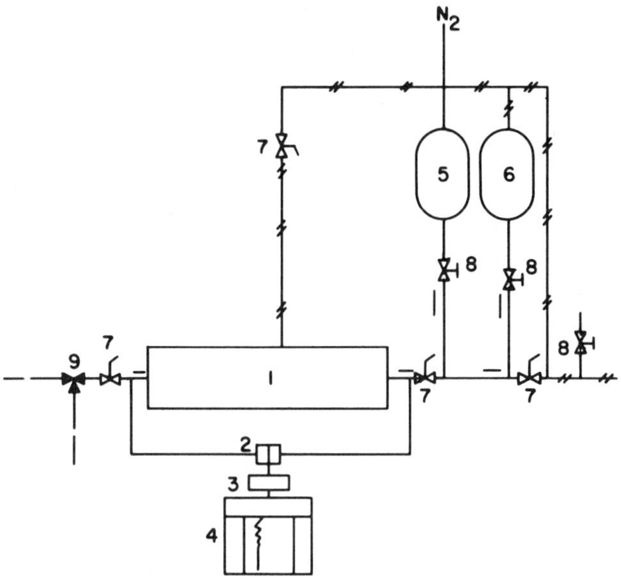

Fig. 3. Auxiliary system designed for the purposes of saturating the core with fluids for measuring the air permeability of the core

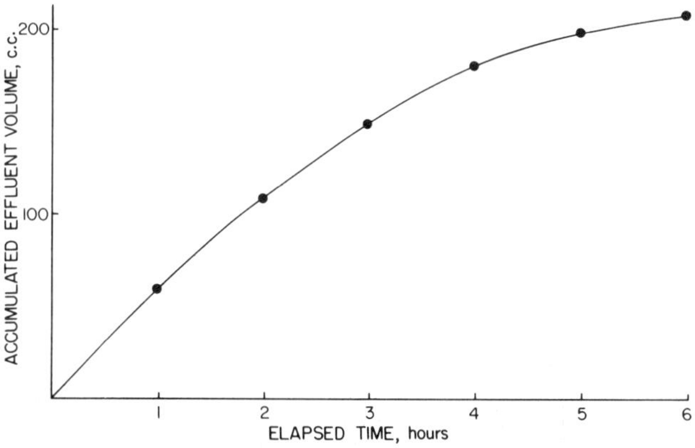

Fig. 4. Accumulated effluent volume vs. elapsed time

SURFACE CHARGE CHARACTERISTICS OF MICROSPHERES

Microspheres were known to possess net negative charges on the surface under neutral and alkaline pH. The surface zeta potential was evaluated with an electrophoretic apparatus according to the equation

$$\zeta \cong \frac{\eta}{\epsilon_r \epsilon_o} \mu \tag{1}$$

where μ is the measured mobility in m/sec per volt/m; ϵ_r is the relative permissivity of aqueous solution, which is around 78.5 at room temperature; ϵ_o is the permittivity of vacuum, having a value of $8.854 \times 10^{-12} J^{-1} C^2 m^{-1}$; and η is the viscosity of aqueous solution. The results are listed in Table 1. Average value of 10 measurements with phosphate buffer at pH 7.8 was taken as the mobility of each species.

The adsorption of negatively charged microspheres onto sand grains might be affected by the ionic environment. Five grams of silica sand and various additives were added to each of the 50-ml aqueous suspensions containing 10^8 microspheres/ml distilled water. The mixture was shaken gently for a period of one hour. The equilibrium concentration of microspheres was determined by measuring the

Table 1. Electrophoretic mobility of bacteria species and microspheres.

Species	Mobility* (μ)	Zeta potential (mV)
Clostridium propionicum	53	− 8.5
Clostridium propionicum spores	74	−12
Clostridium acetobutyricum	48	− 8
Clostridium acetobutyricum spores	67	−11
Negatively charged microspheres	300	−48

*Mobility is expressed in microns per 10 seconds in a field of 10 volts per cm.

turbidity with a visible light spectrophotometer. The original solution (10^8/ml) had an OD value of 0.254 at the wavelength of 500 nm. The results of adsorption experiments are summarized in Table 2. The significance of the table is as follows:

1. The divalent cations increase the adsorption of microspheres onto sand grains to a greater extent than do the monovalent ions, even when the former are at a much lower concentration (runs 2 and 4). Divalent ions such as calcium may serve as bridging agents between negatively charged sites on microsphere surfaces and on silica sand surfaces.
2. The chelating agent $(Na)_4$ EDTA may chelate divalent cations. As a result, the extent to which microspheres adsorb is decreased (runs 3 and 5). Adding $(Na)_4$ EDTA into the solution containing monovalent cations simply increases the ionic strength in the solution. An enhanced adsorption was observed (runs 2 and 4).
3. Adding Tween 80 [polyoxyethylene (20) sorbitan monooleate], a nonionic detergent, into the solution seemed to prohibit adsorption to a certain extent (runs 6 and 7).

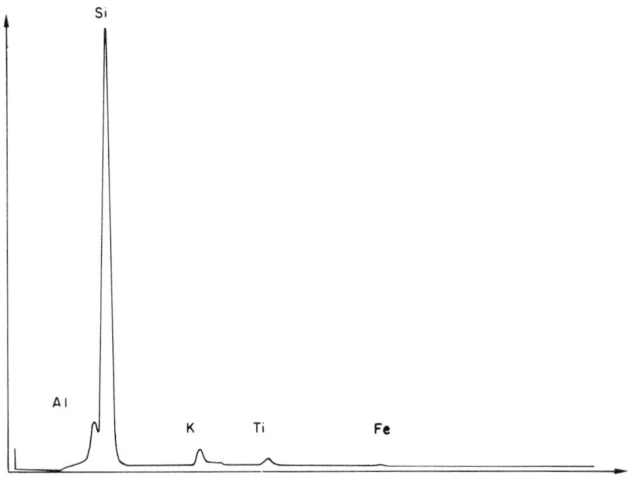

Fig. 5. Energy dispersive X-ray analysis for unbaked Berea sandstone

CHEMICAL CHARACTERIZATION OF UNBAKED BEREA SANDSTONE

The overall elemental composition of Berea sandstone was determined by EDAX. The major elements are Si, Al, K, and Fe (Figure 5). Mineral types were determined by X-ray diffraction (Figure 6). The main mineral types are quartz, cristobolite, orthoclase, and pyrite. The clay is potassium type. A small amount of calcium was detected.

With the mineral composition known, some of the parameters characterizing surface properties in the aqueous environment can be determined experimentally or can be estimated from the already-published data. For example, the Hamaker constants that account for the van der Waals forces between many-molecule bodies were evaluated for several systems.[7] The average surface charge density of the minerals in the sandstone at various pH and ionic strengths remains to be determined.

Table 2. The effect of chemical additives on percent microsphere adsorption onto silica sand.*

Run No.		1	2	3	4	5	6	7 (control)
Additives	NaCl		10,000		10,000			
	$CaCl_2$			2,000		2,000		
	$(Na)_4$ EDTA	2,000	2,000	2,000				
	Tween 80						20,000	
% Adsorption		50%	83%	75%	70%	78%	19%	32%

*The values in this table are the concentrations of additive chemicals in ppm.

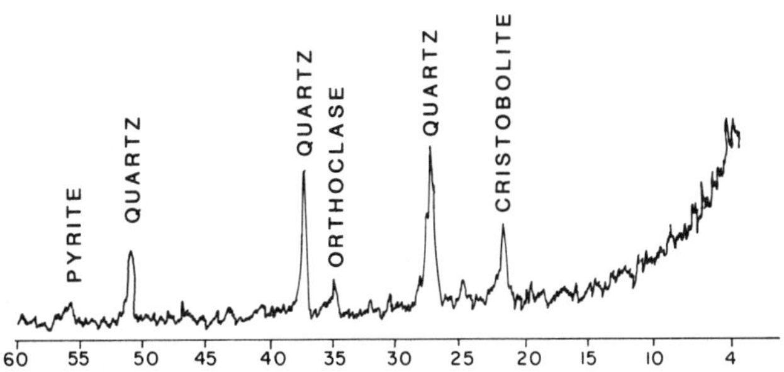

Fig. 6. X-ray diffraction analysis for unbaked Berea sandstone

Along with these factors, a more detailed characterization of porous structure should be carried out. For example, the grain size and pore size distribution, the role of clays in bacterial retention, and the possible changes in the intrinsic porous structure due to the bacterial retention should be studied.

CONCLUSIONS

In the model system comprising negatively charged microsphere suspension and silica sand, divalent ions enhanced adsorption much more strongly than did the monovalent ions. Tween 80, a nonionic surfactant, prohibited adsorption to a certain extent. A chelating agent, $(Na)_4$ EDTA, prohibited adsorption only when divalent ions were present.

The results of core flooding indicated that microspheres were retained strongly even by the Berea sandstone with a high permeability of 400-500 md. Strong adsorption on rock grain surfaces could be one of the major contributions to microsphere retention.

A preliminary effort of characterizing surface properties was carried out.

Other secondary effects, such as porous structure of sandstone rock and the change in intrinsic porous structure due to deposition, must be assessed.

With the background information, the penetration of microspheres and presumably bacterial cells could be achieved through the modification of formation matrix and/or the surface properties.

SUMMARY

The importance of investigating microbial transport in porous media, especially in MEOR applications, is pointed out. Negatively charged microspheres were used to simulate bacterial cells in the preliminary studies. A strong interaction was thought to exist between surfaces of microspheres and those of rock grains. Preliminary study of surface properties was carried out. The effects of chemical additives on microsphere adsorption are reported.

ACKNOWLEDGMENTS

Financial support from the U.S. Department of Energy on Contract No. DE-AS19-81BC10508 is appreciated.

REFERENCES

1 LAZAR, I. 1976. Use of bacteria in the recovery procedures and migration of oil from deposit. Mine, Petrol Si Graze (Romania) *27*(10):475–480 (Trans.).

2 BUBELA, B. 1978. Role of geomicrobiology in enhanced recovery of oil: status quo. APEA J. *18(1)*:161–166.

3 KARASKIEWICZ, J. 1968. Recovery of crude oil from reservoirs by use of bacteria. Nafta (Poland) *24*(7):198–202 (Trans.).

4 HART, R.T., T. FEKETE, and D.L. FLOCK. 1960. The plugging effect of bacteria in sandstone systems. Can. Mining and Metallurgical Bull. *53*:495–501.

5 MYERS, G.E., and R.G.L. McCREADY. 1966. Bacteria can penetrate rock. Can. J. Microbiol. *12*:477.

6 YEN, S.P.S., A. REMBAUM, R.W. MOLDAY, and W. DREYER. 1976. Functional colloidal particles for immunoresearch, p. 236-257. In Emulsion polymerization. ACS Symp. No. 24. Amer. Chem. Soc.

7 VISSER, J. 1972. On Hamaker constants: a comparison between Hamaker constants and Lifshitz-van der Waals constants. Adv. Colloid and Interface Sci. *3*:331–363.

8 BIOSURFACTANTS IN BITUMEN SEPARATION FROM TAR SANDS

J.E. Zajic and Juliana Akit

University of Texas at El Paso

Ninety-eight percent of the tar sands deposits in the United States are located in Utah and are estimated to contain 24–30 billion bbl of bitumen. There are five major Utah tar sands (UTS) deposits: Tar Sands Triangle, P. R. Springs, Sunnyside, Asphalt Ridge and Hill Creek. The largest reserve, Tar Sands Triangle, is estimated to have 14–16 billion bbl. This oil is trapped in White Rim sandstone, which acts as an excellent reservoir. The two methods of enhanced oil recovery systems that are proposed for extracting the bitumen are the electrothermic system developed by Electrothermic Co. and the vapor therm system developed by Carmel.[1]

The electrothermic system has been used commercially to recover heavy asphaltic crudes from wells declining in production in Texas, Utah, and Mexico and has been successfully applied to tar sands in southwest Texas. In this process, an electric current produced by a special electrode is continuously applied to the oil reservoir to maintain an operating temperature of 200–250°F. The heat increases effective permeability and the apparent well-bore radius and decreases viscosity and the amount of pressure drop necessary to move oil through the heated area, thereby enhancing oil recovery.

The vapor therm system is used commercially in Kansas and Missouri. It burns fuel oil to produce hot gases that are quenched with water to give large quantities of steam. The steam is injected in the oil well to heat up the oil-bearing sandstone and, as with the electrothermic system, this improves oil recovery.

Alternative methods of enhanced oil recovery are effective in other areas. In 1922 Fylemann proposed the use of natural soaps (surfactants).[2] Since that time the number of available surfactants has increased to the thousands. Montgomery and Fleet have tested approximately 60 surfactants in a cold water extraction to recover bitumen from Athabasca tar sand, which has a higher water content than Utah tar sands.[3] They found that nonionic surfactants with an HLB value between 6 and 10 gave the highest bitumen yield.

Many bacteria and fungi that grow on hydrocarbons do so because they produce a surface-active agent that enhances the microbes' ability to utilize insoluble substrates such as hydrocarbons. Certain bacteria, such as *Bacillus subtilis,* are able to produce surface-active agents in the absence of hydrocarbons. Zajic and Gerson and Gerson, Zajic and Margaritis evaluated several microbial surfactants in a cold water extraction of bitumen from Athabasca tar sands.[4,5]

This paper evaluates four synthetic surfactants and four bacterial cultures that produce surface-active agents for their ability to extract bitumen in an aqueous system from Utah tar sands. Of the products and cultures tested, Nalco 7710 and *Corynebacterium fascians* grown in hexadecane were most effective in bitumen recovery.

MATERIALS AND METHOD

Cultures and growth media. Acinetobacter calcoaceticus was cultivated in 1 liter of a mineral salts medium (MSM-a; see Table 1 for composition), and 20 ml of hexadecane (Fisher Scientific Co.) *Corynebacterium fascians, Corynebacterium lepus,* and *Arthrobacter* SFC were cultivated in 1 liter of another mineral salts medium (MSM-b; see Table 1), 1.0 g of yeast extract (Difco Laboratories) and 40 ml kerosene (Fisher Scientific Co.) *C. fascians* was also cultivated by substituting 40 ml hexadecane or 40 g of sucrose for the kerosene. All cultures were incubated at room temperature on a New Brunswick gyrotary shaker at 250 rpm for 7 days (5 days for *A. calcoaceticus*). The initial pH was 7.0.

Table 1. Composition (g/l) of two mineral salts media–MSM-a and MSM-b.

	MSM-a	*MSM-b*
NH_4NO_3	—	4
$NaNO_3$	2	—
K_2HPO_4	1	6
NaH_2PO_4	0.5	4
KCl	0.1	—
$MgSO_4 \cdot 7H_2O$	0.5	0.2
$CaCl_2 \cdot 2H_2O$	0.015	0.001
$FeSO_4 \cdot 7H_2O$	0.018	0.001
NaEDTA	—	0.0014

Aqueous surfactants. The following synthetic surfactants were tested at a concentration of 1 g/100 ml: 7710 (Nalco Chem. Co.), 7719 (Nalco Chem. Co.), DB 7946 (BASF Wyandotte Corp.), and Defoamer 267A (Diamond Shamrock). The microbial surfactants were not purified in any way, and the whole broths of the cultures of *C. lepus, C. fascians, A. calcoaceticus,* and *Arthrobacter* SFC were used in the evaluation procedure.

Surface activity. Surface tension and interfacial tension against hexadecane and reciprocal critical micelle concentration (CMC^{-1}) were measured with a Fisher Autotensiomat at room temperature.[6,7] Gibbs Surface Excess (τ) was determined from the equation:

$$\tau = -\frac{1}{RT}\left(\frac{d\gamma}{d\ln c}\right)$$

where

γ = surface tension, erg/cm^2
c = % dilution of surfactant concentration
R = gas constant, $8.3 \times 10^7\ erg\,^\circ K^{-1}\ mol^{-1}$
T = absolute temperature, 298°K

Demulsification procedure. To 4.0 ml of water and 6.0 ml of kerosene, 1.0 ml of a 1.0% aqueous L 92 (BASF Wyandotte Corp.), a synthetic emulsifier, and a known concentration of demulsifier were added. The solution was vortexed for 30 sec. The log percent emulsion present was plotted against time (over a 24-hr period). The half-life ($t_{1/2}$) of the demulsifier was calculated from the following equation, assuming that demulsification rates approximate first-order kinetics:

$$t_{1/2} = \frac{-\ell n\,2}{\ell n\,d(\%\ \text{emulsion})/dt}\ (\text{hr})$$

C. fascians (hexadecane) whole broth was centrifuged at 10,000 rpm for 10 min at 0°C. The top hexadecane bound cells were frozen and removed, while the bottom hexadecane free cells formed a pellet that could be easily separated and studied. Each cell type was washed once, resuspended in distilled water, and tested immediately for demulsification ability. For comparison, the cell free whole broth was also evaluated.

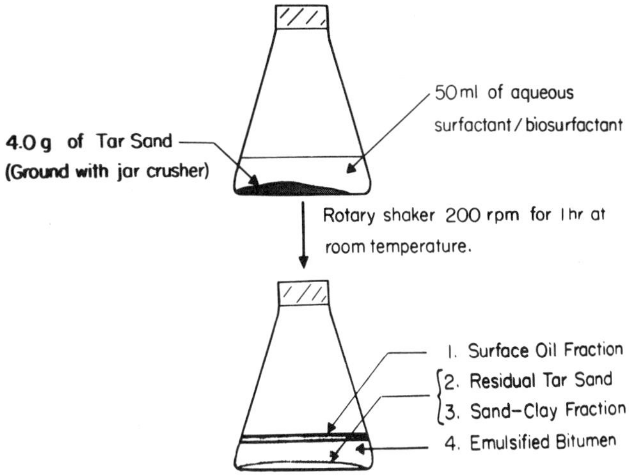

Fig. 1. Cold-water extraction method for recovery of bitumen from Utah tar sands

Bitumen extraction from tar sand. A low shear cold water extraction process was used to evaluate the effectiveness of aqueous surfactants to separate bitumen from Utah tar sands.[8,9] Core samples of UTS were crushed in a Braun Chipmunk crusher (Bico Inc.) and then pulverized in a Bico Inc. Pulverizer unit. Four grams of tar sand were placed in a 500-ml Erlenmeyer flask and were autoclaved separately from the aqueous, synthetic surfactants for 20 min. After cooling, 50 ml of aqueous surfactant were added aseptically. This mixture was shaken on a New Brunswick gyrotary shaker (for 1 hr or 24 hr at room temperature). After shaking, the contents of the flask were allowed to settle and separate in four phases (Figure 1): (*i*) surface oil fraction, which was collected by cleaning the surface with pretreated (prosil-28 from PCR Research Chemicals, Inc.) glass fiber; (*ii*) residual tar sand, which was collected on a 40-mesh screen; (*iii*) sand-clay fraction, which was collected on Whatman #41 filter paper; and (*iv*) emulsified bitumen, which was extracted from the filtrate. The bitumen content of each fraction was determined by extracting with toluene and measuring the optical density at 700 nm (OD_{700}). This value was converted to dry weight of bitumen by the formula:

$$OD_{700}\ \text{value} \times \text{vol of toluene (ml)} \times \text{conversion factor } (g\text{-}ml^{-1}) = \text{dry wt of bitumen (g)}$$

The conversion factor, 0.000479 g/ml, was determined from the dry weight of bitumen extracted from 1 g of UTS in 10 ml of toluene.

RESULTS AND DISCUSSION

Table 2 and Figure 2 show the surface active properties of the aqueous surfactants that were tested for their ability to extract bitumen from UTS after 1 hr of treatment. The results of this experiment are presented in Table 2. The most surface-active synthetic product was 7719 (Nalco), which reduced the surface tension to 28.0 mN/m and had an interfacial tension against hexadecane of 0.2 mN/m. However, the highest concentration of surfactant was found in the whole broths of *Arthrobacter* SFC and *C. fascians* (hexadecane), which gave CMC^{-1} values of greater than 830.

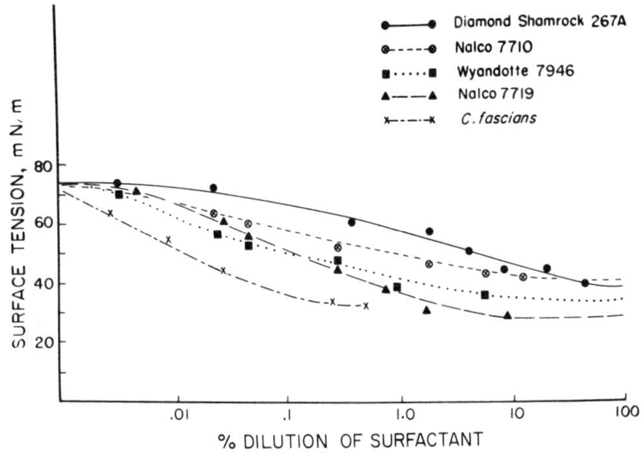

Fig. 2. Surface tension reduction of four synthetic and one microbial surfactant

The ability to lower surface tension or interfacial tension was not a reflection on the ability of these surfactants to extract bitumen, as seen in Table 2. The product 7719 (Nalco) extracted the lowest concentration of free bitumen, while *C. fascians* (hexadecane) and 7710 (Nalco) released the most bitumen of those products tested. The negative numbers indicate a decrease in improvement. *C. fascians* (hexadecane) and 7710 (Nalco) had an improvement factor [(test − control)/control] of 17.4 and 14.5, respectively, in the surface oil.

The microbial surfactants were similarly analyzed after 24 hr of treatment. Those results are expressed in Table 3 and 4. *C. fascians* (hexadecane) was the only culture to show an improvement factor of 0.42. This lower recovery rate was a reflection of the control against which it was compared. The control contained hexadecane, which by itself was able to extract a considerable amount of free bitumen.

Although a carbohydrate substrate would be a cheaper carbon source than hexadecane, *C. fascians* was poor at extracting bitumen from Utah tar sands when grown on sucrose (Table 4). It had a surface tension of 41 mN/m, an interfacial tension of 7, and a CMC^{-1} value of 15.

The demulsification half-lives of the surfactants are shown in Table 5. DB 7946 (Wyandotte) was the most potent; it was 5190 times better than the control while 7719 (Nalco), the least potent, was only 5.8 times better than its control. The microbial products compare well with the synthetic surfactants; the hexadecane bound cells, the free cells, and the cell-free whole broth were 306, 41, and 30 times better than their controls, respectively. Roughly one-tenth less microbial product was needed to achieve these differences when compared to the synthetic products.

The microbial surfactants compare favorably with the synthetic surfactants for bitumen recovery and demulsification.

CONCLUSION

Of the four microbes evaluated, only *C. fascians* produced surfactant(s) that by the fermentation of hydrocarbons were effective in extracting free bitumen from Utah tar sands. This compared favorably with the four synthetic

Table 2. Surface properties of surfactants tested.

Surfactant	*Surface Tension* (mN/m)	*Interfacial Tension (Hexadecane)* (mN/m)	CMC^{-1}	*Gibb's Surface Excess* τ (moles/cm^2)
DB 7946 (Wyandotte)	34.0	1.0	16	0.48
7719 (Nalco)	28.0	0.2	250	0.28
7710 (Nalco)	39.0	8.5	11	0.84
Defoamer 267 A (Diamond Shamrock)	34.0	10.0	0.6	0.009
C. fascians (hexadecane)	31.0	2.0	830	0.05
C. fascians (kerosene)	29.0	1.75	110	—
C. fascians (sucrose)	47.0	7.0	15	—
Arthrobacter SFC	28.0	1.75	1,340;60	—
A. calcoaceticus	49.0	20.0	110	
Control (water)	73.0	34.0	—	—

Table 3. Weight and improvement of bitumen extracted after 1 hr cold-water extraction of Utah tar sands (UTS).

	Wt (g) of bitumen extracted from 4 g ore			
Aqueous surfactant	*Surface Oil*	*Residual US*	*Sand-Clay*	*Emulsified Fraction*
Control (water)	0.0033 ± 0.0006	0.076 ± 0.027	0.0024 ± 0.0004	$1.95 \times 10^{-5} \pm 9.3 \times 10^{-6}$
C. fascians (hexadecane)	0.0612 ± 0.0070 (17.4)[a]	0.051 ± 0.005 (−0.32)	0.0215 ± 0.0005 (10)	$1.1 \times 10^{-5} \pm 0$ (54)
C. fascians (sucrose)	0.0010 ± 0.0003 (−0.69)	0.061 ± 0.003 (−0.20)	0.0023 ± 0.0008 (−0.04)	$0.9 \times 10^{-5} \pm 0$ (−0.5)
7710 (Nalco)	0.0512 ± 0.0043 (14.5)	0.042 ± 0.017 (−0.45)	0.0703 ± 0.0123 (28)	$2.7 \times 10^{-3} \pm 0$ (144)
7719 (Nalco)	0.0001 ± 0.00005 (−0.96)	0.112 ± 0.014 (0.48)	0.0024 ± 0.0007 (N.I.)	$5.6 \times 10^{-5} \pm 1.9 \times 10^{-5}$ (2)
DB 7946 (Wyandotte)	0.0026 ± 0.0002 (−0.22)	0.060 ± 0.018 (−.2)	0.0516 ± 0.0033 (12)	$1.6 \times 10^{-3} \pm 4.3 \times 10^{-5}$ (84)
Defoamer 267 A (Diamond Shamrock)	0.0034 ± 0.0006 (0.03)	0.070 ± 0.014 (0.05)	0.131 ± 0.002 (4.4)	$1.8 \times 10^{-3} \pm 4.6 \times 10^{-5}$ (8.5)

[a] improvement factor = (test – control)/control
(NI) = No improvement

surfactants that were also evaluated. Nalco 7710 was the best synthetic surfactant tested. Although the sample size was small, the data indicate that biosurfactants are just as effective as synthetic surfactants in the water extraction of bitumen from UTS. In these studies no attempt was made to optimize either biosurfactant production or the bioextraction process for bitumen. The prime objective was to examine microbes which are known to produce surfactants to see if they could achieve a separation of bitumen from tar sand. As noted above, only *C. fascians* was effective.

Table 4. Weight and improvement of bitumen extracted after 24 hr cold-water extraction of Utah tar sands (UTS) using microbial surfactants.

	Mass Recovered (g) From 4 g Ore			
Microbial Surfactant	*Surface Oil*	*Residual UTS*	*Sand-Clay*	*Emulsified Fraction*
Sucrose control	0.0034	0.150	0.0278	4.7×10^{-5}
C. fascians (sucrose)	0.0005 (−0.85)[a]	0.111 (−0.26)	0.0425 (0.53)	2.8×10^{-4} (5.0)
Hexadecane control	0.118 ± 0.007	0.090 ± 0.030	0.016 ± 0.001	N.D.
C. fascians (hexadecane)	0.167 ± 0.046 (0.42)	0.043 ± 0.003 (−0.52)	0.021 ± 0.003 (0.31)	N.D.
A. calcoaceticus (hexadecane)	0.028 ± 0.004 (−0.76)	0.061 ± 0.002 (−0.32)	0.128 ± 0.004 (7)	$6.0 \times 10^{-4} \pm 4.6 \times 10^{-5}$ (N.D.)
Kerosene control	0.143 ± 0.026	0.018 ± 0	0.018 ± 0.005	0.001 ± 0
C. lepus (kerosene)	0.095 ± 0.002 (−0.34)	0.001 ± 0.0005 (−0.39)	0.023 ± 0.003 (0.28)	$1.7 \times 10^{-3} \pm 2.8 \times 10^{-4}$ (0.7)
Arthrobacter SFC (kerosene)	0.100 ± 0.043 (−0.30)	0.048 ± 0.008 (1.67)	0.010 ± 0.007 (−0.44)	$5.9 \times 10^{-3} \pm 3.7 \times 10^{-4}$ (4.9)

[a] ( ) = improvement factor = (test − control)/control
ND = not determined

ACKNOWLEDGMENTS

This research was sponsored by Petroleum Sciences of Houston, Texas. We would also like to acknowledge the technical assistance of Delfina Cisneros and Jaime Pinales.

Table 5. Demulsification of L-92 stablized emulsion (O/W emulsion).

Surfactant	*Concentration Tested g/10 ml Emulsion*	*Half-Life (hr) Control*	*Surfactant*
DB 7946 (Wyandotte)	0.004	1,038	0.2
7710 (Nalco)	0.005	789	29
7719 (Nalco)	0.009	644	110
Defoamer 267A (Diamond Shamrock)	0.004	349	3
Corynebacterium fascians			
hexadecane bound cells	0.0007	245	0.8
hexadecane free cells	0.0002	245	6
cell-free whole broth	0.8	151	5

REFERENCES

1 U.S. DEPARTMENT OF ENERGY. 1980. Alternative fuel feasibility study, oil recovery in the Utah Tar Sands Triangle, vol. 1. Technical proposal. U.S. Department of Energy, Office of Resource Application, Washington, D.C.

2 GERSON, D.F., J.E. ZAJIC, D.G. COOPER, and M.D. OUCHI, 1977. Microbial separation of bitumen from Athabasca tar sand. Part II. Biochemical Engineering, Faculty of Engineering Science, University of Western Ontario, London, Ontario, Canada.

3 MONTGOMERY, D.S., and M.P. FLEET. 1958. Mines Branch Report IR-58-41. Canada Department of Mines and Technical Survey.

4 ZAJIC, J.E., and D.F. GERSON. 1978. Oil Sand and Oil Shale Symposium, Am. Chem. Soc. Preprints *22*:195.

5 GERSON, D.F., J.E. ZAJIC, and A. MARGARITIS. 1976. Microbial separation of bitumen from Athabasca Tar Sand. University of Western Ontario, London, Ontario, Canada.

6 COOPER, D.G., and J.E. ZAJIC. 1980. Adv. Appl. Microbiol. p. 227. *26*:229.

7 COOPER, D.G., J.E. ZAJIC, and D.F. GERSON. 1979. Proceedings, microbial degradation of pollutants in marine environments. National Technical Information Service, Springfield, Va.

8 GERSON, ZAJIC, COOPER, and OUCHI, op cit.

9 ZAJIC and GERSON, op cit.

9

PRODUCTION AND CHEMICAL CHARACTERIZATION OF SURFACTANTS FROM *RHODOCOCCUS ERYTHROPOLIS* AND *PSEUDOMONAS* SP. MUB GROWN ON HYDROCARBONS

Fritz Wagner, Ulrich Behrendt, Hans Bock, Axel Kretschmer, Siegmund Lang, and Christoph Syldatk

Technical University of Braunschweig, West Germany

The microbial oxidation or cooxidation of sparingly water-soluble carbon substrates is almost always associated with the production of surfactants. The formation of ionic and nonionic surfactants from growing microbial cultures and the effect of these surfactants on the initial interaction of the lipophilic carbon substrates with the microbial cell have been reviewed recently[1-5]. The most commonly isolated and characterized surfactants are glycolipids, such as trehalose, rhamnose, and sophorose lipids. Trehalose-6,6′-dimycolates, the so-called cord factors, are well-known cell wall constitutents in *Corynebacterium, Mycobacterium,* and *Nocardia.*[6]

In cultures of *Arthrobacter paraffineus* grown on n-alkanes, trehalose dicorynomycolates were enriched with corynomycolic acid residues of unknown substitution pattern.[7] An overproduction of trehalose-6,6′-dicorynomycolates by *Rhodococcus erythropolis* grown on n-alkanes was described in which the number of carbon atoms in the corynomycolic acid residues ranged from C_{32} to C_{38}.[8] The majority of the nonionic trehalose dicorynomycolates produced are associated with the cell wall of *Rhodococcus erythropolis* and are not excreted into the culture broth. In contrast, the rhamnolipids formed by *Pseudomonas* sp. when grown on n-alkanes are extracellular anionic metabolites,[9,10,11] in which the hydrophilic part of the molecule contains one or two rhamose units and the hydrophobic part contains one or two covalently bound β-hydroxy fatty acids. Glycolipids are involved in the uptake of strongly hydrophobic carbon substrates by growing cultures of microorganisms.[12-16] Two different mechanisms for the initial interaction of n-alkanes in the presence of trehalose or rhamnose lipids with the microbial cell can be postulated:

(*i*) The anionic, extracellular rhamnose lipids produce a microemulsion of liquid n-alkanes in the water phase, leading to facilitated transport of the microdroplets or micelles into the cell.

(*ii*) The nonionic, cell-wall-associated trehalose lipids impede the wetting of the cell surface by n-alkanes. This increases the direct microorganism-hydrocarbon contact at the aqueous-hydrocarbon interface and improves diffusion through the cell wall.

Because of these two different proposed mechanisms of n-alkane uptake, an investigation was undertaken to examine growth and formation of glycolipids of two microorganisms that occur outside the cell or associate with the cellular envelope. The first part of this paper describes the production of trehalose lipids from *Rhodococcus erythropolis* using n-alkanes for in situ extraction of the cell-wall-associated glycolipids as well as the chemical characterization of the isolated and purified products. The remainder of the text discusses the production of extracellular rhamnose lipids by *Pseudomonas* sp. MUB under conditions where the nitrogen content of the medium was limited.

RESULTS AND DISCUSSION

Improved formation of trehalose lipids by Rhodococcus erythropolis using an in situ extraction method. The kinetics of growth and production of α,α-trehalose dicorynomycolates from *Rhodococcus erythropolis* DSM 43215 on a mineral salt medium with 2% (by weight) n-alkanes C_{12} to C_{18} was published recently.[17,18] The unusual growth profile can be divided into several different phases. In the first phase the organism exhibited a growth rate of 0.22 hr^{-1}

within the given hydrocarbon droplets (ϕ about 50 μm). After 10 hr of incubation in a defined proportion and concentration of n-alkanes/cell mass/trehalose lipids, large clumps were observed. The following second phase, lasting about 20 hr, was marked by a strong retardation of the growth and production rate. After the concentration of n-alkanes was decreased further, the large clumps were disintegrated into much smaller cell aggregates without any emulsifying effect. This was followed by a second increase in the growth and production rate. Final yields of 19 g/l dry cell mass and 2.1 g/l of trehalose lipids were obtained after consuming more than 90% of the carbon source. The content of trehalose lipids was measured by the colorimetric anthrone method.[19]

These results were used to investigate further the optimization of trehalose lipid production. One possible approach to increasing the production rate of trehalose lipids is to repress the strong aggregation phase that is caused by a defined ratio of cell mass/n-alkanes/trehalose lipids. Using this approach the starting concentration of the n-alkanes was increased up to 10% by weight. The results are shown in Figure 1.

The formation of the normally cell-wall-associated trehalose lipids was influenced markedly by these culture conditions. During the exponential growth phase with a

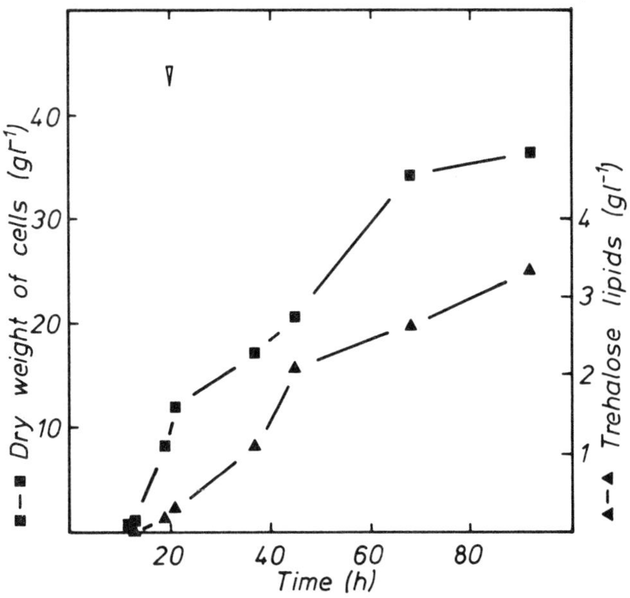

Fig. 1. Time course of trehalose lipids formation by *Rhodococcus erythropolis* DSM 43 215 in an 80-l bioreactor on a modified mineral salt medium with 10% by weight of n-alkanes (1.7% C_{13}, 89.4% C_{14}, 9.2% C_{15}, 0.7% C_{16}). The composition of the mineral salt medium is based on the elemental composition of the cell mass of *Rhodococcus erythropolis.* Aeration rate = 1.5 v/v/min. Temperature = 30°C and constant pH at 5.8. Working pressure is 2 bar

growth rate of 0.34 hr^{-1}, the organism was located in the hydrocarbon droplets as in the case with 2% n-alkanes. Since an emulsion is formed, a second insoluble phase containing 2% n-alkanes was not present. In the presence of the high content of n-alkanes, the aggregation was significantly reduced. This was due to the continuous extraction of the trehalose lipids, leading to the formation of a stable oil-water emulsion (droplet size: ϕ 2.4–5 μm). The microorganism furthermore grew with a linear growth rate within the nonemulsified oil droplets. The synthesis of trehalose lipids then paralleled growth, and the degradation of the trehalose lipids during the stationary growth phase in the presence of excess emulsified hydrocarbons was repressed. The product yield could be doubled by means of this in situ extraction method, bearing in mind that in this case the yield was based on the extracted and isolated trehalose lipid fraction.

Characterization of the main components in the organic crude extract. The whole cell suspension of *Rhodococcus erythropolis* obtained in the stationary growth phase was extensively extracted at pH 3.0 with dichloromethane/methanol (2:1,v/v). The combined organic phases were evaporated at 40°C under reduced pressure. The residue was separated by repeated column chromatography on silica gel, and each fraction was purified further by preparative thin layer chromatography. The structures of the pure substances were determined by chemical and spectroscopic methods.[20,21,22] From their recorded hydrophobic behavior, the following categories of substances were observed:

Group 1: n-alkanes; methyl-branched, unsaturated hydrocarbons

Group 2: triacylglycerols, 2-alkyl-3-keto fatty acids, fatty acids, fatty alcohols, corynomycolates (2-alkyl-3-hydroxy fatty acids)

Group 3: α,α-trehalose-6,6′-dicorynomycolates, α,α-trehalose fatty acid diesters, α,α-trehalose-6-monocorynomycolates, α,α-trehalose fatty acid monoesters

Group 4: Phosphatidylethanol amines and phosphatidic acids.

In addition to these substances, the organic crude extract contained other metabolites, comprising 5% by weight, that have not been characterized.

Group 1 The recovered n-alkanes were identical with the n-alkanes that were used as carbon source by the growing culture. Also, the chain length proportions of the remaining n-alkanes were the same as the initial proportions. The spectroscopic data of the methyl-branched, unsaturated hydrocarbons suggest that these hydrocarbons are isoprenoids.

Group 2 The profile of the free fatty acids, bound fatty acids in the triacylglycerols, and fatty alcohols showed a good correlation with the chain length of the n-alkanes used as substrates. However, the free fatty acids contained about 25% by weight cis-monounsaturated fatty acids, whereas in the triacylglycerols only 2% by weight could be detected. In addition significant amounts of long chain fatty acids in the range of C_{19} to C_{22} were observed in the free fatty acid fraction. These long chain fatty acids are proposed intermediates of corynomycolate synthesis.[23]

The 2-alkyl-3-keto fatty acids and the 2-alkyl-3-hydroxy fatty acids were found as a homologous series of the following general structures:

$$CH_3-(CH_2)_y-C(=O)-CH[(CH_2)_x-CH_3]-COOH$$

and

$$CH_3-(CH_2)_y-CH(OH)-CH[(CH_2)_x-CH_3]-COOH$$

$x = 7\text{–}12;\ y = 15\text{–}24$

The substitution profiles obtained in the free corynomycolates were identical to those of the corynomycolic acids esterified with α,α-trehalose.

Group 3 The chemical structure of α,α-trehalose-6,6′-dicorynomycolate from *Rhodococcus erythropolis* (fraction 3.1.) has recently been elucidated.[24] From a bulk of 1 kg organic crude extract, two fractions (3b and 3d) could be separated as byproducts. The yields are less than 10% by weight with respect to the total amount of trehalosecorynomycolates (fractions 3a and 3c). The product α,α-trehalose-6-monocorynomycolate seems to be a biosynthetic intermediate of the corresponding diesters (fraction 3a).[25] The following structures of trehalose lipids were determined:

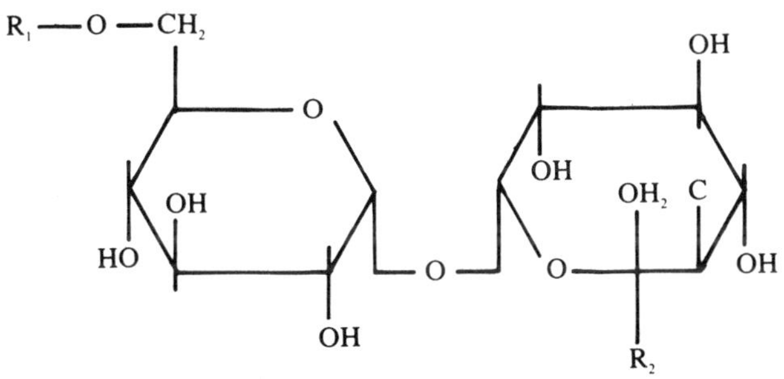

3a:

$$R_1 = R_2 = -C(=O)-CH[(CH_2)_x-CH_3]-CH(OH)-(CH_2)_Y-CH_3$$

3b:

$$R_1 = -C(=O)-CH[(CH_2)_x-CH_3]-CH(OH)-(CH_2)_Y-CH_3$$

$$R_2 = -C(=O)-CH[(CH_2)_x-CH_3]-CH_2-(CH_2)_y-CH_3$$

or:

$$R_2 = -C(=O)-CH[(CH_2)_6-CH=CH-(CH_2)_3-CH_3]-CH(OH)-(CH_2)_y-CH_3$$

or:

$$R_2 = -C(=O)-CH_2-(CH_2)_z-CH_3$$

3c:

$R_1 = H$

$$R_2 = -C(=O)-CH[(CH_2)_x-CH_3]-CH(OH)-(CH_2)_y-CH_3$$

3d:

$R_1 = H$

$R_2 =$ with 3.2

$x = 7 - 12; y = 15 - 24; z = 12 - 20$

The quantitative composition of the Group 3 trehalose lipids can be greatly influenced by variation of the growth conditions when *Rhodococcus erythropolis* is grown on n-alkanes as the sole carbon source.[26] However, in contrast to the results obtained with species of *Arthrobacterium, Corynebacterium,* and *Nocardia* grown on sucrose or fructose, no substitution of the trehalose moiety in the Group 3 trehalose lipids by *Rhodococcus erythropolis* could be observed.[27,28] Also, no substitution was seen when a carbohydrate was utilized as a cosubstrate.

Group 4 Unlike the free fatty acids and the acyl groups in the triacylglycerols (see Group 2), the acyl moieties in the phosphatidylethanolamines and phosphatidic acids contain considerable amounts of methyl branched C_{13}–C_{20} fatty acids. For example, when *Rhodococcus erythropolis* was grown on n-tetradecane as the sole carbon source, the isolated phosphatidylethanolamines were mainly 1,2-di-tetradecanoyl-sn-glycero-3-phospho-ethanolamine and 1,2-tetradecanoyl-(15-methyl-nonadecanoyl)-sn-glycero-phospho-ethanolamine.[29]

Formation of extracellular glycolipids by Pseudomonas *sp. MUB under nitrogen limitation.* The organism used in this study was isolated by enrichment culture techniques from a natural water sample. Its morphological characteristics and physiological tests indicate that the organism belongs to the genus *Pseudomonas.*

The formation of rhamnolipids by different strains of *Pseudomonas* from various carbon sources such as glycerol or fructose and hydrocarbons has been reported.[30-34] Very little has been published about the controlled process conditions for the production of glycolipids, or about the possibility of alteration of the product composition with *Pseudomonas* sp. grown on hydrocarbons as the sole carbon source. *Pseudomonas* sp. MUB grown on a mineral salt medium with 2% by weight of either glucose, glycerol, citrate, ethanol, or a mixture of n-alkanes C_{13}–C_{16} as sole carbon sources showed a significant formation of a mixture of glycolipids. However, n-alkanes were generally superior to the other carbon substrates in product yield. One component (glycolipid II) of the glycolipid mixture was determined as the described rhamnolipid from *Pseudomonas aeruginosa.*[35,36] In contrast it was demonstrated that *Pseudomonas aeruginosa* produced no rhamnolipid with citrate as the carbon source.[37]

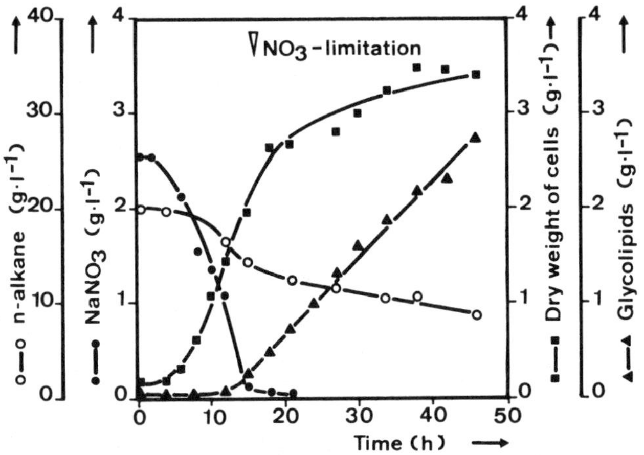

Fig. 2. Time course of glycolipids formation by *Pseudomonas* sp. MUB in a 30-l bioreactor on a mineral salt medium at 30°C and constant pH of 6.5.
N source: 0.25% $NaNO_3$
C source: 2% n-alkanes
(0.7% C_{13}, 98.4% C_{14}, 9.2% C_{15}, 0.7% C_{16})

The accumulation of glycolipids in the culture broth of *Pseudomonas* sp. MUB was markedly increased by nitrogen limitation. In Figure 2 the results of a typical experiment with nitrate limitation are represented. Corresponding to the exponential growth phase, a decrease of the nitrate content was detected. During this time no formation of glycolipids was observed. The low content of about 20 mg per liter glycolipids in the culture supernatant was transferred with the inoculum. No visible emulsion of the hydrocarbon droplets in the water phase could be observed and the organism showed no orientation to the hydrocarbon phase. These observations are surprising when compared with the initial hydrocarbon uptake mechanism postulated in the literature.[38,39,40] It seems that the mass transfer of the hydrocarbons is not growth limiting in the presence of 20 mg of glycolipids per liter. In this connection it is noteworthy that the critical micelle concentration of the glycolipids for lowering the interfacial tension was also in the range of 20 mg/l.

A linear phase of glycolipid production began after consumption of the nitrate in the culture broth. This interval was associated with a decrease of the growth rate and with a sharp drop in QO_2 and QCO_2. The concentration of n-alkanes decreased quite slowly. After reaching a glycolipid level of about 40 mg/l, the total contents of the bioreactor became strongly emulsified. A final yield of more than 1.5 g of glycolipids per gram dry cell mass was obtained. In further experiments it was demonstrated that ammonium ions as well as nitrate ions also repressed the biosynthesis of glycolipids from *Pseudomonas* sp. MUB.

Most of the whole emulsion obtained in the stationary growth phase was extracted at pH 3.0 with dichloromethane/methanol (2:1, v/v). The organic extract was separated by repeated column chromatography on silica gel. The organic crude extract contained five fractions besides the unconsumed alkanes. A comparison of the chromatographic behavior is shown in Table 1.

Table 1. Comparison of chromotographic behavior.

Fraction	*Solvent system*[a] *A*	*B*	*C*
F I	0.71	0.78	—
Gl II	0.43	0.52	0.10
F II	0.31	0.10	—
Gl III	0.20	0.29	0.02
F III	0.15	0.10	—
2-0-α-L-Rhamnopyranosyl-α-L-rhamnopyranosyl-β-hydroxy-decanoyl-β-hydroxy-decanoic acid[b]	0.43	0.52	0.10

[a]A: $CHCl_3/CH_3OH/H_2O$ (65:15:2)
B: $CHCl_3/CH_3OH/H_2O/CH_3COOH$ (65:15:2:2)
C: $CHCl_3/CH_3OH/H_2O/NH_3$, 25% (65:15:2:2)
[b]Rhamnose lipid derived from *Pseudomonas fluorescens* ATCC 15 453

The observed chromatographic behavior suggested that the fraction G1 II was identical to the anionic rhamnolipid from *Pseudomonas fluorescens* ATCC 15 453 and *Pseudomonas aeruginosa.*[41,42] After acid hydrolysis of G1 II, only rhamnose was detected.

Fraction G1 III was also an anionic glycolipid. However, this glycolipid was not identical to the rhamnolipids described in the literature. The structure of F I is unknown. F II and F III were characterized as phosphatidylethanolamines and phosphatidic acids, respectively.

REFERENCES

1 COOPER, D.G., and J.E. ZAJIC. 1980. Surface active compounds from microorganisms. Adv. Appl. Microbio. *26*:229–259.

2 RATLEDGE, C. 1980. Microbial lipids derived from hydrocarbons. p. 133–154. *In* D.E.F. Harrison, U.J. Higgins, R. Watkinson (ed.), Hydrocarbons in biotechnology. Heyden and Son Ltd., London.

3 SPENCER, J.F.T., D.M. SPENCER, and A.P. TULLOCH. 1979. Extracellular glycolipids of yeasts. *In* A.H. Rose (ed.), Economic microbiology, vol. 3. 523–540. Academic Press, London.

4 WAGNER, F., H. BOCK, and A. KRETSCHMER. 1981. Gewinnung von tensiden mit n-alkan-oxidierenden mikroorganismen p. 181–192. In R.M. Lafferty (ed.). Fermentation. Springer Verlag, Vienna.

5 WATKINSON, R.J. 1980. Interaction of microorganisms with hydrocarbons. p. 11–24. In D.E.F. Harrison, I.J. Higgins, R. Watkinson (ed.), Hydrocarbons in biotechnology. Heyden and Son Ltd., London.

6 LEDERER, E. 1976. Cord factor and related trehalose esters. Chemistry and Physics of Lipids *16*:91–106.

7 SUZUKI, T., K. TANAKA, I. MATSUBARA, and S. KINOSHITA. 1969. Trehalose lipid and α-branched-β-hydroxy fatty acid formed by bacteria grown on n-alkanes. Agr. Biol. Chem. *33*:1619–1627.

8 RAPP, P., H. BOCK, V. WRAY, and F. WAGNER 1979. Formation, isolation and characterization of trehalose dimycolates from *Rhodococcus erythropolis* grown on n-alkanes. J. Gen. Microbiol. *115*:491–503.

9 HISATSUKA, K., T. NAKAHARA, N. SANO, and K. YAMADA. 1971. Formation of rhamnolipid by *Pseudomonas aeruginosa* and its function in hydrocarbon fermentation. Agr. Biol. Chem. *35*:686–692.

10 ITOH, S., and T. SUZUKI. 1972. Effect of rhamnolipids on growth of *Pseudomonas aeruginosa* mutant deficient in n-paraffin utilizing ability. Agr. Biol. Chem. *36*:2233–2235.

11 YAMAGUCHI, M., A. SATO, and A. YUKUYAMA. 1976. Microbial production of sugar lipids. Chem. and Ind. *4*:741–742.

12 HISATSUKA, NAKAHARA, SANO, and YAMADA, op cit.

13 NAKAHARA, T., K. HISATSUKA, and Y. MINODA. 1981. Effect of hydrocarbon emulsification on growth and respiration of microorganisms in hydrocarbon media. J. Ferment. Technol. *59*:415–418.

14 SCHÖMER, U., and F. WAGNER. 1980. Production of steroid-ring-B-seco acid from β-sitosterol and influence of surfactants on sterol uptake in *Nocardia* sp. M 29-40. Eur. J. Appl. Microbiol. Biotechnol. *10*:99–106.

15 WAGNER, BOCK, and KRETSCHMER, op cit.

16 WATKINSON, op cit.

17 KRETSCHMER, A., H. BOCK, and F. WAGNER. 1982. Formation and characterization of interfacial-active lipids from *Rhodococcus erythropolis* grown on n-alkanes. (In preparation).

18 RAPP, BOCK, WRAY, and WAGNER, op cit.

19 HISATSUKA, NAKAHARA, SANO, and YAMADA, op cit.

20 KRETSCHMER, A., and F. WAGNER. 1982. Biosynthesis of interfacial-active trehalose-dicorynomycolates from *Rhodococcus erythropolis* grown on n-alkanes. (In preparation).

21 KRETSCHMER, BOCK, and WAGNER, op cit.

22 RAPP, BOCK, WRAY, and WAGNER, op cit.

23 KRETSCHMER and WAGNER, op cit.

24 RAPP, BOCK, WRAY, and WAGNER, op cit.

25 KRETSCHMER and WAGNER, op cit.

26 RISTAR, E. 1982. Dissertation TU Braunschweig. (In preparation).

27 SUZUKI, T., H. TANAKA, and S. ITOH. 1974. Sucrose lipids of *Arthrobacteria*, *Corynebacteria* and *Nocardia* grown on sucrose. Agr. Biol. Chem. *38*:557–563.

28 ITOH, S., and T. SUZUKI. 1974. Fructose lipids of *Arthrobacteria, Corynebacteria, Nocardia* and *Mycobacteria* grown on fructose. Agr. Biol. Chem. *38*:1443–1449.
29 KRETSCHMER, BOCK, and WAGNER, op cit.
30 HAUSER, G., and M.L. KARNOVSKY. 1954. Studies of the production of glycolipid by *Pseudomonas aeruginosa.* J. Bacteriol. *68*:645–654.
31 HISATSUKA, NAKAHARA, SANO, and YAMADA, op cit.
32 ITOH and SUZUKI, op cit.
33 JARVIS, F.G., and M.J. JOHNSON. 1949. A glycolipid produced by *Pseudomonas aeruginosa.* J. Am. Chem. Soc. *71*:4124–4126.
34 YAMAGUCHI, SATO, and YUKUYAMA, op cit.
35 HISATSUKA, NAKAHARA, SANO, and YAMADA, op cit.
36 JARVIS and JOHNSON, op cit.
37 HAUSER and KARNOVSKY, op cit.
38 HISATSUKA, NAKAHARA, SANO, and YAMADA, op cit.
39 NAKAHARA, HISATSUKA, and MINODA, op cit.
40 WATKINSON, op cit.
41 SUZUKI, T., and S.I. ITOH. 1972. Verfahren zur herstellung rhamnosehaltiger glykolipide. DE-OS 21 50 375
42 HISATSUKA, NAKAHARA, SANO, and YAMADA, op cit.

10 RHODOCHROUS BACTERIA: BIOSURFACTANT PRODUCTION AND DEMULSIFYING ABILITY

B.A. Ramsay and D.G. Cooper
McGill University, Montreal

A. Margaritis
University of Western Ontario

J.E. Zajic
University of Texas at El Paso

Microbial processes have been implicated in oil recovery for a long time.[1,2,3] Whether microbes produced surfactants was not evaluated. It has been shown that microbes produce biosurfactants, which compare favorably with synthetic surfactants in aiding cold-water extraction of bitumen from tar sand.[4] Cost estimates place the production of biosurfactants at 30% of that of synthetic surfactants.[5] It has also been shown that the addition of biosurfactants produced by *Mycobacterium rhodochrous*, which is a member of the rhodochrous complex of bacteria, enhances oil recovery by the water-flooding technique.[6] This paper focuses on the *Mycobacterium rhodocrous* group of bacteria. In addition to surface properties, the demulsification ability of the whole culture broths of these microbes was examined. Primary and secondary production of oil result in emulsion formation. Oil containing significant quantities of water increases transportation costs and is unacceptable for refining.[7] Recently, microbial broths have been tested for their ability to demulsify standard oil-in-water emulsions.[8]

Although the genus *Rhodococcus* is not represented in the eighth edition of Bergey's manual, there is now little doubt that the rhodochrous group forms a recognizable taxon.[9,10] Goodfellow and Alderson have proposed that the genus *Rhodococcus* be reintroduced to accommodate bacteria that have been classified as *Gordona, Jensenia, Mycobacterium rhodochrous*, the rhodochrous complex and the rhodochrous taxon.[11,12] One of the most important sources of chemotaxonomic features in distinguishing between members of the rhodochrous complex and other actinomycetes is their characteristic lipids since the lipids of actinomycetes have many unique structures.[13,14] Some of the unusual lipid structures found in members of the rhodochrous complex and other actinomycetes are long apolar alkyl chains, which are a common constituent of biosurfactants.[15] Indeed, some members of the rhodochrous complex produce biosurfactants when grown on hydrocarbons.[16,17,18]

This paper deals with 13 different strains belonging to the rhodochrous complex and investigates their ability to produce biosurfactants and to demulsify emulsions.

MATERIALS AND METHODS

The mineral salts medium used for growth of cultures had the following w/v composition: NH_4NO_3, 0.2%; K_2HPO_4, 0.1%; KH_2PO_4, 0.05%; $MgSO_4$, 0.05%; KCl, 0.01%; $CaCl_2$, 0.001%; $FeSO_4$, 0.001%; EDTA, 0.0002% (pH 7.0). Difco yeast extract was added in a 0.006% (w/v) concentration as a source of vitamins and growth factors. The carbon sources were 2% (v/v) n-hexadecane (Humphrey, 99% pure) or 2% (w/v) D-glucose (Fisher Scientific Co.), as specified. All shake-flask studies were conducted with 50 ml of medium in 500-ml Erlenmyer flasks using a New Brunswick gyrotory shaker at 150 rpm and 25° ± 2°C. All cultures were grown in duplicate for eight days, after which time the contents of the flasks were sampled and analyzed. Results were averaged. A 5% v/v inoculum of medium-acclimatized exponential growth phase cells was used for all shake-flask experiments.

A solvent extraction method of biomass determination was used.[19] A standard curve, used in the solvent extraction method to compensate for extraction of cellular material, was not included in this study. Thus, reported biomass concentration data are lower than actual dry weight concentrations. However, this was a convenient and necessary method of comparison in cases when no other biomass determination method could be employed due to problems associated with biomass concentration determinations of cultures still containing residual hydrocarbon.

The surface and interfacial tensions were measured with a Fisher Autotensiomat, which is a modified de Nuöy surface tensiometer equipped with a motorized sample stage and a sensitive strain gauge connected to a platinum ring. Critical micelle concentration (CMC) was measured as the concentration of whole broth needed to reduce the surface tension of 40 ml of distilled deionized water in a 50-ml Pyrex beaker to the minimum value possible. Therefore, an X^{-1} CMC value indicates the number of times that the culture broth can be diluted in water and still retain its minimum surface tension. It is the reciprocal of the dilution required to reach the CMC.

Spreading tension was calculated as:

$$\gamma_s = \gamma_a - \gamma_b - \gamma_{ab} \qquad (1)$$

where:

γ_s = spreading tension, mN/m
γ_a = surface tension of aqueous phase, mN/m
γ_b = surface tension of hydrocarbon, mN/m
γ_{ab} = interfacial tension of water against hydrocarbon, mN/m

To determine foaming ability, bacterial whole broth was first diluted to 10% v/v in distilled deionized water. Twenty-five ml of this dilution was placed into a vertical glass tube with a 27-mm inside diameter. Air was pumped through a glass sparger at the base of the tube at a rate of 0.4 l/min. Foaming ability was measured as the maximum height of foam produced.

Demulsifying ability was tested by the addition of whole bacterial broth to test emulsions that contained just enough synthetic surfactant to allow formation of a stable emulsion. Two oil-in-water emulsion types were used.[20] The L-92 stabilized emulsion contained 5 ml of kerosene (Imperial Oil #9 refined oil) and 4 ml of 0.068% (w/v) L-92 pluronic surfactant (BASF, Wyandotte Corporation) in deionized distilled water. The Tween-Span stabilized emulsion contained 6 ml of kerosene and 4 ml of 0.072% (w/v) Tween 60 (ICI United States Inc.) and 0.028% (w/v) Span 60 (J.T. Baker Chemical Company) in deionized distilled water. Both of these systems were vigorously vortexed in 16-mm × 150-mm glass test tubes until a stable emulsion was formed. One ml of test broth from a bacterial culture was added to the test emulsions, while 1 ml of mineral salts medium was added to the control emulsions. The tubes were vortexed for approximately 5 sec and then were left to stand vertically. A plot was made of the natural logarithm of the percent of the total test volume occupied by the emulsion versus time. The slope generated over the linear portion of the curve was used to calculate emulsion half-lives as shown below:

$$\text{half-life (hr)} = \frac{\ell n 2}{\text{slope (hr}^{-1})} \qquad (2)$$

Rhodococcus aurantiacus 80001, *R. terrae* 70006, *R. rubropertinctus* 60003, *R. rhodochrous* 40001, *R. reseus* 65001, and *R. lentifragmentus* 23002 were obtained from Dr. M. Tsukamura. *Nocardia amarae* LL-Se-6, *N. pellegrini* SN 5101, *N. rubra* SN 5201, and *N. rhodochrous* LL-100-22 were obtained from Dr. Mary P. Lechevalier. *N. rhodochrous* UWO 837 is an isolate of Dr. J. Rozanis and was obtained from the University of Western Ontario Culture Collection. *N. amarae* 27808 was obtained from the American Type Culture Collection.

RESULTS AND DISCUSSION

Thirteen different organisms belonging to or closely related to the rhodochrous complex of bacteria were tested for the production of surface-active compounds. The evaluation process involved testing shake-flask-grown cultures for their ability to lower the whole broth surface, interfacial, and spreading tensions; to destabilize standard oil-in-water emulsions; and to produce foam.

Table 1 contains the results of shake-flask fermentations using glucose as the sole carbon and energy source. Some of these cultures were able to produce surface-active compounds when growing on glucose. The biomass concentration values were low because it was necessary to use the solvent extraction method of biomass determination. However, these values can be compared to each other and do show that all of the bacteria except *N. rhodochrous* L-100-22 grew well in the glucose medium. In all cases the surface and interfacial tensions were lower than the controls. This was expected since most microbially produced substances have some degree of surface activity in water. However, surface activity was particularly evident in the cases of *R. aurantiacus* 80001, *R. lentifragmentus* 23002, and *N. rubra*

Table 1. Surface properties of Rhodochrous bacteria grown on 2% w/v glucose

Organism	*Biomass* (g/l)	*Surface Tension* (mN/m)	*Interfacial Tension* (mN/m)	*Spreading Tension* (mN/m)	*Biosurfactant Concentration* ($X\ CMC^{-1}$)	*Half-life of L92 Emulsion* (hr)	*Half-life of Tween-Span Emulsion* (hr)
R. aurantiacus 80001	1.8	40	5	7	3	9	<0.08
R. rubropertinctus 60003	1.6	63	16	19	3	7	<0.08
R. lentifragmentus 23002	1.0	38	4	6	4	<0.08	>100
R. reseus 65001	1.1	52	6	18	5	20	>100
R. terrae 70006	0.6	65	9	25	<1	10	100
R. rhodochrous UWO 837	0.9	66	26	12	<1	30	>100
R. rhodochrous 40001	1.9	65	17	20	<1	8	40
N. rhodochrous L-100-22	0.3	65	29	8	2	80	>100
N. rubra SN 5201	1.3	42	5	9	2	<0.08	40
N. pellegrini SN 5101	1.0	66	13	25	2	10	>100
N. erythropolis V49	0.7	66	26	12	<1	>100	>100
N. amarae 27808	1.3	60	11	21	<1	0.5	>100
N. amarae LL-SE-6	1.6	56	18	10	4	>100	>100
Control	0	74	41	5	—	>100	>100

SN 5201. Unfortunately, the concentration of biosurfactant produced was very low in glucose-grown cultures, reaching a maximum value of only five times the CMC in the case of *R. reseus* 65001.

All of the test cultures except *N. amarae* LL-Se-6 were able to grow in the mineral salts medium using n-hexadecane as their sole carbon and energy source, as shown in Table 2. All of the organisms which did grow in this medium showed a large increase over glucose-grown cultures in their ability to lower the surface and interfacial tensions of the whole broth, with the exception of *R. lentifragmentus* 23002. This strain produced considerable surface activity using either D-glucose or n-hexadecane.

In most cases, the concentrations of biosurfactant produced also increased when these bacteria were grown using n-hexadecane instead of glucose. These results were expected since it is to the advantage of the organisms to produce biosurfactants for efficient growth when using hydrocarbon substrate.[2] Hydrocarbon emulsification and bacterial surface energies are affected by biosurfactants, and these factors influence hydrocarbon assimilation. In all cases when the bacteria grew well on n-hexadecane, they reduced the spreading tension of the whole broth to a much greater extent than when they were grown on glucose. A low or negative spreading tension value favors the stability of emulsions.[22] Bacteria tend to create the emulsion condition

Table 2. Evaluation of Rhodochrous bacteria grown on 2% v/v hexadecane

Organism	*Biomass* (g/l)	*Surface Tension* (mN/m)	*Interfacial Tension* (mN/m)	*Spreading Tension* (mN/m)	*Biosurfactant Concentration* ($X\ CMC^{-1}$)	*Half-life of L92 Emulsion* (hr)	*Half-life of Tween-Span Emulsion* (hr)
R. aurantiacus 80001	1.0	28	0.35	−0.35	8	5	30
R. rubropertinctus 60003	1.0	31	2	1	2	9	1
R. lentifragmentus 23002	0.5	32	6	−2	30	30	>100
R. reseus 65001	0.2	33	5	0	4	20	>100
R. terrae 70006	0.3	30	3	−1	4	30	>100
R. rhodochrous UWO 837	0.3	35	3	4	20	1	40
R. rhodochrous 40001	0.6	32	3	1	30	5	20
N. rhodochrous L-100-22	0.4	33	2	3	1	>100	100
N. rubra SN 5201	0.2	30	1	1	30	5	20
N. pellegrini SN 5101	0.6	30	5	−3	30	140	90
N. erythropolis V49	1.6	34	3	3	20	8	>100
N. amarae 27808	0.3	33	2	3	5	20	>100
N. amarae LL-SE-6	0.1	65	28	9	2	>100	>100
Control	0	68	38	2	—	>100	>100

necessary for efficient uptake of hydrocarbon substrate when cells are grown on such substrates.

Whether grown on glucose as shown in Table 1 or on n-hexadecane as shown in Table 2, the addition of bacterial whole broth decreased the half-lives of test emulsions. Normally these oil-in-water emulsions had half-lives greater than 100 hr, but the addition of a standard volume of whole broth in most cases caused the emulsions to separate much more quickly. Results show that the test emulsion half-lives were reduced to less than 5 min (0.08 hr) in some cases after the addition of bacterial whole broth. Although it was shown that in most cases the concentration of biosurfactant responsible for lowering the whole broth surface tension increased in hexadecane-grown cultures, the ability to destabilize oil-in-water emulsions was more dependent on the amount of growth than on the carbon source. This result indicates that the major factor(s) responsible for demulsification is different than that which causes reduction of surface tension. It also points to the possible involvement of the cells themselves rather than any extracellular product in the destabilization of emulsions. Both types of test emulsions were oil-in-water emulsions. The difference between them lay in the surfactant used to give them stability. Therefore, any difference in the bacterial whole broth's ability to demulsify the two emulsion types must be related to its interaction with the stabilizing surfactant(s). This is particularly evident in light of the fact that some bacterial broths had a greater effect on the Tween-Span stabilized emulsion, while others had a greater ability to destabilize the L-92 stabilized emulsion, although in general the L-92 emulsion type was the most sensitive. No significant foaming ability was found with any of the sample broths. However, this may be due to the time of sampling as well as the culture conditions.

Most of the rhodochrous complex bacteria tested showed several types of surface activity. Demulsification by these bacteria was observed whether the carbon source was glucose or hexadecane. Surfactant production, as indicated by the lowering of surface and interfacial tensions as well as the X CMC^{-1} values, was observed for most strains grown on n-hexadecane. Several of the strains tested also produced surfactants when grown on D-glucose, but in general the amount was less than observed for the hydrocarbon-grown cultures.

SUMMARY

Thirteen different microbial strains belonging to or closely related to the rhodochrous complex of bacteria were screened for their ability to produce biosurfactants and to demulsify. Although *Rhodococcus aurantiacus* 80001, *Rhodococcus lentifragmentus* 23002, and *Nocardia rubra* SN 5201 reduced whole broth surface tension values to 40, 38, and 42 mN/m, respectively, when grown on a D-glucose carbon source, the highest biosurfactant concentration observed was a value of only five times the CMC that was produced by *Rhodococcus reseus* 65001. When grown using n-hexadecane as the carbon source, all of the cultures except *Nocardia amarae* LL-SE-6 lowered the whole broth surface tension to 35 mN/m or less, and biosurfactant concentrations of up to 30 times the CMC were recorded. When grown using n-hexadecane, *Rhodococcus aurantiacus* 80001 was able to lower the interfacial tension between the whole bacterial broth and n-hexadecane to 0.35 mN/m. Whether grown using n-hexadecane or D-glucose, many of the microorganisms produced a culture broth that greatly shortened the half-lives of surfactant-stabilized oil-in-water emulsions when added to them.

ACKNOWLEDGMENTS

This work was supported by the Natural Science and Engineering Research Council of Canada.

REFERENCES

1 FORBES, A.D. 1980. Microorganisms in oil recovery, p. 169–180. *In* D.E.F. Harrison, I.J. Higgins, and R. Watkinson (ed.), Hydrocarbons in biotechnology. Heyden and Sons Ltd., London.

2 ZOBELL, C.E. 1946. Bacteriological process for treatment of fluid-bearing earth formations. United States patent No. 2, 413,278.

3 ZOBELL, C.E. 1947. Bacterial release of oil from oil-bearing materials. World Oil. *127*:35–41.

4 COOPER, D.G., and J.E. ZAJIC. 1980. Surface-active compounds from microorganisms. Adv. Appl. Microbiol. *26*:229–253.

5 GERSON, D.F., and J.E. ZAJIC. 1977. Bitumen extraction from tar sands with microbial surfactants. The Oil Sands of Canada-Venezuela. CIM Special Publication No. *17*:705–710.

6 WAGNER, F., W. LINDORFER, and W. SCHULTZ, 1976. Verfahren zur Verbesserung des Ausbeute bei Gewinnung von Erdol durch Wasserflutten. German patent No. 24:19–267.

7 LISSANT, K.J. 1977. Emulsification and demulsification in oil recovery, p. 93–100. *In* Improved oil recovery surfactant polym. flooding. Pap. AIChE Symp.

8 AKIT, J., D.G. COOPER, K.I. MANNINEN, and J.E. ZAJIC. 1981. Investigation of potential biosurfactant production among phytopathogenic *Cornyebacteria* and related soil microbes. Current Microbiol. *6*:145–150.

9 BOUSFIELD, I.J., and M. GOODFELLOW. 1976. The "Rhodochrous" complex and its relationship with allied taxa, p. 39-65. *In* M. Goodfellow, G.H. Brownell, and J.A. Serrano (ed.), The Biology of the *Nocardiae*. Academic Press, New York.

10 BUCHANAN, R.E., and N.E. GIBBONS. 1974. Bergey's manual of determinative bacteriology, 8th ed. The Williams and Wilkins Co., Baltimore.

11 GOODFELLOW, M., and G. ALDERSON. 1977. The Actinomycetegenus Rhodococcus: a home for the "Rhodochrous" complex. J. Gen. Microbiol. *100*:99–122.

12 GOODFELLOW, M., and D.E. MINNIKIN 1977. Nocardioform bacteria. Ann. Rev. Microbiol. *31*:159–180.

13 MINNIKIN, D.E., and M. GOODFELLOW. 1976. Lipid composition in the classification and identification of *Nocardiae* and related taxa, p. 160–219. *In* M. Goodfellow, G.H. Brownell and J.A. Serrano (ed.). The biology of the *Nocardiae*. Academic Press, New York.

14 ASSELINEAU, J. 1966. The bacterial lipids. Herman, Paris.

15 COOPER and ZAJIC, op cit.

16 HOLDOM, R.S., and A.G. TURNER. 1968. Growth of *Mycobacterium rhodochrous* on n-decane: a new growth factor and emulsifying agent. J. Appl. Bact. *34*:448–456.

17 MARGARITIS, A., K. KENNEDY, J.E. ZAJIC, and D.F. GERSON. 1979. Biosurfactant production by *Nocardia erythropolis*. Dev. Ind. Microbiol. *20*:623–630.

18 RAPP, P., H. BOCK, V. WRAY, and F. WAGNER. 1979. Formation isolation and characterization of trehalose dimycolates from *Rhodococcus erythropolis* grown on n-alkanes. J. Gen. Microbiol. *115*:491–503.

19 HUG, H., and A. FIECHTER. 1973. Assimilation of aliphatic hydrocarbons by *Candida tropicalis*. I. Analytical problems in hexadecane batch experiments. Arch. Mikrobiol. *88*:77–86.

20 AKIT, COOPER, MANNINEN, and ZAJIC, op cit.

21 BARNETT, S.M., S.K. VALENKAR, and C.W. HOUSTON. 1974. Mechanism of hydrocarbon uptake by microorganisms. Biotechnol. Bioeng. *16*:863–865.

22 GERSON, D.F., and J.E. ZAJIC. 1978. Surfactant production from hydrocarbons by *Corynebacterium lepus* sp. nov. and *Pseudomonas asphaltenicus* sp. nov. Dev. Ind. Microbiol. *19*:577–599.

11

EFFECT OF NITROGEN SOURCES ON SURFACTANT PRODUCTION BY *ARTHROBACTER PARAFFINEUS* ATCC 19558

Z. Duvnjak, D.G. Cooper, and N. Kosaric

University of Western Ontario

Although it is known that microbes synthesize surfactants, little is known about the best production media and optimization.

It has been reported that *Arthrobacter paraffineus* KY403 produced surface active agents when grown on hydrocarbons.[1] In that study, anthrone positive lipids were found in the culture broth layer, which also contained bacterial cells and the residual hydrocarbon. The lipids were characterized as α-branched-β-hydroxy-fatty acid trehalose esters. When using the same organism, a primary fatty alcohol was found with the carbon skeleton corresponding to n-paraffin, which was used as the carbon source. This product appeared rapidly in the early lag phase of the growth.[2] A surface active agent was also detected in the culture broth after growth of an *Arthrobacter* species RAG-1 on either n-hexadecane or crude oil.[3]

Nitrogen sources played an important role in surfactant production. The highest surfactant concentrations were noticed when *Corynebacterium* sp. SFC-D was grown in a medium containing ammonium sulfate as the nitrogen source. Surfactant production was 75% lower when this compound was replaced with sodium nitrate.[4] The nitrate form of nitrogen was superior to ammonium nitrogen in production of extracellular fatty acids by *Nocardia* and *Pseudomonas* species.[5]

Yeast extract and nutrient broth had a profound effect on surfactant production. When these two materials were added to cultures of *Corynebacterium*, the surfactant production was increased by ten and eight times, respectively.[6] This increase in surfactant production with *C. lepus* was fifty and one hundred times, respectively.[7]

The effect of various nitrogen sources on surfactant production by *Arthrobacter paraffineus* ATCC 19558 is presented in this work.

MATERIALS AND METHODS

Arthrobacter paraffineus ATCC 19558 was grown on a slant medium containing Difco nutrient broth (0.8%) and agar (2%) for 7 days at room temperature.

For inoculum preparation, a medium containing nutrient broth (0.8%) and glucose (1%) was used. Growth of inoculum was carried out for 3 days at $23 \pm 1°C$. Five percent (v/v) of inoculum was used for inoculation.

A modified mineral salts medium (MMSM) was used for surfactant production and as a control for all experiments except for testing yeast extract and peptone.[8] It contained NH_4NO_3 (0.4%), KH_2PO_4 (0.4%), Na_2HPO_4 (0.6%), $MgSO_4 \cdot 7H_2O$(0.02%), $CaCl_2 \cdot 2H_2O$(0.001%), $FeSO_4 \cdot 7H_2O$ (0.001%), and Na_2EDTA (0.0014%). The medium used in testing the effect of yeast extract and peptone had the same composition except that the concentrations of $FeSO_4 \cdot 7H_2O$ and Na_2EDTA were 0.0001% and 0.00014%, respectively. Hexadecane (5% v/v) was used as a source of carbon. Surfactant production in these media served as a control broth if not stated otherwise. The different materials containing nitrogen were added to this medium at concentrations shown. Initial pH of the media was 7.0. The microorganism was grown for 5 days.

Both inoculum preparation and production of surfactant were carried out at $23 \pm 1°C$ in 500-ml Erlenmeyer flasks containing 100 ml of the medium, placed in a rotary shaker operated at 200 rpm.

Biomass was determined by the extraction method according to Hug and Fiechter, which uses a mixture of ethanol-chloroform-n-butanol for extraction of residual hydrocarbons.[9]

Surface tension and interfacial tension were measured by the Fisher autotensiomat model 215. The surfactant concentration was expressed in terms of the critical micelle concentration (CMC). The CMC is defined as the concentration of surfactant necessary to initiate micelle formation.[10] The CMC could be estimated from a plot of surface tension vs. the log of the percent dilution as the highest dilution at which the surface tension reaches the lowest value. The reciprocal of this dilution is proportional to the amount of surfactant present in the broth. Results are presented in terms of CMC^{-1}, which actually represents how the dilution factor for a culture broth should be diluted to reach the critical micelle concentration.

The demulsification ability of the culture was measured by a method described previously.[11] This ability is expressed in terms of emulsion half-lives ($t_{1/2}$) when aliquots of whole broth were added to oil-in-water (o/w) and water-in-oil (w/o) emulsions stabilized by synthetic surfactants.

RESULTS

As CMC^{-1} values are generally known to be subject to variations among different groups of experiments, due to minor differences in environmental conditions, controls were included with each group. In this way a more accurate assessment of the effect within each group of experiments could be made. Uninoculated media contained no measurable amount of surfactant and showed a surface tension of about 68 mN/m.

Yeast extract in the MMSM had a positive effect on surfactant production by *A. paraffineus* (Table 1). When concentrations of yeast extract were between 0.01% and 0.1% (w/v), CMC^{-1} in the broths was about 600. This value was approximately 6 times higher in comparison to the control medium that contained no yeast extract. Yeast extract concentrations higher than 0.1% gave CMC^{-1} values similar to that in the control medium.

Table 1. Effect of yeast extract on surfactant biosynthesis by *A. paraffineus*.

Yeast Extract % (w/v)	*Surface Tension (mN/m)*	*CMC^{-1} × 100*
Control	31	96
0.01	31	606
0.02	32	606
0.04	35	625
0.05	30	595
0.06	34	613
0.08	34	678
0.10	41	434
0.20	39	79
0.30	37	72
0.40	41	100
0.05*	33	5
0.20*	38	1

Control — MMSM containing NH_4NO_3 and no yeast extract
*Yeast extract as a sole source of nitrogen.

Surface tensions of the whole broth containing yeast extract ranged from 31 to 41 mN/m. The higher surface tensions corresponded to higher concentrations of yeast extract. The effect of yeast extract concentration on surface tension of broth has been noticed previously in the study of surfactant production by *Corynebacterium lepus*.[12] When the broths were diluted with distilled water to a ratio of 1:10, surface tensions dropped to 30–31 mN/m. The same phenomenon was noticed for interfacial tensions. The interfacial tensions after dilution were in the range of 1–2 mN/m.

When NH_4NO_3 was omitted from the medium and was replaced by yeast extract, the production of surfactant dropped substantially, as is shown in Table 1.

It is evident from these results that yeast extract cannot be used as a sole source of nitrogen for surfactant production by this microorganism

Peptone, bacto-tryptone, and nutrient broth in the MMSM had a similar effect on surfactant production as the yeast extract (Figure 1). Their optimum concentrations were found to be 0.2%, 0.8%, and 0.4%, respectively. Higher than optimum concentrations of peptone and nutrient broth caused a decrease in surfactant production.

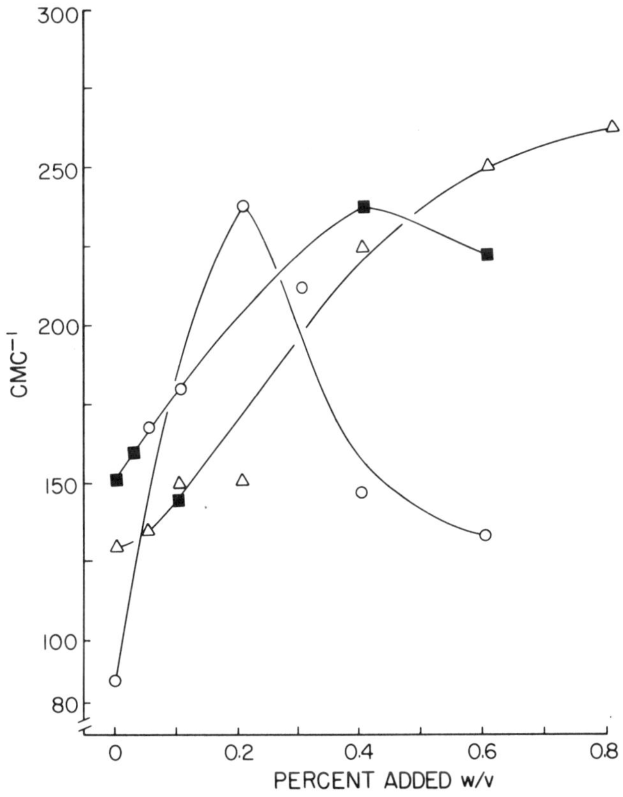

Fig. 1. Effect of peptone, bacto-tryptone and nutrient broth on surfactant biosynthesis in *A. paraffineus*.
○—peptone; △—bacto-tryptone; ■—nutrient broth

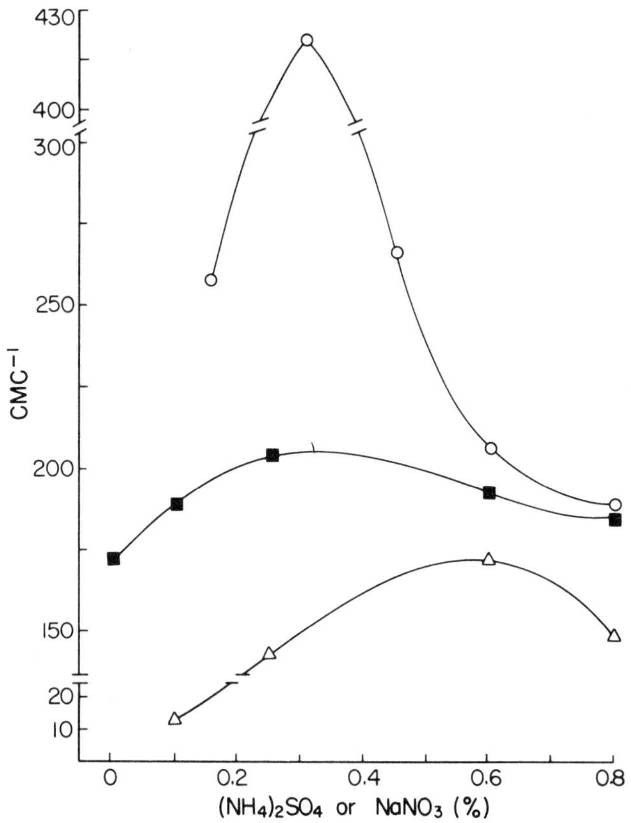

Fig. 2. Effect of ammonium sulfate and sodium nitrate on surfactant biosynthesis in *A. paraffineus*.
○—$(NH_4)_2SO_4$; △—$NaNO_3$; ■—0.4% NH_4NO_3 + varying concentrations of $NaNO_3$

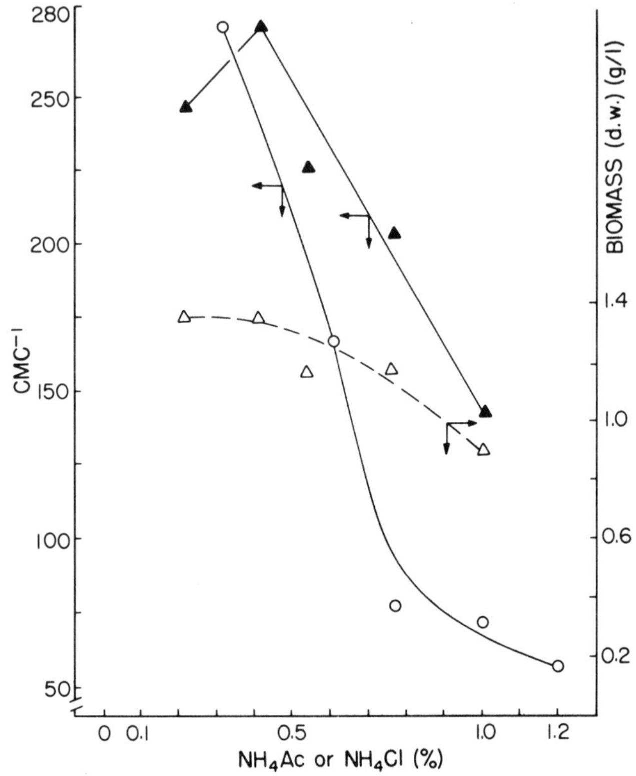

Fig. 3. Effect of ammonium acetate and ammonium chloride on surfactant biosynthesis in *A. paraffineus*.
△,▲—NH_4Cl; ○—NH_4-acetate

When 0.6% and 0.8% bacto-tryptone were used in the MMSM from which ammonium nitrate was omitted, CMC^{-1} values were around 2 and 5 and surface tensions were 38 and 36 mN/m, respectively. CMC^{-1} values were negligible when nutrient broth in a concentration of 0.05 or 0.60% was utilized as a sole nitrogen source in the medium. Surface tensions of the cultures containing 0.05% and 0.60% of nutrient broth were 34 and 45 mN/m, respectively. According to these results, neither bacto-tryptone nor nutrient broth can serve as the sole nitrogen source for surfactant production.

When ammonium sulfate was used instead of ammonium nitrate, its optimum concentration was found to be 0.3%, and a CMC^{-1} value of 422 was obtained (Figure 2). This surfactant concentration is about 2.2 times higher than the CMC^{-1} value obtained with ammonium nitrate. Surface tensions of broth with various concentrations of ammonium sulfate were in the range of 30–32 mN/m.

Sodium nitrate also appears to be a good nitrogen source for surfactant production. Its optimum concentration was 0.6% in the MMSM from which ammonium nitrate was omitted.

When the MMSM containing 0.4% of ammonium nitrate was supplemented by 0.25% (w/v) of sodium nitrate, about 18% higher yield of the surfactant was obtained as compared to the control broth (Figure 2).

When NH_4NO_3 was tested, its optimum concentration was found to be 0.8%, yielding a CMC^{-1} value of 220. This is about 20% higher than the value obtained in the control broth, which contained 0.4% of this compound.

The optimum concentration of ammonium chloride was found to be 0.4%. At concentrations tested, the maximum biosurfactant production was achieved with 0.3% of ammonium acetate (Figure 3). In both cases, surfactant yields were 42% higher in comparison to the control. Surface tensions varied and were in the range of 29–32 mN/m. The maximum biomass concentration achieved in the broth with NH_4Cl was the same as in the control broth. A decrease in biomass concentration was observed at higher concentrations of ammonium chloride. The decrease in surfactant production is directly proportional to the decrease in the yield of biomass.

The effect of ammonium bicarbonate on surfactant production is shown in Table 2. The highest surfactant concentration was obtained in the broth containing 0.6% of ammonium bicarbonate. A 0.2% increase in bicarbonate concentration caused about a 50% drop in the CMC^{-1} value. No biosurfactant was produced when the concentration of this salt was further increased to 1.1%, even though some biomass was produced at the same concentration. This can also be seen from the value of surface tension, which was reduced only to 53 mN/m. It is also evident that increasing concentrations of ammonium bicarbonate above optimum had more effect on surfactant production than on the biomass yield.

Table 2. Influence of ammonium bicarbonate on biosynthesis of surfactant by *A. paraffineus*.

NH_4HCO_3, % (w/v)	Surface Tension (mN/m)	CMC^{-1} × 100	Biomass (dw) (g/l)
0.3	31	263	2.0
0.6	32	305	0.84
0.8	32	146	0.70
1.1	53	—	0.44
1.4	50	—	0.06
Control	30	183	1.12

Control: MMSM with NH_4NO_3.

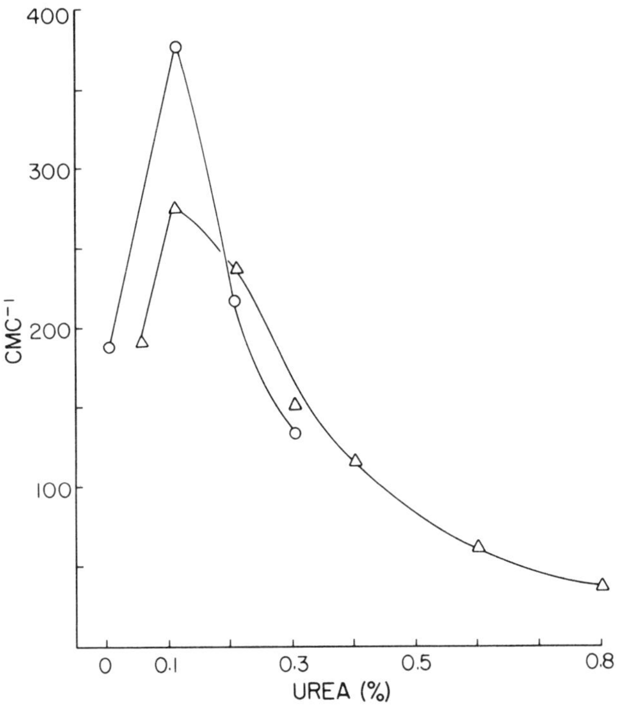

Fig. 4. Effect of urea on surfactant biosynthesis in *A. paraffineus*. △—urea; ○—0.4% NH_4NO_3 + varying concentrations of urea

Urea can be used as a sole nitrogen source both for growth and for surfactant production. With 0.1% of this compound in the MMSM from which NH_4NO_3 was omitted, the yield of surfactant was about 47% higher than the control (Figure 4). It can be seen from the same figure that the MMSM containing NH_4NO_3 and supplemented with 0.1% of urea yielded double the amount of surfactant in respect to broth that contained no urea.

The addition of aspartic acid or asparagine to the MMSM caused an increase in surfactant production (Figure 5). The yield of surfactant at optimum concentration of asparagine (0.4%) was 43% higher than the control. The amount of surfactant increased with increasing concentration of aspartic acid. Biomass also increased with addition of aspartic acid and reached a value of 2.5 g/l dry weight (dw) at the maximum tested concentration of this amino acid. This is more than twice the amount obtained in the control broth.

The optimum concentration of glycine and glutamic acid in the MMSM containing ammonium nitrate was found to be around 1% (Figure 6).

Experiments were also performed with the MMSM in which ammonium nitrate was omitted. In this case, asparagine could serve as a nitrogen source, as shown in Table 3. The yield of the surfactant was about the same as in the control broth that contained NH_4NO_3 as the only nitrogen source. However, aspartic acid, glutamic acid, and glycine do not appear to be good nitrogen sources for this process.

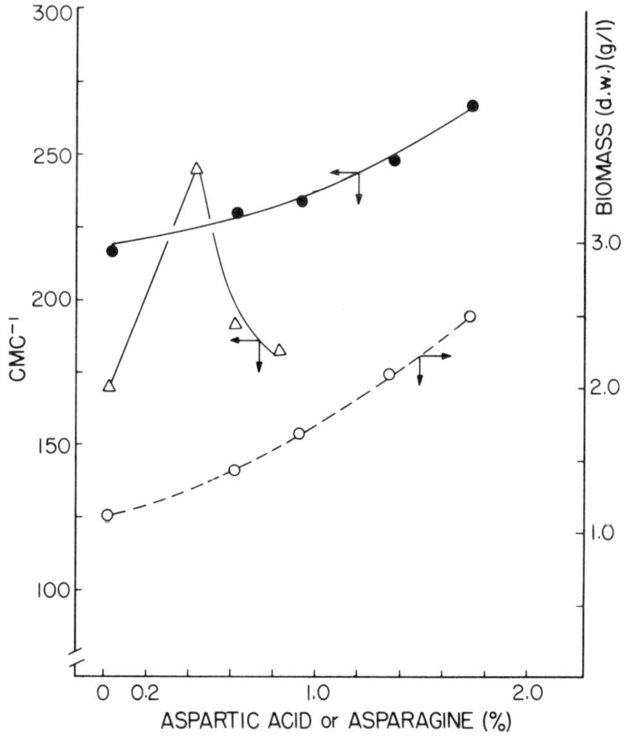

Fig. 5. Effect of asparagine and aspartic acid on surfactant biosynthesis in *A. paraffineus*;
△—asparagine; ○,●—aspartic acid

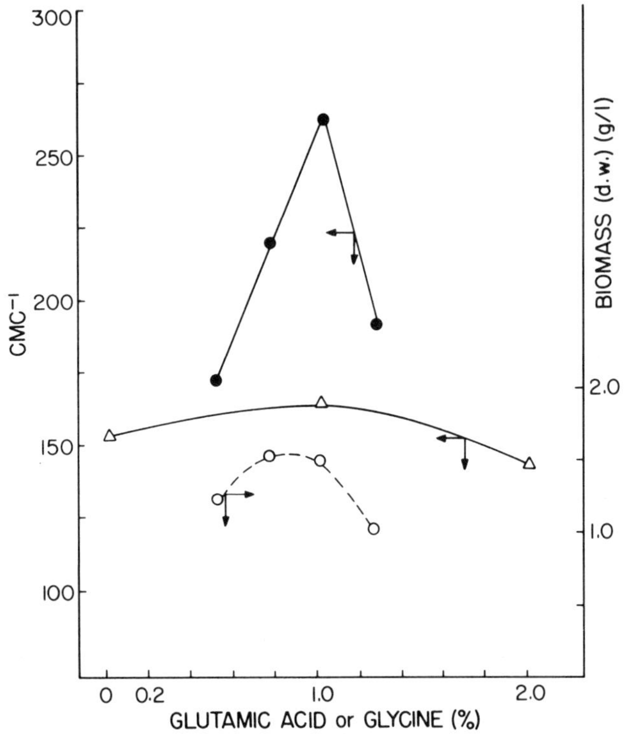

Fig. 6. Effect of glycine and glutamic acid on surfactant biosynthesis in *A. paraffineus*;
●,○—glycine; △—glutamic acid

The whole culture broth also acted as a good demulsifier. Table 4 shows some of the results obtained. As shown in the table, the half-lives of w/o and o/w emulsions were affected by the source of nitrogen that was used for growth. When

Table 4. Effect of nitrogen source on demulsification of o/w and w/o emulsions by *A. paraffineus*.

Nitrogen Source	*Salt Concentration in Broth (%)*	*Half-Lives* ($t_{1/2}$) (min)* W/O	O/W
NH_4NO_3 (control)	0.4	29	48
Urea	0.1	12	38

*0.5 ml of whole broth was added to 10 ml of emulsion.

urea was used, the half-lives ($t_{1/2}$) of emulsions were shorter in comparison to the control, which contained NH_4NO_3 as a nitrogen source. The demulsification effect did not appear when the media were added to emulsions.

DISCUSSION

It is evident from the obtained results that yeast extract in low concentrations enhanced surfactant production. This observation is in agreement with the positive effect of yeast extract on surfactant production as shown for *Corynebacterium* sp. SFC-D, *C. lepus*, and *C. hydrocarboclastus*.[13,14,15] With *A. paraffineus* the results were not as dramatic as with *C. lepus* and *C.* sp. SFC-D. Surfactant production was up to 50 times higher when *C. lepus* was grown in a medium containing yeast extract in comparison to the control.

It is known that yeast extract has a growth-promoting function, but it was found that surfactant production by *Corynebacterium* sp. SFC-D was much more enhanced (10

Table 3. Amino acids as a sole nitrogen source in surfactant biosynthesis by *A. paraffineus*.

Amino Acid	*Concentration, % (w/v)*	*Surface Tension (mN/m)*	*CMC^{-1} × 100*	*Biomass (dw) (g/l)*
Aspartic acid	0.6	35	21	1.12
	1.0	32	14	0.90
Control	—	32	217	1.11
Asparagine	0.2	30	143	
	0.6	33	169	
Control	—	31	171	
Glutamic acid	0.5	31	9.0	
	1.5	34	11	
Control	—	30	154	
Glycine	0.25	41	—	
	1.00	30	2.0	
Control	—	32	217	

Control — MMSM containing NH_4NO_3 as a nitrogen source.

times) with addition of the nutrient than the biomass was increased (1.8 times).[15] This shows a high stimulating effect of yeast extract on surfactant production when added to a medium that contains a mineral source of nitrogen.

A similar effect on surfactant production was noticed when bactotryptone, peptone, and nutrient broth were added to the MMSM.

Growth occurred when these materials were used as the sole nitrogen source in the MMSM from which ammonium nitrate was omitted. Under these conditions the surfactant production was very poor. The same effect was caused by amino acids (Table 3). This observation may also explain why the tested complex nitrogen sources (yeast extract, peptone, bacto-tryptone, and nutrient broth) cannot support surfactant production. Nitrogen in these sources is mainly in the form of amino nitrogen, which does not seem to be preferred for production of biosurfactant by this organism.

A considerable amount of surfactant was produced when an amino form of nitrogen was used as a sole nitrogen source. Table 3 shows that this form of nitrogen is superior to amino nitrogen for surfactant production.

When inorganic sources of nitrogen were used in the MMSM, the best results were obtained with ammonium salts (Table 2, Figure 2). Urea also appeared to be a good source of nitrogen in surfactant production. These results are in agreement with findings reported for *Corynebacterium* sp. SFC-D grown in a hydrocarbon medium.[15] However, they differ from the results obtained in production of extracellular fatty acids by *Nocardia* and *Pseudomonas* species since nitrate was superior to an ammonium form of nitrogen.[16] In addition, it was reported that sodium nitrate was superior to other sources of nitrogen in flocculant production by *A. hydrocarboclastus*.[17]

It was also found in this study that surfactant production followed biomass increase (Figures 3, 5, and 6 and Table 2). Similar observations were previously reported.[17,18,19]

The whole culture broth also showed a good demulsification capability for both w/o and o/w emulsions. This capability could have practical applications in breaking various kinds of industrial emulsions (see Gutnick & Rosenberg and Zajic & Panchal).

SUMMARY

Arthrobacter paraffineus ATCC 19558 increased surfactant production when yeast extract, peptone, bacto-tryptone, nutrient broth, aspartic acid, asparagine, glycine, or glutamic acid was present in modified mineral salts medium containing 0.4% NH_4NO_3. This effect was especially pronounced with yeast extract, which yielded six times more surfactant than the control. Of these materials only asparagine can serve as a sole nitrogen source for surfactant production. When inorganic salts were used as a nitrogen source, the microorganism preferred ammonium to nitrate forms of nitrogen.

Urea could be utilized both as a sole source of nitrogen and in combination with an inorganic nitrogen salt, yielding relatively high surfactant concentration in the broth.

REFERENCES

1 SUZUKI, T., K. TANAKA, I. MATSUBARA, and S. KINOSHITA. 1969. Trehalose lipids and α-branched-β-hydroxy fatty acid formed by bacteria grown on n-alkanes. Agr. Biol. Chem. *33*:1619–1627.

2 SUZUKI, T., and K. OGAWA. 1972. Transient accumulation of fatty alcohols by n-paraffin-grown microorganisms. Agr. Biol. Chem. *36*:457–463.

3 REISFIELD, A., E. ROSENBERG, and D.L. GUTNICK. 1972. Microbial degradation of crude oil: factors affecting the dispersion in seawater by mixed and pure cultures. Appl. Microbiol. *24*:363–368.

4 DE PAUW, S.B. 1981. Biosurfactant production from hydrocarbons by *Corynebacterium* sp. SFC-D. Master of Engineering Science thesis, University of Western Ontario, London, Canada.

5 LINDAY, E.M., and M.B. DONALD. 1961. Small scale studies of microbiological oxidation of hydrocarbons to long-chain fatty acids. Biotech. Bioeng. *3*:219–233.

6 DE PAUW, op cit.

7 GERSON, D.F., and J.E. ZAJIC. 1978. Surfactant production from hydrocarbons by *Corynebacterium lepus*, sp. nov., and *Pseudomonas asphaltenicus*, sp. nov. Developments in Industrial Microbiology *19*:577–600.

8 COOPER, D.G., J.E. ZAJIC, W.L. CAIRNS, and N. KOSARIC. 1980. Demulsification and microbes. Biochemical engineering research reports, Vol. VI. Biochemical Engineering, Faculty of Engineering, University of Western Ontario, London, Canada.

9 HUG, H., and A. FIECHTER. 1973. Assimilation of aliphatic hydrocarbons by *Candida tropicalis*. 1. Analytical problems in hexadecane batch experiments. Arch. Microbiol. *88*:77–86.

10 COOPER, D.G., J.E. ZAJIC, and D.F. GERSON. 1979. Production of surface active lipids by *Corynebacterium lepus*. Appl. Environ. Microbiol. *37*:4–10.

11 COOPER, D.G., C.R. MacDONALD, J. AKIT, Z. DUVNJAK, J.E. ZAJIC, and N. KOSARIC. 1980. Evaluation of surface active compounds produced by microorganisms. Part 1. Biochemical engineering research reports, Vol. VII. Biochemical Engineering, Faculty of Engineering, University of Western Ontario, London, Canada.

12 GERSON and ZAJIC, op cit.

13 KNETTIG, E., and J.E. ZAJIC. 1972. Flocculant production from kerosene. Biotech. Bioeng. *14*:379–390.

14 GERSON and ZAJIC, op cit.

15 DE PAUW, op cit.

16 LINDAY and DONALD, op cit.

17 KNETTIG and ZAJIC, op cit.

18 PANCHAL, C.J., and J.E. ZAJIC. 1978. Isolation of emulsifying agents from a species of *Corynebacterium*. Developments in Industrial Microbiol. *19*:569–576.

19 SUZUKI, TANAKA, MATSUBARA, and KINOSHITA, op cit.

ADDITIONAL REFERENCES

ERICKSON, L.E., and T. NAKAHARA. 1975. Growth in cultures with two liquid phases: hydrocarbon uptake and transport. Proc. Biochem. *10*:9–13.

GUTNICK, D.L., and E. ROSENBERG. 1977. Oil tankers and pollution: a microbiological approach. Annual Rev. Microbiol. *31*:379–396.

ZAJIC, J.E., and C.J. PANCHAL. 1976. Bio-emulsifiers. CRC Crit. Rev. Microbiol. *5*:39–66.

12 SOME CHARACTERISTICS OF THE BACTERIAL INOCULUM USED FOR OIL RELEASE FROM RESERVOIRS

I. Lazar

Institute of Biologica Sciences, Bucharest

The theoretical and partly practical bases of the use of bacteria for oil release from reservoirs were elaborated as a result of investigations carried out by Zobell between 1943–1955. As a consequence of the field investigations carried out in the 1957–1972 period in Czechoslovakia, Hungary, the USSR, and Poland, the method initially designed by Zobell was developed and improved.[1-6] At present the idea of using bacteria or some of their products to stimulate oil release from porous media has become generally accepted.

After injection into the reservoirs, bacteria act both on the oil and on the rock where the oil is located as well as on the injected nutrient support system. This results in the formation of several products (gases, organic or mineral acids, biopolymers, biosurfactants, etc.) that stimulate the oil release from reservoirs. This shows that the bacterial inoculum represents the major component injected into the formation. The effect of the inoculum depends upon the treatment applied to wells. This also explains why Zobell paid special attention to the bacterial inoculum.

Zobell used strains of *Desulfovibrio* on nutrient media with salts, calcium carbonate, or some other organic acids.[7] He also introduced the idea of using other heterotrophic bacteria having the ability to ferment carbohydrates anaerobically with the production of gases and other useful products.[8] Updegraff and Wren as well as Beck and Coty used bacterial species belonging to genera *Clostridium* and *Pseudomonas* for the same purpose.[9-13] In the U.S. in the 1962–1967 period, Hitzman used both suspensions of bacterial spores and methanogenic bacteria as well as living hydrocarbon-utilizing organisms to dislodge and upgrade oil in situ.[14,15,16]

In addition to the utilization of bacteria for oil release from reservoirs, research has been conducted in related areas such as the effects of microorganisms on a reservoir's fluids, oil-water interactions, composition of crude oil, and the physicochemical consequences of microbial action in the reservoir. This also includes investigations on the influence of different factors such as temperature, pH, organic and inorganic nutrient limitations, and rock permeability on bacterial activity.[17-27]

As mentioned earlier, an intensification of investigations took place in the 1957–1972 period in some European countries such as Czechoslovakia, Hungary, Poland, and the USSR where special attention was given to field trials. Dostalek and associates used bacterial inocula taken from species belonging to genera *Desulfovibrio, Pseudomonas,* and *Clostridium* and developed a nutrient support system involving glucose (2%) or molasses (4%).[28,29,30] Field trials for stimulating oil release from reservoirs were conducted in Czechoslovakia. In the Soviet Union, Kuznetsov et al., Ekzertsev, and Senyukov et al. used bacterial cultures (which were injected into the formation) along with such nutrients as molasses; gases such as methane were observed.[31,32,33]

In Hungary, the papers published between 1967–1973 give details about the field trials and the use of bacterial populations adapted to the reservoir's conditions.[34,35,36] The predominant bacteria in these adapted bacterial populations were the species belonging to genera *Pseudomonas, Clostridium,* and *Desulfovibrio*. The inoculum formed from such kinds of bacteria was injected into the formation along with a nutrient system containing molasses, mineral salts, a source of inorganic nitrogen, and phosphorus.

The field trials carried out in Poland between 1961–1971 were based on the use of bacterial inoculum represented by adapted bacterial populations formed from the species belonging to genera *Arthrobacter, Clostridium, Mycobacterium, Peptococcus,* and *Pseudomonas.*[37–42] Such bacterial populations were injected into the formation with the nutrient substrate of 4% molasses.

In contrast to the field trials in Czechoslovakia, Hungary, and the Soviet Union, the studies from Poland utilized collector models for establishing the oil-releasing capability from porous media.[42] The investigations carried out in Hungary and Poland have contributed to the development of the possibility of using bacterial populations adapted to the oil reservoir's conditions. The idea is to select populations belonging to mesophilic heterotrophic bacteria, which have the capacity to ferment carbohydrates as well as degrade crude oil.

After 1971 studies similar to those carried out in Hungary and Poland were started in Romania. The research was conducted by a group of microbiologists from the Institute of Biological Sciences, Bucharest, in collaboration with specialists from the Oil and Gas Research Institute, Cimpina, Romania, and from some oil and gas economical units. In the framework of these investigations, special attention was given to isolating bacterial populations for the reservoirs selected for microbiological treatment. The bacterial populations were characterized by the number of bacteria per milliliter, the main types of bacteria, oil-releasing capacity from a collector, and utilization of crude oil. Only selected bacterial populations with high oil-releasing activity were kept for multiplication in the quantities necessary for well injection. Between 1975–1980, seven wells from reservoirs with different physicochemical and petrophysical characteristics were submitted to microbiological treatment. In the work published during 1976–1982 are details concerning the sources used for the isolation of the bacterial population, the process for obtaining the adapted bacterial populations, and the characterization of bacterial inocula used in laboratory and field trials.[43–47] This paper focuses on some of the methods and processes for obtaining effective bacterial inocula used for well injection as well as the practical value and the possibilities of improving the qualities of such inocula.

SOURCES FOR THE ISOLATION OF THE BACTERIAL POPULATIONS

In the framework of the research carried out by Lazar (Romania) as well as that carried out mainly in Hungary by Dienes and Jaranyi (Hungary) and by Karaskiewicz (Poland), special attention was given to the category of sources of bacterial flora that had been partly adapted to the reservoir conditions yet still possessed a high capacity of fermentation. The first category of the sources refers to formation water from the reservoir that will be submitted to microbiological treatment, well mud resulting from the cleaning of the bottom of the well, soil around the well, and sewage mud from purification stations for formation water. The second category of sources refers to residual water mainly from the food industry, anaerobically fermented sludge from biogas stations, sewage collected from the canalization of petrochemical plants, residual water or the scum (defecation foam) from sugar processing plants, and mud from the bottom of the therapeutic lakes. Several papers published during the last five years by Lazar present the results concerning the possibility of getting adapted bacterial populations from such kinds of sources.[48–54] The sources that proved to be the most adequate for our purpose were the formation water from the first category of sources and the scum (defecation foam) from sugar processing plants from the second category. The bacterial populations from these two sources adapt more easily for molasses metabolization with high gas production in the presence of water and oil and at the temperature of the reservoir for which these populations are adapted. Due to these reasons the bacterial inocula used in Romania for well injection between 1975–1980 was mainly represented by bacterial populations obtained from these two sources.

INSTALLATION OR SYSTEMS USED FOR THE ADAPTATION OF BACTERIA

For obtaining adapted bacterial populations from any source, it is necessary to use some adequate installations or systems. The first papers that give indications on the use of a special system or installation in the process of obtaining adapted bacterial populations by simple selection are those of Jaranyi, Jaranyi et al., and Karaskiewicz. Investigations carried out in Romania between 1975–1980 were based both on the use of the installations adopted by Jaranyi et al. and then by Karaskiewicz and on three other installations or systems described in detail in several patents and published papers by Lazar.[55–59] The inoculum used for well injection was prepared using these four installations, schematically represented in Figures 1, 2, 3, and 4 and conventionally designated in these published papers as A, B, C, and D. Of these, system B offered aerobic conditions, system A offered partly anaerobic conditions, and systems C and D offered strictly anaerobic conditions. System D is that used by Jaranyi et al. and then by Karaskiewicz. Some of the installations, for instance D, were designed with variations in accordance with the quantity of the inoculum needed, and consisted of vessels of 2-, 10-, and 50-1 capacity. The high capacity variations were used only for the preparation of the inoculum for well injection. The C installation was

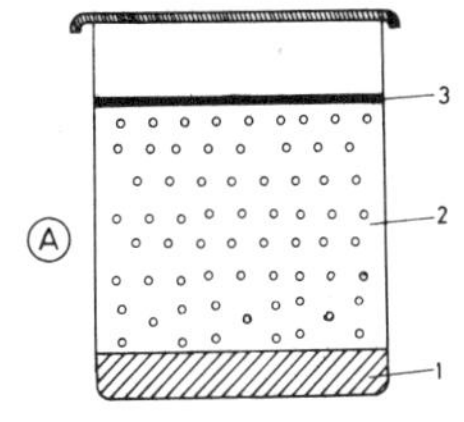

Fig.1

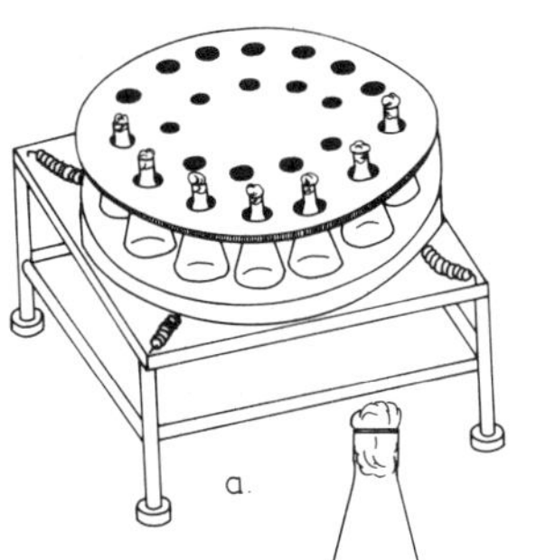

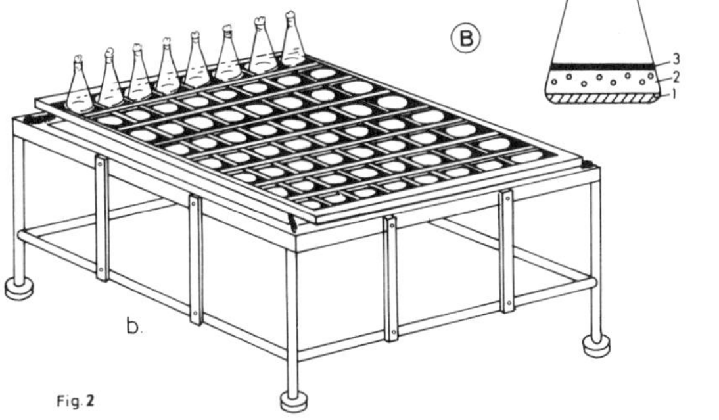

Fig.2

Fig. 1. Installation A for bacterial adaptation
1 = fermented scum (foam) from sugar processing plants
2 = formation water with 4% molasses
3 = crude oil

Fig. 2. Installation B for bacterial adaptation
a, b = types of rotary shakers used
1, 2, 3 = similar to Figure 1

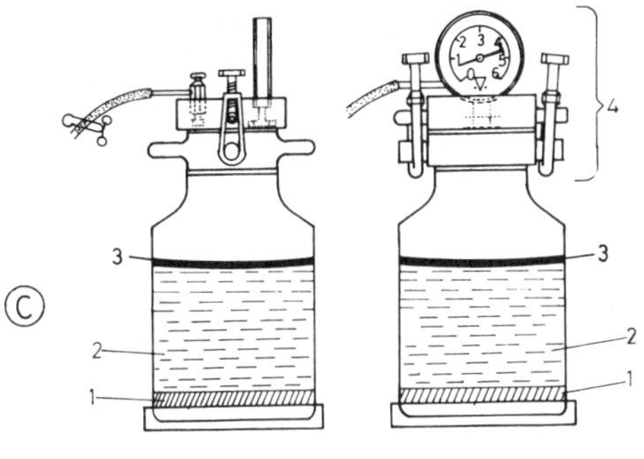

Fig.3

Fig. 3. Installation C for bacterial adaptation
1, 2, 3 = similar to Figure 1

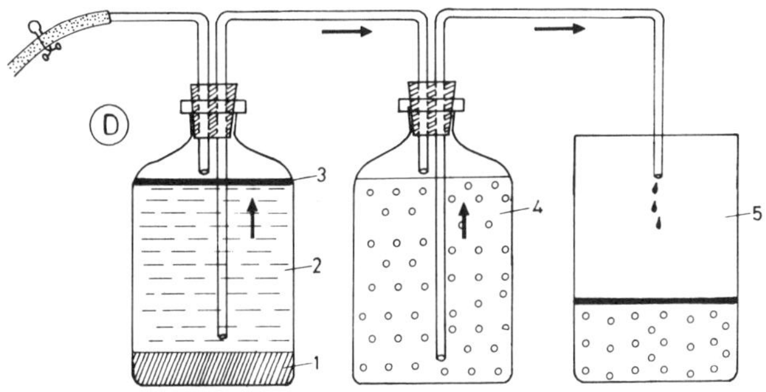

Fig.4

Fig. 4. Installation D for bacterial adaptation
1, 3 = similar to Figure 1
2 = formation water with 5% molasses
4 = formation water with 4% molasses
5 = vessel for collection of bacterial inoculum

represented by aluminum containers of 25-1 capacity, and the B installation was represented by two types of rotary shakers. One may hold 26 Erlenmeyer flasks of varying capacity (250 ml–1,000 ml), while the second type may hold 64–80 flasks of 750 ml each.

METHODS FOR ADAPTATION TO RESERVOIR CONDITIONS

On the basis of the work carried out by Jaranyi et al. and by Karaskiewicz, the bacterial inoculum used in the field trials was obtained by a simple method that did not involve mutagenesis or other special techniques of selection. Mainly, the work carried out in Romania in 1975–1980 was based on the same method. Probably there is little chance to modify this method before 1985. The main idea of the method is to obtain bacterial populations from several sources, mentioned above. Before multiplying such bacterial populations in the quantities necessary for field trials, the populations are submitted to a simple process of selection. During this process, the main test refers to establishing the capacity of the adapted bacteria to release the oil from porous media of a collector in conditions as much similar to those existing in the reservoir where the inoculum is to be injected as possible.

The investigations carried out in Romania after 1973 by the Institute of Biological Sciences concentrated on obtaining an inoculum that represented both the adapted bacterial strains or mixture of strains initially isolated from the several sources in pure culture and by adapted bacterial populations. The temperature at which the adaptation process of bacterial strains, the mixture of the bacterial strains, and the growth of the bacterial population were conducted was similar to that of the reservoir for which the inoculum was prepared. The resulting bacterial populations had the capacity to use molasses with high production of gases as well as organic or mineral acids, biopolymers, and biosurfactants. At the same time some of the bacteria from the inoculum were able to utilize some fractions of crude oil; for example, the heavy fractions were utilized, with a resulting decrease in oil viscosity which possibly improved oil recovery.

The laboratory experiments on collector models showed that the bacterial populations were more active in releasing oil than the strains or mixture of strains initially isolated in pure cultures.[60]

Thus it was established that in collector models the strains isolated in pure culture from several sources (mainly from formation water of the reservoir for which adaptation is made) yielded an oil release up to 3.5%; the mixture of strains, initially isolated in pure cultures, yielded an oil release up to 14.5%; while the bacterial population determined an 18–32.5% oil release. The high percentages of oil release were obtained by the adapted bacterial populations isolated from formation water, fermented scum from the sugar processing plants, and sludge or sediment from the plant used to purify injection water. These adapted bacterial populations were used as an inoculum for well injection during 1976–1980. The populations were prepared using the four above-mentioned installations of Lazar, and the predominant bacterial species belonged to the following genera: *Pseudomonas, Escherichia, Arthrobacter, Mycobacterium, Micrococcus, Peptococcus, Bacillus,* and *Clostridium*. In the inoculum injected into the formation, the bacteria from genera *Aerobacter, Flavobacterium, Cellulomonas*, and *Desulfovibrio* were identified with a low frequency. The methods for obtaining bacterial populations, adapted to reservoir conditions, are presented in detail in several patents and the aforementioned papers of Lazar and consist of the following three steps:

Step One. In the recipient vessels of 10 l, the following ingredients were introduced:

1. Scum from the sugar processing plants (1 kg). The scum was collected from the places very near to the processing plants where it was evacuated, during warm months of the year, when the maximum fermentation was observed with a noticeable, intense gas emanation.
2. Crude oil and formation water from the wells of the reservoirs from which the adaptation was made (50–70 ml oil and 7 l coproduced water).
3. Molasses, 200–250 g.

After the homogenization of the above-mentioned ingredients, the vessel was kept at a temperature similar to that of the reservoir for which the inoculum was prepared. The bacteria from the sources introduced into the vessel adapted to the formation water mineralization, the crude oil, and the reservoir's temperature and multiplied vigorously, causing the medium to become cloudy with a heavy frothing on the surface. This was observed after 24 hr for reservoirs with a low degree for formation water mineralization, while the water with a high degree of mineralization took 48–72 hr or even more. In this step the process for obtaining the first adapted bacterial population lasted from 4–7 days, depending upon the peculiarities of the formation reservoir water, crude oil, and the sources of the bacteria introduced in vessel. Scum from sugar processing plants always quickened the process and gave it maximum intensity. Evaluation of the adaptation process indicated that it was possible to obtain adapted bacterial populations for the reservoir from which the formation water and used crude oil were obtained.

Step Two. Small quantities of inoculum taken from the bacterial populations adapted in the first step were simultaneously introduced into the following four adaptation installations or systems:

Installation A (Figure 1) was represented by a vessel in which the following ingredients were introduced:

1. Bacterial inoculum obtained in step 1, 250–300 ml
2. Fermented scum from the sugar processing plants, 250–300 g
3. Crude oil and coproduced water from the wells of the reservoir for which the inoculum was prepared (50–70 ml crude oil + 7 l coproduced water)
4. Molasses, 200–250 g

Installation B (Figure 2) was represented by a rotary shaker that could hold Erlenmeyer flasks of varying capacities, 250–1,000 ml; usually flasks with 750-ml capacity were used. In these flasks 150–200 ml of nutrient support and inoculum were introduced as in Installation A.

Installation C (Figure 3) was represented by aluminum containers of 25-l capacity, fitted with a system for hermetic sealing, a manometer for reading the pressure as a result of bacterial activity, and valves for evacuating or collecting gases. The containers were loaded with 20 l of the same material used for Installations A and B.

Installation D (Figure 4) was represented by the installation previously used in Hungary and Poland. The first vessel was filled with nutrient support with 5% molasses and inoculum. The ingredients were in the same proportion as used for Installation A. The second vessel was filled with formation water and 4% molasses. The migration of the medium from the first two vessels to the third was an indication of the bacterial growth and activity, developed as a consequence of the adaptation to the formation water, crude oil, and reservoir temperature.

As already mentioned, all four installations were kept in a room with a temperature similar to that of the reservoir where the inoculum was to be injected. The process for obtaining the adapted bacterial population lasted from 2–4 to 7–10 days, depending upon the characteristics of the reservoir fluids (water and crude oil) and the ecological environments of the sources of bacteria.

The bacterial populations obtained from each installation were characterized on the basis of bacterial concentration per milliliter, main types of bacteria, oil-releasing ability, and utilization of the crude oil. Data concerning these aspects have been presented in previously published papers.[57–64] Mainly, a good bacterial growth in the four enumerated installations led to an inoculum with a cell concentration of 10^6–10^9 ml^{-1}, composed most frequently of bacteria from the genera *Pseudomonas, Escherichia, Arthrobacter, Mycobacterium, Peptococcus, Micrococcus, Clostridium*, and *Bacillus* and less frequently of bacteria from the genera *Flavobacterium, Nocardia, Aerobacter, Brevibacterium, Cellulomonas, Corynebacterium, Desulfovibrio*, and some yeasts.

The ability of adapted bacterial populations to mobilize petroleum was tested on three types of basic model collectors. The basic model collector and three variations (Figures 5, 6, and 7) were described in previous papers by Lazar.[62–66]

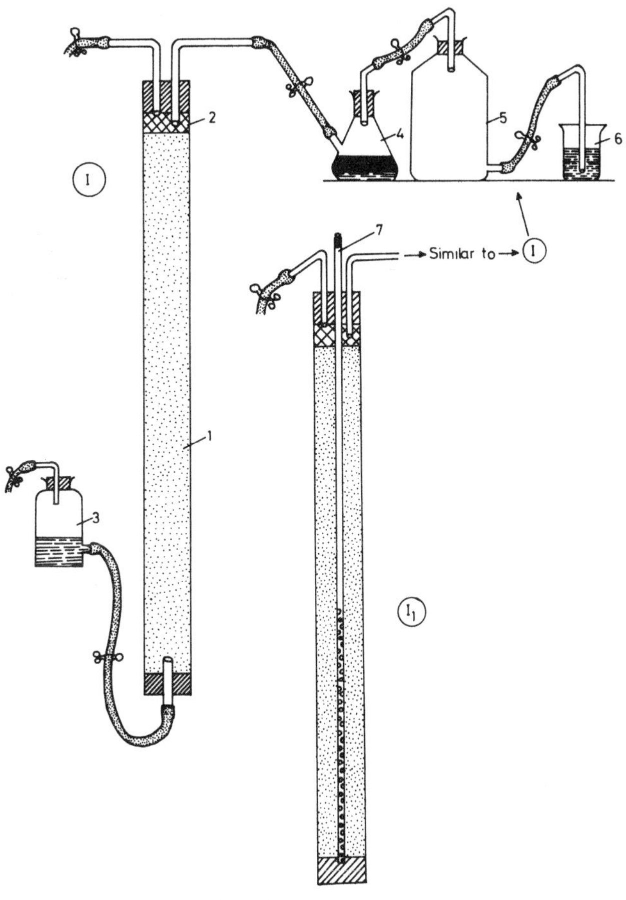

Fig. 5. Collector model I with variant I_1 for testing the oil releasing ability of adapted bacterial populations
1 = sand saturated with crude oil and then injected with nutrient support on molasses base and adapted bacterial populations
2 = glass wool for the filtration of the mobilized fluids
3 = vessel with nutrient support and inoculum
4 = vessel for collecting crude oil, mobilized due to bacterial activity
5 = vessel for collecting mobilized gases
6 = vessel with saturated NaCl solution for eliminating the surplus gases
7 = glass tube with lower part perforated, used for introducing nutrient support and inoculum in the collector

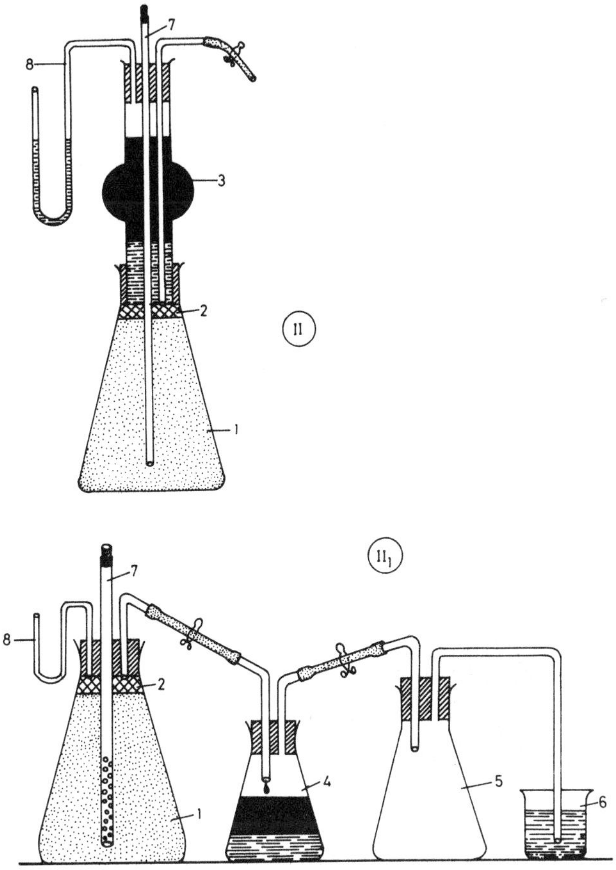

Fig. 6. Collector model II with variant II_1 for testing the oil releasing ability of adapted bacterial populations
1 = sand saturated with crude oil and then injected with nutrient support on molasses base and adapted bacterial populations
2 = glass wool for the filtration of the mobilized fluids
3 = special glass tube for collecting the gases and fluids from collector, mobilized due to bacterial activity
4 = vessel for collecting crude oil, mobilized due to bacterial activity
5 = vessel for collecting mobilized gases
6 = vessel with saturated NaCl solution for eliminating the surplus gases
7 = glass tube with lower part perforated, used for introducing nutrient support and inoculum in the collector
8 = capillary glass tube for indication of gases formation in the collector

Several adapted bacterial populations could be tested in series, in groups of ten, over a very short period of time on this collector model. Due to this advantage, the third collector model was most frequently used (Figure 8, III, III_1). The bacterial populations were selected on the basis of their highest oil-releasing ability.

Parallel with the testing of oil-releasing ability, tests for crude oil utilization by the adapted bacterial populations were also conducted. The degree of oil utilization varied

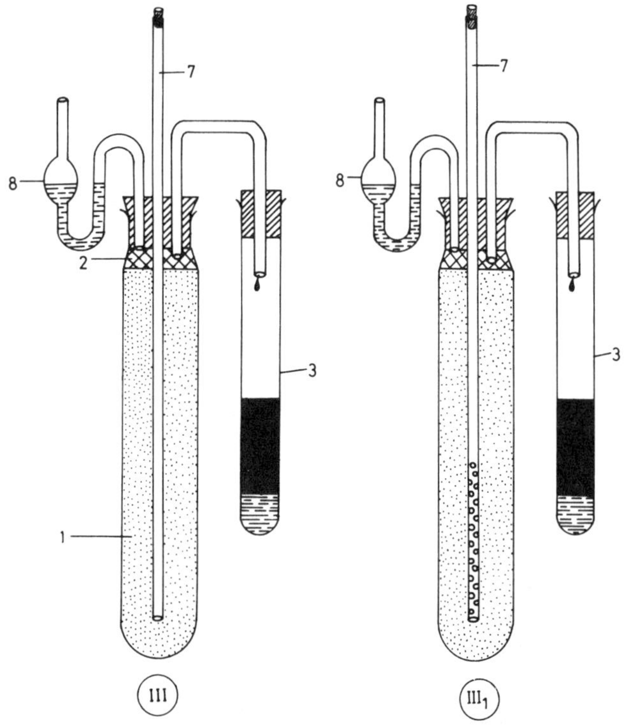

Fig. 7. Collector model III with variant III_1 for testing the oil releasing ability of adapted bacterial populations
1 = sand saturated with crude oil and then injected with nutrient support on molasses base and adapted bacterial populations
2 = glass wool for the filtration of the mobilized fluids
3 = special glass tube for collecting the gases and fluids from collector, mobilized due to bacterial activity
4 = vessel for collecting crude oil, mobilized due to bacterial activity
5 = vessel for collecting mobilized gases
6 = vessel with saturated NaCl solution for eliminating the surplus gases
7 = glass tube with lower part perforated, used for introducing nutrient support and inoculum in the collector
8 = capillary glass tube for indication of gases formation in the collector

Fig. 8. Three types of collector models

from 2–39%, depending upon the nutrient support, adapted bacterial populations, and crude oil.

The adapted bacterial populations obtained on Installations A, B, and C were more active in the mobilization of crude oil from the collector than the populations obtained on Installation D. Also, the mixture of bacterial populations obtained from these four installations was more active in oil release than individual populations.[67-71] During some experiments it was noticed that when the migration of oil from collector models was stopped, the addition of nutrient medium caused the bacteria to renew their activity and brought about the release of an additional quantity of crude oil.

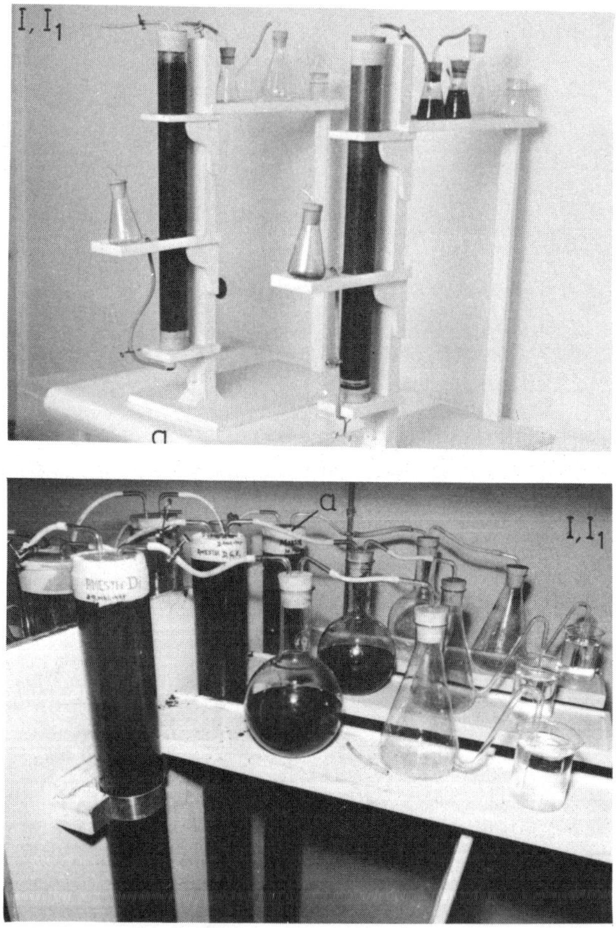

Fig. 9. Testing oil release ability of several adapted bacterial populations on collector model type I, I_1, a = control

Finally, the adapted bacterial populations that proved to be the most active in crude oil release (Figure 8–10) were retained for multiplication in the third step of adaptation in the quantities required for well injections.

Step Three. The adapted bacterial populations selected in the second step were multiplied on a large scale in A, B, C, and D installations. On each installation about 100 l inoculum was prepared for well injection. The period for multiplying the inoculum in this third step was about 4–6 days. The resulting bacterial inoculum was again characterized, individually as well as in mixture, before injection into the well. Data concerning the characteristics of the bacterial inoculum injected through seven wells into reservoirs having different physicochemical and petrophysical characteristics has already been published.[71,72] The bacterial inoculum obtained by the adaptation process described above had the following main characteristics:

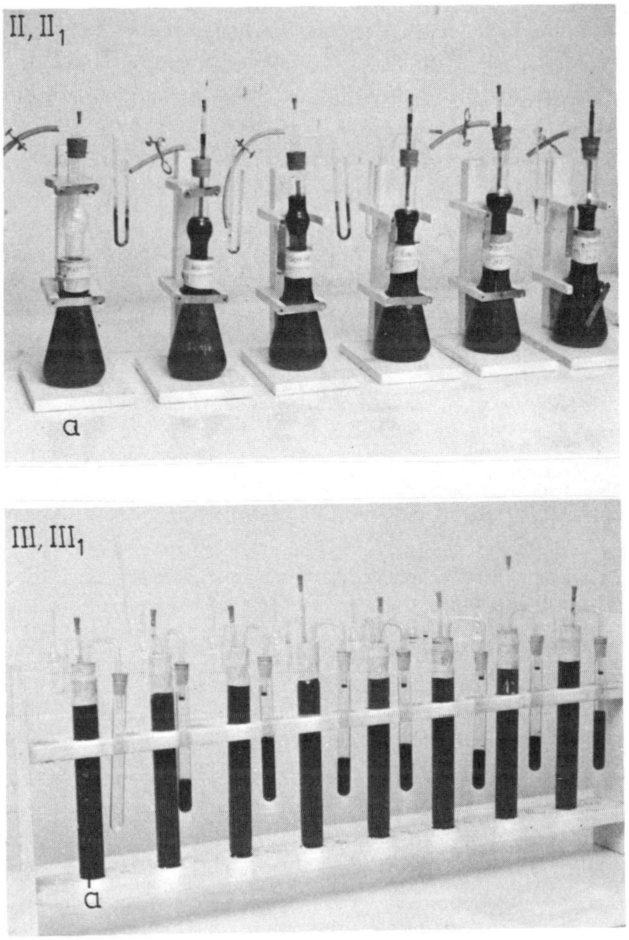

Fig. 10. Testing oil release ability of several adapted bacterial populations on collector models types II, II_1, and III, III_1

1. concentration of bacterial cells per milliliter reaching values of 10^7–10^9; composed of bacteria from genera previously mentioned
2. ability to mobilize the petroleum from collector models up to a percentage between 19.5–48.8 in a period of 5–10 days
3. utilization of crude oil in ideal laboratory conditions varying from 2–51%.

THE RESULTS OF FIELD TRIALS

The results of field trials carried out in Romania during 1975–1979 are presented in a recent paper.[73] In these field trials the adapted bacterial populations obtained after the method mentioned above were used.

Out of seven reservoirs, positive results were only obtained from the wells of two reservoirs. In some low-production (0.2 tons/day) wells from these two reservoirs, an increase of oil production up to 200% was registered, while in other wells increases from 1.0–1.9 tons/day (90%), from 1.0–1.5 tons/day (50%), and from 8.8–10.2 tons/day (16%) were registered for 1–5 years. Some of these wells still show increased oil production, while in others the oil production has returned to the initial level.

The failure of five wells to respond to the microbiological treatment was due to insufficient knowledge about their ecological environments. The unsatisfactory results from these five reservoirs show once again that the ecology of the reservoir or the corresponding physicochemical and petrophysical characteristics play a very important role in microbially enhanced oil recovery.

THE FUTURE OF WELL INJECTION WITH ADAPTED BACTERIAL POPULATIONS

Although the field trials carried out in several countries have yielded encouraging results, it is still difficult to say that all aspects concerning the bacterial inoculum used for well injection are satisfactorily resolved.[73–88] For example, the bacterial inoculum used so far in field trials was represented mainly by mesophilic heterotrophic bacteria whose development is not possible in the reservoirs due to high temperatures.

As has already been pointed out in some papers as well as in the Final Report of the Conference on Microbiological Processes Useful in Enhanced Oil Recovery, an intensive laboratory and field research is necessary in at least three directions to overcome present limitations:[89,90]

1. using genetic engineering techniques to obtain desired microorganisms

2. obtaining more information about the reservoir ecology and those physicochemical, petrophysical, and biological parameters that would permit the reservoir to serve as a large-scale bioreactor or fermentor
3. gaining more knowledge about the transformations occurring in the reservoirs, rocks, crude oil, and formation water due to direct bacterial action or due to the action of bacterial products (gases, mineral and organic acids, biopolymers, biosurfactants, etc.)

Until we are able to get the desired super bacteria by genetic engineering techniques, the laboratory and field studies should be conducted using adapted, mixed, natural populations, in view of the experience accumulated during the last 20–30 years.

In the future it will be necessary to improve the adaptation process according to the reservoir conditions and to stimulate the activity of adapted bacterial populations at the formation level where they are injected.

Accumulated data are necessary regarding the bioproducts resulting from metabolization of nutrient support, usually based on molasses. Studies are also necessary for a better understanding of the mechanism of oil release and its migration as a result of direct bacterial activity or bioproducts. The factors that positively influence the stimulation of oil release have been insufficiently studied. Experiments conducted on fermented scum from sugar processing plants showed that it contains some elements (of vitamin and growing-factor type) that stimulate bacterial activity, including their ability to release oil from porous media.[91]

All of these aspects concerning improvement of the method for obtaining adapted bacterial populations with high oil-releasing ability and utilization of heavy fractions from crude oil are in progress.

It is emphasized that this paper refers only to aspects concerning the bacterial inoculum used for stimulation of oil release. Other aspects, such as the use of biopolymers and biosurfactants as well as other possibilities of using microorganisms in enhanced oil recovery, are also in progress and will be presented in other papers in the future.[92]

CONCLUSION

1. Field trials carried out in Romania during 1975–1980 were based on the use of an inoculum containing bacterial populations adapted to the reservoir's conditions. Such adapted bacterial populations were isolated mainly from the formation water of the reservoir, scum from sugar processing plants, and sediments from plants used to purify injection water. These specialized bacteria possessed an optimum growth temperature in the mesophilic range, degraded crude oil, and intensively metabolized the molasses-based nutrient with a high production of gases as well as other bioproducts such as organic and mineral acids, biopolymers, biosurfactants, etc. The predominant species in these bacterial populations belonged to the genera *Pseudomonas, Escherichia, Arthrobacter, Peptococcus, Micrococcus, Bacillus,* and *Clostridium*; less numerous were bacteria from the genera *Aerobacter, Flavobacterium, Brevibacterium, Corynebacterium, Nocardia, Cellulomonas, Desulfovibrio*, and yeasts.

2. The adapted bacterial populations produced a 19.5–48.8% oil release in laboratory experiments on collector models. In field trials out of seven reservoirs injected with bacterial populations, only in two reservoirs was an evident increase of oil production registered. The negative response from the other five reservoirs was in part due to the physicochemical and petrochemical characteristics of the reservoir, which negatively influenced the spread and growth of bacteria in formations.

3. Until a reservoir's ecology and those transformations taking place due to bacterial activity at the level of rocks, crude oil, and formation water are better understood, and as long as practical applications of a super bacteria are not possible, the use of adapted bacterial populations is recommended. This requires intensive research to improve the selection process, obtain bacterial populations with nutrient preferences for heavy fractions of the crude oil, and explore factors that can stimulate the growth of these populations at formation levels.

REFERENCES

1 DIENES, M., and I. JARANYI. 1973. Kőolaj és Földgáz, 6, (106), Evral, p. 205–208.

2 DOSTALEK, M., and M. SPURNY. 1957. Ceska slov. microbiol. 2:300–317.

3 JARANYI, I. 1968. M. All.Földtani Intézet Évi jelentése az 1968-as évröl, 423–426.

4 JARANYI, I., L. KISS, and G. SZALANCZY. 1967. M.All.Földtani Intézet évi jelentèse az 1968-as évröl, 345–349.

5 KARASKIEWICZ, J. 1974. Zastosowanie metod mikrobiologeznyck w intensyfikacji eksploatacji Karpackick zloz ropy naftowej; widawnictwo "Slask" Katowice, 66 p.

6 LAZAR, I., A. GRIGORIU, A. MIHOC, I. ZAMFIRESCU, L. DUMITRU, and L. TRIFU. 1982. Lucrarile celui de al III-lea Simpozion de Microbiol. ind., Bucureşti 12–13 iunie 1981.

7 ZOBELL, C.E. 1946. U.S. Patent 2,413,278.

8 ZOBELL, C.E. 1947. World Oil *126*(13):36–47;*127*(1): 35–41.

9 UPDEGRAFF, D.M. 1957. U.S. Patent 2,807,570.

10 UPDEGRAFF, D.M., and G.B. WREN. 1953. U.S. Patent 2,660,550.

11 UPDEGRAFF, D.M., and G.B. WREN. 1954. Appl. Microbiol. 2:309.

12 BECK, J.V. 1947. Producer's Monthly *11*(11):13.

13 COTY, V.F. 1975. The role of microorganisms in the recovery of oil. National Science Foundation, Easton, Maryland, 9-14 XI. 1975, 77.
14 HITZMAN, D.O. 1962. U.S. Patent 3,032,472.
15 HITZMAN, D.O. 1965. U.S. Patent 3,185,216.
16 HITZMAN, D.O. 1967. U.S. Patent 3,340,930.
17 EKZERTSEV, V.A. 1958. Mikrobiologiya. *27*:626.
18 EKZERTSEV, V.A. 1960. Geokhimiya, p. 362.
19 FOSTER, J.W., R.M. COWAN, and T.A. MAAG. 1962. J. Bacteriol *83*:330.
20 KNOSEL, D., and W. SCHWARTZ. 1954. Arch. Mikrobiol. *20*:362.
21 KOLESNIK, Z.A., and N.I. SKMONOVA. 1957. Doklady akad. Nauk., S.S.S.R. *115*:1197.
22 MYERS, G.E., and W.D. SAMIRODEN. 1967. Producers Monthly *4*:22.
23 NORENKOVA, I.K., I.V. BREZHNEVA, R.M. SVECHINA, M.N. KARPENKO, and E.M. KULIKOVA. 1976. Mikrobiologiya *45*:724.
24 SIMAKOVA, T.L., Z.A. KILESNIK, N.V. STRIGLEVA, I.K. NORENKOVA, and N.I. SHMONOVA. 1968. Mikrobiologyia *37*:233.
25 TRAXLER, R.W., and W.L. FLAMMERY. 1968. p. 44-54. *In* H. Walters and J. Elphick (ed.), Biodeterioration of materials. Elsevier Publ. Co., New York.
26 WALHAUSSER, K.H. 1969. Erdol-Erdgas Zeit. *85*:14.
27 ZAJIC, J.E. 1969. Microbial biogeochemistry. Academic Press, New York.
28 DOSTALEK, M. 1965. Abh. Dent. Akad. Wiss. Berlin, Kl. Chem. Geol. Biol. 1965 (2):81 (Publ. 1966).
29 DOSTALEK and SPURNY, op cit.
30 DOSTALEK, M., M. SPURNY, and A. ROSYPALOVA. 1958. Prace ustavu pro naftovy vyzkum. Publ. 9:29.
31 EKZERTSEV, 1960, Geokhimiya, op cit.
32 KOLESNIK and SKMONOVA, op cit.
33 SENYUKOV, V.M., E.M. YULBARISOV, N.N. TALBUKINA, and E.P. SHISHNEVA. 1970. Mikrobiologiya *39*:705-710.
34 DIENES and JARANYI, op cit.
35 JARANYI, op cit.
36 JARANYI, KISS, and SZALANCZY, op cit.
37 KARASKIEWICZ, J. 1964. Nafta (Katowice) *20*:61.
38 KARASKIEWICZ, J. 1965. Abh. Dent. Akad. Wiss. Berlin, Kl. Chem. Geol. Biol. 1965 (2):63 (Publ. 1966).
39 KARASKIEWICZ, J. 1968. Nafta (Katowice) *24*:198.
40 KARASKIEWICZ, J. 1977. Isvestia, Akad. Nauk. SSSR, Ser. Biol. *5*:790.
41 KARASKIEWICZ, J. 1975. Nafta (Katowice) *31*:144.
42 KARASKIEWICZ, 1974, op cit.
43 LAZAR, I. 1978. Certificat de inovator No. 44, eliberat de M.E.I. la 6.III. 1978.
44 LAZAR, GRIGORIU, MIHOC, ZAMFIRESCU, DUMITRU, and TRIFU, op cit.
45 LAZAR, I., I. ZAMFIRESCU-BALINSCHI. 1977. Lucrarile I-liu Simpozion de Microbiol. ind., Iași 18-19 XII. 1976. p.284-289.
46 LAZAR, I., I. ZAMFIRESCU, L. DUMITRU, A. MIHOC, and A. GRIGORIU. 1979. Brevet de invenție No. 73892 acordat de OSIM la 28.XII.1979.
47 ZAMFIRESCU-BALISCHI, I., A GRIGORIU, and I. LAZAR. 1980. Lucrarile celui de al II-lea simpozion de Microbiol. ind., Iasi, 25-26 Mai, 1979, p.460-465.
48 LAZAR, 1978, Certificat de inovator No. 44, op cit.
49 LAZAR, I. 1978. Certificat de inovator No. 46, eliberat de M.E.I. la 6.III.1978.
50 LAZAR, I. 1979. St. și Cerc. Biol. Seria Biol. Veget., 31, *2*:123-128.
51 LAZAR, GRIGORIU, MIHOC, ZAMFIRESCU, DUMITRU, and TRIFU, op cit.
52 LAZAR and ZAMFIRESCU-BALINSCHI, op cit.
53 LAZAR, I. and I. ZAMFIRESCU. 1979. Brevet de inventie No. 73891 acordat de OSIM la 28.XII.1979.
54 LAZAR, I., I. ZAMFIRESCU, L. DUMITRU, A. GRIGORIU, and A. MIHOC. 1978. Certificat de inovator No. 77 eliberat de M.E.I. la 28.X.1978.
55 LAZAR, 1978, Certificat de inovator No. 44, op cit.
56 LAZAR, 1978, Certificat de inovator No. 46, op cit.
57 LAZAR, ZAMFIRESCU-BALINSCHI, op cit.
58 LAZAR and ZAMFIRESCU, op cit.
59 LAZAR, I., I. ZAMFIRESCU-BALINSCHI, A. GRIGORIU, L. DUMITRU, and A. MIHOC. 1977. Lucrarile I-lui Simpozion de Microbiol ind. Iași, 18-19.XII.1976. p.290-296.
60 LAZAR, I., L. DUMITRU, I. BALINSCHI-ZAMFIRESCU, A. GRIGORIU, and A. MIHOC. 1977. Lucrarile I-lui Simpozion de Microbiol. ind. Iași, 18-19.XII.1976. p.297-303.
61 LAZAR, I. 1980. Lucrarile celui de al II-lea Simpozion de Microbiologie industriala Iași, 25-26.V.1979. p.435-438.
62 LAZAR, GRIGORIU, MIHOC, ZAMFIRESCU, DUMITRU, and TRIFU, op cit.
63 LAZAR, ZAMFIRESCU, DUMITRU, GRIGORIU, and MIHOC, op cit.
64 LAZAR, ZAMFIRESCU, DUMITRU, MIHOC, and GRIGORIU, op cit.
65 LAZAR, I. 1978. Certificat de inovator No. 45, eliberat de M.E.I. la 6.III.1978.
66 LAZAR, I., I. ZAMFIRESCU, and L. DUMITRU. 1978. Certificat de inovator No. 47 eliberat de M.E.I. la 6.III.1978.
67 LAZAR, 1979, St. si Cerc. Biol. Seria Biol. Veget., op cit.
68 LAZAR, DUMITRU, BALINSCHI-ZAMFIRESCU, GRIGORIU, and MIHOC, op cit.
69 LAZAR, ZAMFIRESCU-BALINSCHI, op cit.
70 LAZAR, ZAMFIRESCU, DUMITRU, GRIGORIU, and MIHOC, op cit.
71 LAZAR, ZAMFIRESCU-BALINSCHI, GRIGORIU, DUMITRU, and MIHOC, op cit.
72 LAZAR, GRIGORIU, MIHOC, ZAMFIRESCU, DUMITRU, and TRIFU, op cit.
73 LAZAR, I., A. MIHOC, A GRIGORIU, I. ZAMFIRESCU, L. DUMITRU, and L. TRIFU. 1982. Lucrarile celui de al III-lea Simpozion de Microbiol. ind. București 12-13.VI.1981.

74 DIENES and JARANYI, op cit.
75 DOSTALEK, op cit.
76 DOSTALEK and SPURNY, op cit.
77 JARANYI, op cit.
78 JARANYI, KISS, and SZALANCZY, op cit.
79 JOHNSON, C.A. 1979. *In* Final Report of the Conference on Microbiological Processes Useful in Enhanced Oil Recovery, San Diego, California.
80 KARASKIEWICZ, 1964, Nafta (Katowice), op cit.
81 KARASKIEWICZ, 1965, Abh. Dent. Akad. Wiss. Berlin, op cit.
82 KARASKIEWICZ, 1968, Nafta (Katowice), op cit.
83 KARASKIEWICZ, 1974, op cit.
84 KARASKIEWICZ, 1975, Nafta (Katowice), op cit.
85 KARASKIEWICZ, 1977, op cit.
86 LAZAR, I. 1978. European Symposium on Enhanced Oil Recovery, Herriot-Watt University, Edinburgh, July 5-7, 1978.
87 LAZAR, ZAMFIRESCU, DUMITRU, GRIGORIU, and MIHOC, op cit.
88 SENYUKOV, YULBARISOV, TALBUKINA, and SHISHNEVA, op cit.
89 Final Report of the Conference on Microbiological Processes Useful in Enhanced Oil Recovery, August 29–September 1, 1979. San Diego, California.
90 MOSES, V. 1978. European Symposium on Enhanced Oil Recovery, Herriot-Watt University, Edinburgh, July 5–7, 1978.
91 LAZAR, 1979, op cit.
92 FORBES, A.D. 1980. p. 169–180. *In* Hydrocarbons Biotechnol. Proc. Meet. Canterbury 1979, London, 1980.

ADDITIONAL REFERENCES

DAVIS, J.B. 1967. Petroleum microbiology. N.Y. Elsevier Publ. Co., New York.

HEYER, J., and W. SCHWARTZ. 1970. Z. Allgem. Mikrobiol. *8*:545.

HEYER, J., and W. SCHWARTZ. 1970. Z. Allgem. Mikrobiol. *8*:565.

KUZNETSOV, S.I., M.V. IVANOV, and N.N. LYALIKOVA. 1963. Introduction to geological microbiology. McGraw-Hill, New York.

13 MICROBIAL DESULFURIZATION AND DENITROGENATION OF HYDROCARBONS

W.R. Finnerty, K. Shockley, and H. Attaway

University of Georgia

The metabolism of hydrocarbons is distributed ubiquitously throughout the microbial world. Numerous studies document the oxidation of alkanes to fatty acids and simple aromatic and polynuclear aromatic hydrocarbons to cis-diols with subsequent ring cleavage reactions to produce acidic products.[1,2,3] The oxidation of hydrocarbons by microorganisms represents an obligately oxygen-dependent reaction process. Literature concerning the metabolism of hydrocarbon-like molecules with covalently bonded heteroatoms such as sulfur or nitrogen is scant.

A number of *Pseudomonas* species were isolated that oxidize dibenzothiophene to water-soluble acidic products.[4-7] Similar observations have been reported for the oxidation of dibenzothiophene by a mixed culture.[8] *Pseudomonas aeruginosa* ERC-8 has been reported to oxidize dibenzothiophene dissolved in alkanes with the accumulation of 4-[2-(3-hydroxy)-thionaphthenyl]-2-hydroxy-3-butenoic acid.[9] A recent report has described the oxidation of dibenzothiophene by a *Beijerinckia* species to (+) cis-1,2-dihydroxy-1,2-dihydrodibenzothiophene and dibenzothiophene-5-oxide.[10] Sagardia et al. demonstrated the oxidation of benzothiophene solubilized in oil by bacteria.[11]

A review of the microbiological literature shows a limited data base concerning the metabolism of nitrogen-containing heterocyclic organic compounds. Several reports exist in the very early literature concerning the mineralization of pyridine to NH_3, CO_2, and H_2O. Recently, soil isolates have been studied with respect to pyridine degradation by a ring hydroxylation and cleavage mechanism. A few reports have appeared concerning the bacterial metabolism of such compounds as quinoline and indole, with no supporting data on products or mechanisms of dissimilation. In recent years numerous reports have appeared concerning the bacterial metabolism of xenobiotics and as acylanilides, phenylcarbamates, and phenylureas.[12-15] More recent studies have investigated the comparative metabolism of aniline and chloroanilines by *Ps. multivorans* AN-1.[16,17] A current study has demonstrated the plasmid-determined utilization of 2-hydroxypyridine by *Arthrobacter crystallopoietes*.[18]

This report concerns the isolation and characterization of microorganisms that oxidize the model sulfur- and nitrogen-containing aromatic compounds, dibenzothiophene and carbazole. The rationale for this work was the isolation of microorganisms that selectively oxidize specific sulfur- and nitrogen-containing aromatic compounds present in petroleum or derived fractions of petroleum, coal liquid, and shale oil.

MATERIALS AND METHODS

Cultures. All bacterial isolates were obtained by enrichment techniques as described in the text.

Media. The basal salts medium for dibenzothiophene studies consisted of (in g/l): KH_2PO_4, 4; $Na_2HPO_4 \cdot 12H_2O$, 4; $(NH_4)_2SO_4$, 2; $MgSO_4 \cdot 7H_2O$, 0.2; $CaCl_2 \cdot 2H_2O$, 0.001; and $FeSO_4 \cdot 7H_2O$, 0.001; pH 7.4. Substrates were added to the basal salts medium as specified. Cultures were grown on a New Brunswick gyrotory shaker at 28°C (200 rpm).

The basal salts medium for carbazole studies consisted of (in g/l): K_2HPO_4, 10; $Na_2HPO_4 \cdot 12H_2O$, 5; $(NH_4)_2SO_4$, 2; $MgSO_4 \cdot 7H_2O$, 0.2; $CaCl_2 \cdot 2H_2O$, 0.001; $FeSO_4 \cdot 7H_2O$, 0.001; and $MnCl_2 \cdot 4H_2O$, 0.001; pH 7.0. Recrystallized carbazole was added as an acetone solution to hot medium, facilitating a fine dispersion of the substrate. All agar-based media contained 1.6% agar.

Oxidation of dibenzothiophene. Cultures were pregrown on a 0.8% nutrient-0.5% yeast extract medium, harvested, and suspended in a buffer consisting of 29 mM KH_2PO_4, 11 mM $Na_2HPO_4 \cdot 12H_2O$, and 15 mM $(NH_4)_2SO_4$; pH was 7.4. This cell suspension was added at predetermined cell densities to a medium in the experimental reaction vessel containing the same phosphate buffer, pH 7.4, supplemented with 3.5 μM $FeSO_4 \cdot 7H_2O$. Reactions were performed in a New Brunswick 2-liter multigen fermentor operated at a 1,200-ml working volume, and samples (30 ml) were obtained at appropriate intervals of time. Phenanthrene was added to the samples as an internal standard and was extracted for residual dibenzothiophene with petroleum ether, with a 95% or better efficiency of recovery. Dibenzothiophene was quantified with a Tracor Model 560 gas chromatograph equipped with a flame ionization detector and a coiled glass column (½ in. × 8 ft) containing 20% SP2100 (Analabs Inc.). Operating conditions were oven temperature, 200°C; detector temperature, 250°C; injection point, 225°C; flow rate, 22 cc nitrogen/min.

Dibenzothiophene was also monitored by the formation of characteristic water-soluble products with a Varian Cary 210 spectrophotometer.

Detection of plasmids. Plasmid screens were performed essentially as described by Kado and Liu.[19] Cells were grown in nutrient broth-yeast extract (NBYE) medium; 1.5 ml of this cell suspension were centrifuged and the cell pellet was suspended in 30 μl of 40 mM TRIS containing 20 mM sodium acetate and 2 mM Na_2 EDTA, pH 7.0. This was followed by the addition of 130 μl of 50 mM TRIS, pH 12.6, containing 20% sodium dodecyl sulfate. The reaction mixture was mixed and heated at 50°C for 20 min followed by extraction with chloroform:phenol (1:1,v/v). This mixture was then centrifuged for 7 min at 4°C, and a sample (25 μl) was placed in the well of a 0.8% agarose gel prepared in TRIS-acetate buffer, pH 8.3, and electrophoresced at 100 volts for 3 hr. Plasmids were visualized by staining the gels with ethidium bromide (5 μg/ml). Plasmid standards were *Escherichia coli* containing 102.5-, 62-, and 19-megadalton plasmids; *E. coli* AB2463 (pAS8T^S rep-1::Tn7), containing a 50-megadalton plasmid; and *E. coli* J53-2(R826), containing a 174-megadalton plasmid. Plasmids were cured from selected isolates by growth in the presence of novobiocin over a concentration range of 15–400 μg/ml.

Oxidation of carbazole. Cultures were pregrown on a 0.8% nutrient broth-0.05% yeast extract medium containing 0.05% carbazole. The culture served as the inoculum to the multigen fermentor containing the complete basal salts medium and 0.05% carbazole. Samples (50 ml) were obtained throughout the growth of the organism and were extracted with ethylacetate. Carbazole was quantified by gas chromatography, as described for dibenzothiophene quantification. Spent growth media was extracted for metabolic products by acidification of the bacteria-free medium and extraction with ethylacetate.

RESULTS

Isolation of dibenzothiophene-oxidizing microorganisms. Soil isolates were obtained through conventional enrichment techniques that oxidized dibenzothiophene but were unable to grow at the expense of dibenzothiophene as a sole source of carbon and energy. Soil samples (5 g) were moistened with a standard basal salts solution, supplemented with 0.1% dibenzothiophene final concentration, and incubated at room temperature for 4 days. The moist soil slurry was transferred to 250-ml Erlenmeyer flasks containing 100 ml of 0.8% NBYE growth medium supplemented with 0.05% dibenzothiophene. This culture flask was incubated with shaking and was subcultured every 24 hr to fresh medium. After 3 days of serial subculture, the final culture was streaked onto NBYE-agar plates containing 0.01% dibenzothiophene. Colonies that exhibited a red or intense yellow halo were identified and transferred to NBYE-agar slants for further analysis. As might be expected, the population density was extremely high, with a number of apparently diverse microorganisms present as determined visually by colonial characteristics. The number of isolates present per plate possessing the halo scoring characteristic was small, usually 1–2 colonies. Three isolates that exhibited the halo characteristics were selected for further study.

Characteristics of dibenzothiophene-oxidizing isolates. The three selected cultures were examined for their ability to oxidize dibenzothiophene in liquid shake culture. All cultures oxidized dibenzothiophene but were unable to grow on dibenzothiophene as a sole source of carbon and energy. Culture #1 (DBT-1) consisted of two separate bacterial species (*a* and *b*) that individually did not oxidize dibenzothiophene, whereas cultures #3 (DBT-3) and #4 (DBT-4) were pure cultures.

The mixed culture (DBT-1) was studied further to determine the role for each organism in the oxidation of dibenzothiophene. The hypothesis was considered that each organism was enzymatically deficient in specific but different enzymes involved in the partial oxidation of dibenzothiophene. This hypothesis was tested by growing each organism individually in NBYE medium supplemented with dibenzothiophene. The spent growth medium obtained from each culture was reestablished with respect to initial growth conditions, filter sterilized, supplemented with dibenzothiophene, and inoculated with the opposite organism. Only under conditions where both organisms were grown together in the same growth medium was dibenzothiophene oxidized (Table 1). These results suggested that the organisms were not enzyme deficient with respect to oxidation of specific intermediates. This conclusion was confirmed by examining the spent growth medium of the pure cultures for accumulation of products resulting from the oxidation of dibenzothiophene. No products were detected nor was dibenzothiophene oxidized, as determined by solvent extraction of the spent growth medium and quantification of dibenzothiophene by gas chromatography.

Table 1. The oxidation of dibenzothiophene by culture DBT-1.[a]

	Dibenzothiophene Oxidation in Spent Culture Medium Derived From:		
Organism	*a*	*b*	*a and b*
a	–	–	–
b	–	–	–
a and b	+	+	+

[a]Organisms *a* and *b* are the pure cultures derived from DBT-1.
– = no growth
+ = growth

It was considered at this point that only one organism in the mixed culture was involved in the oxidation of dibenzothiophene. The possible requirement for one organism was the production of a metabolite that served as an inducer for the enzymes involved in dibenzothiophene oxidation in the second organism. First, the growth kinetics of the mixed culture system were examined to determine precisely whether equal cell numbers existed throughout the growth cycle or if disproportionate growth kinetics characterized this mixed culture system. These studies proved highly interesting and instructive in that grossly disproportionate growth kinetics occurred with the mixed culture system. A differential of 7–9 logarithms of colony-forming units existed between organism *a* and organism *b*. This fact indicated that organism *a* was growth restricted, while organism *b* exhibited normal growth kinetics characteristic of a wide variety of microorganisms. The same growth characteristics prevailed when the growth kinetics of *a* and *b* were determined from pure culture.

These results suggested the possibility that organism *a* produced specific extracellular metabolite(s) that served as inducers for dibenzothiophene oxidation in organism *b*. The restricted growth of organism *a* appeared to provide a threshold concentration and constant supply of specific metabolites to organism *b*, which were necessary for maintaining the capacity for dibenzothiophene oxidation. This conclusion was supported by the fact that spent growth medium derived from organism *a* would not induce organism *b* to oxidize dibenzothiophene. The hypothesis was evaluated by dialysis culture.

Dialysis culture allowed for the separation of organism *a* from organism *b* in the same culture flask. The essential feature was that organism *a* would grow in a restricted manner, producing extracellular metabolite(s) that would diffuse through the dialysis membrane. The continued production of metabolites by organism *a* would provide a constant supply to organism *b*, which in turn would become induced for dibenzothiophene oxidation. The external medium, containing water-insoluble dibenzothiophene and organism *b*, could then be readily assessed for dibenzothiophene oxidation by the formation of characteristic products resulting from the oxidation of dibenzothiophene. This experimental design allowed organism *b* to oxidize dibenzothiophene, indicating that organism *a* produced diffusable products that induced the enzyme systems for dibenzothiophene oxidation in organism *b*.

The isolation and identification of the putative inducer(s) produced by organism *a* was accomplished by growing 10 l of organism *a* and concentrating this spent growth medium to 1 l. This concentrated growth medium was filtered, adjusted to 70% ethanol final concentration, and filtered again. The filtrate then was reduced to dryness and the residue was extracted with 70% ethanol. This alcoholic extract was strongly ninhydrin positive, weakly positive for carbohydrate, and contained a small amount of chloroform-soluble lipoidal material. Further analyses established the presence of leucine (50%), tryptophan (30%), tyrosine (15%), and phenylalanine (5%), plus a number of other amino acids at trace levels. Total amino acid, assuming 100% recovery, was 0.49 mg/l.

These amino acids were evaluated as possible inducers of dibenzothiophene oxidation in organism *b*. A chemically defined medium with 0.1% succinate as the sole source of carbon and energy was prepared and supplemented with 0.05% of L-leucine, L-tryptophan, L-tyrosine, or L-phenylalanine, and 0.05% dibenzothiophene. Only the culture flask containing L-leucine exhibited dibenzothiophene oxidation, as evidenced by the formation of characteristic water-soluble products. Control cultures lacking amino acids did not oxidize dibenzothiophene. We have simplified a mixed culture system to a pure culture and identified an inducer effective in promoting dibenzothiophene oxidation. Organism *b* derived from the mixed culture DBT-1 will hereafter be designated DBT-2.

The concentration of leucine with respect to optimal dibenzothiophene oxidation was determined for DBT-2 (Table 2). A 100-mM concentration of L-leucine appeared optimal for obtaining the maximum oxidation of diben-

Table 2. Leucine induction of dibenzothiophene oxidation.

Leucine Concentration (mM)	*Dibenzothiophene Induction (%)*
0	0
50	44
100	100
150	100
200	100

zothiophene. No explanation for the role of leucine as a gratuitous inducer in the oxidation of dibenzothiophene is offered at this time. It is noteworthy that the leucine requirement is in the range of substrate concentrations with possible implications for the metabolic transformation of leucine to other product(s) with inducer function. Further studies are in progress to determine the role of leucine in dibenzothiophene oxidation.

The further characterization of isolates DBT-2, DBT-3, and DBT-4 concerned the growth substrate spectrum in relationship to dibenzothiophene oxidation (Table 3). These data revealed a rather restricted growth substrate spectrum with fewer substrates initiating the induction of dibenzothiophene oxidation. All isolates are gram-negative rods, do not produce diffusable pigments, and will grow at temperatures up to 41°C. DBT-2 grows anaerobically with nitrate as a final electron acceptor but does not oxidize dibenzothiophene under these conditions.

Plasmids and dibenzothiophene oxidation. The presence of plasmids and their involvement in dibenzothiophene oxidation was determined for DBT-2, DBT-3, and DBT-4. DBT-2 harbored a 55-megadalton plasmid plus 2 plasmids less than 10 megadaltons. It was impossible to detect plasmid DNA in DBT-3, indicating that the genes coding for dibenzothiophene oxidation were chromosomal localized. DBT-4 exhibited a 58-megadalton plasmid. The 55-megadalton plasmid of DBT-2 was cured with either ethidium bromide or by growth in the presence of novobiocin. The phenotype characteristics of wild type and cured strains are shown in Table 4. The cured strain lost the enzymatic capacity to oxidize dibenzothiophene while retaining the ability to grow at the expense of leucine as a sole source of carbon and energy, suggesting that leucine is probably not the direct

Table 3. Growth characteristics of dibenzothiophene-oxidizing microbial isolates.

	DBT-2		*DBT-3*		*DBT-4*	
Carbon Source	*Growth*	*DBT Oxidation*	*Growth*	*DBT Oxidation*	*Growth*	*DBT Oxidation*
NBYE	+++	+	+++	+	+++	+
Leucine	+++	+++	+++	+	+++	+
Tryptophan	−	−	−	−	−	−
Tyrosine	−	−	−	−	−	−
Phenylalanine	−	−	−	−	−	−
Histidine	−	−	++	−	++	−
Serine	−	−	+	−	+	−
Norleucine	−	−	−	−	−	−
Glycine	−	−	−	−	−	−
Sucrose	−	−	−	−	−	−
Lactose	−	−	−	−	−	−
Glucose	++	−	+	−	+	−
Xylose	−	−	+	−	+	−
Acetate	++	−	+	−	+	−
Succinate	+++	−	++	−	++	−
Benzoate	++	++	+	+	−	−
m-hydroxybenzoate	−	−	−	−	+	−
p-hydroxybenzoate	++	++	+	−	+	−
Dibenzothiophene	−	−	−	−	−	−
Carbazole	−	−	−	−	−	−
Naphthalene	−	−	+	−	+	−
Phenol	−	−	−	−	−	−
n-alkanes (C_{10}–C_{20})	−	−	−	−	−	−

− = No growth or no dibenzothiophene oxidation
+++ = Excellent growth or dibenzothiophene oxidation
++ = Fair growth or dibenzothiophene oxidation
+ = Poor growth or dibenzothiophene oxidation

Table 4. Phenotypic characteristics of the plasmid cured isolate DBT-2.

	Wild Type		*Cured Strain*	
Carbon Source	*Growth*	*Dibenzothiophene Oxidation*	*Growth*	*Dibenzothiophene Oxidation*
Leucine	+++	+++	+++	–
Benzoate	+++	+++	–	–
p-hydroxybenzoate	++	++	++	–
Mandelate	++	++	++	–
Catechol	++	–	++	–

– = No growth or no dibenzothiophene oxidation
+++ = Excellent growth or dibenzothiophene oxidation
++ = Fair growth or dibenzothiophene oxidation
+ = Poor growth or dibenzothiophene oxidation

Table 5. Biocatalytic oxidation of sulfur heterocycles by DBT-2.

Time (hr)	*Dibenzothiophene (0.2% w/v)*	*Benzothiophene (0.2% w/v)*	*Thianthrene (0.2% w/v)*	*Thioxanthine (0.2% w/v)*
	Percent Conversion to Water-Soluble Products			
0	0	0	0	0
12	83.7	76.4	68.1	71.5
24	97.4	98.2	93.1	95.2

inducer of dibenzothiophene oxidation. Plasmid screens of the cured strain demonstrated the loss of the 55-megadalton plasmid, indicating the plasmid localization of genes coding for dibenzothiophene oxidation.

Oxidation of sulfur-containing heterocyclic compounds by DBT-2. The oxidation of related sulfur-containing aromatic substrates was evaluated with DBT-2. In all experiments DBT-2 was pregrown in an NBYE medium, harvested, and suspended in a defined, complete basal salts medium lacking a carbon source. These cells were established in a nongrowing condition to be used strictly as biocatalysts. A select number of sulfur heterocyclic compounds were tested for conversion to water-soluble products (Table 5). The oxidation of these substrates was determined by quantifying the disappearance of the substrate with time. Dibenzothiophene, benzothiophene, thianthrene, and thioxanthine were converted to water-soluble products at greater than 90% efficiency within 24 hr. Controls with heat-killed cells established the enzymatic basis for this oxidation. In addition, attempts to grow DBT-2 at the expense of the water-soluble products were routinely negative, indicating that dibenzothiophene oxidation was a strictly gratuitous reaction process from which the cells derived no discernible physiological advantage. The reaction conditions were optimized for DBT-2 with respect to cell density, pH, aeration, dibenzothiophene concentration, rate of agitation, and temperature to achieve a 98–100% conversion of dibenzothiophene to water-soluble products within a 5-hr reaction time.

The application of this bacterial biocatalytic reaction process was evaluated with crude oil. A low-sulfur crude oil was supplemented with 10 g of dibenzothiophene. The dibenzothiophene-oxidizing isolates converted 100% of this oil-soluble dibenzothiophene to water-soluble products within 5 hr (Table 6). Concentration limits have not been investigated at this time. These analyses were extended to the treatment of a high-sulfur crude oil containing 2.9% nascent dibenzothiophene, as per the analysis. This treatment utilized DBT-2 with 100% conversion of dibenzothiophene to water-soluble, sulfur-containing products in 5 hr (Table 7). These results indicate that the biocatalysts were functionally effective in the oxidation of dibenzothiophene

Table 6. Oxidation of dibenzothiophene dissolved in low-sulfur oil.

Time (hr)	*DBT-2*	*DBT-3*	*DBT-4*
	Percent Dibenzothiophene Removal		
0	0	0	0
1	23	46	9
2	47	71	51
3	63	96	73
4	90	100	85
5	100	—	100

Table 7. Treatment of high-sulfur crude oil with DBT-2.

Time (hr)	*Percent Depletion Dibenzothiophene*
0	0
1	43
2	76
3	94
4	99
5	100

2.9% nascent dibenzothiophene present in crude oil.

Table 8. Growth characteristics of four carbazole-utilizing microbial isolates.

	Microbial Isolate			
Substrate	*C-1*	*C-2*	*C-3*	*C-4*
Glucose	–	–	–	+
Succinate	+	+	+	+
Carbazole	+	+	+	+
Acetate	+	–	+	+
Alanine	+	+	+	+
Malate	+	+	+	+
Salicylate	–	–	–	–
Protocatechuate	ND	ND	ND	+
Catechol	+	+	+	–
Indole	–	–	+	–
Tryptophan	+	+	+	+
Kynurenate	+	+	+	+
Anthranilate	+	+	+	+
p-aminobenzoate	ND	ND	ND	–
m-hydroxybenzoate	ND	ND	ND	–
p-hydroxybenzoate	ND	ND	ND	+
Benzoate	–	–	–	–
1-naphthol	–	–	–	–
Naphthalene	–	–	–	–
Fluorene	ND	ND	ND	–
Dibenzothiophene	–	–	–	–
Hexane	–	–	–	–
Hexadecane	–	–	–	–
NBYE	+	+	+	+

+ = growth in 48 hr – = no growth in 48 hr

when dissolved in a chemically complex medium such as crude oil.

Isolation of carbazole-utilizing microorganisms. Soil samples (5 g) were incubated in 100 ml of NBYE medium containing 0.05% carbazole for 72 hr at 30°C with shaking. This culture was subcultured to fresh NBYE-carbazole medium three times at 72-hr intervals and was streaked onto NBYE-0.01% carbazole agar plates for single-colony isolation. Colonies that produced a clear halo were selected, indicating the utilization of carbazole. Carbazole, a water-insoluble compound, was incorporated into agar plates as a finely divided suspension producing a uniform turbid background. Four bacterial isolates were obtained that grow on carbazole as a sole source of carbon and energy, designated hereafter as C-1, C-2, C-3, and C-4.

Characteristics of carbazole-utilizing isolates. All isolates were gram-negative bacteria, exhibited pleotrophic morphology, and grew on carbazole as a sole source of carbon and energy. The ability of these carbazole-utilizing isolates to assimilate the nitrogen atom of carbazole as the sole nitrogen source was determined. Carbazole nitrogen served as the sole nitrogen source for all isolates. Nitrogen fixation was eliminated as a source of nitrogen by demonstrating the absence of growth in a nitrogen-free medium with either glucose or succinate as the sole source of carbon and energy.

Selected growth characteristics of the carbazole isolates are shown in Table 8. A number of paraffinic and aromatic hydrocarbons and dibenzothiophene failed to support growth. The more conventional growth substrates elicited maximum growth in 24 hr or less. The carbazole isolates exhibited varying rates of growth on carbazole as a sole source of carbon and energy, with C-4 metabolizing 0.05% carbazole to completion within 24 hr. The growth of all isolates on carbazole was a strictly aerobic reaction. The growth kinetics were similar with either carbazole or ammonium sulfate as the sole source of nitrogen and carbazole as the sole source of carbon and energy.

The ability of C-4 to utilize alternative nitrogen-containing aromatic compounds as the sole source of carbon and energy or of nitrogen is shown in Table 9. Carbazole, pyrrole, kynurenate, and anthranilate served as sole sources of nitrogen, with succinate as a source of carbon and energy. Only carbazole was able to support excellent growth of C-4 as a sole source of carbon, energy, and nitrogen, although tryptophan, kynurenate, and anthranilate supported marginal growth as sole sources of carbon and energy.

Rate studies on carbazole utilization were conducted since carbazole was able to support the growth of C-4 as the sole source of carbon, energy, and nitrogen. C-4 consumed 4.95 millimoles of oxygen per hour in the metabolism of 12 moles of carbazole to completion. These initial rates of oxidation were quite rapid but decreased at subsequent time intervals, possibly due to product inhibition.

The pathway of carbazole dissimilation by C-4 is presently unknown. However, a number of water-soluble products were present in the spent culture medium, as visualized by thin-layer chromatography (Table 10). A minimum of 6–7 products were present in small concentrations, and these are currently under structural analysis.

Plasmids and carbazole utilization. The carbazole-utilizing isolates were screened for plasmids. C-1 harbored

Table 9. Growth of carbazole culture C-4 on nitrogen heterocyclics.

	0.01% Succinate		*Nitrogen Heterocyclic*	
Nitrogen Heterocyclic Substrate	*$-(NH_4)_2SO_4$*	*$+(NH_4)_2SO_4$*	*$-(NH_4)_2SO_4$*	*$+(NH_4)_2SO_4$*
Carbazole	+++	+++	+++	+++
Pyrrole	++	++	–	–
Pyridine	ND	ND	ND	ND
Quinoline	–	–	–	–
Indole	–	–	–	–
Tryptophan	–	+	–	+
Kynurenate	++	++	–	+
Anthranilate	++	+++	–	+
Aniline	+	++	–	–

+++ = excellent growth
++ = fair growth
+ = poor growth
– = no growth

Table 10. Chromatographic identification of products of carbazole dissimilation.

R_f Values	
Ethylacetate:Hexane:Chloroform (20:40:40, v/v)	*Chloroform:Acetone (80:20, v/v)*
0.75 (blue)	0.81 (blue)
0.58 (carbazole)	0.78 (carbazole)
0.55 (red)	0.77 (red)
0.50 (yellow)	0.63 (aqua)
0.25 (aqua)	0.56 (yellow)
0.16 (violet)	0.50 (violet)
0.08 (yellow)	0.38 (dark yellow)
	0.11 (yellow)

Table 11. Growth of carbazole culture C-4 on carbazole-supplemented crude oils.

Crude Oil	*Plus Carbazole*	*Minus Carbazole*
West Texas crude	+	–
Venezuela Cerro Negro crude	++	–
Venezuela Monagas crude	+++	–
Alaskan crude	+++	–
Monterey crude	–	–
Mineral oil	+++	–

+++ = excellent growth
++ = fair growth
+ = poor growth
– = no growth

a 150-megadalton and a 40-megadalton plasmid, C-2 a 60-megadalton plasmid, C-3 a 55 megadalton plasmid, and C-4 a 103-megadalton plasmid.

Removal of carbazole from oil. The metabolic ability of C-4 to grow at the expense of various crude oils was evaluated with and without carbazole supplementation (Table 11). Carbazole was dissolved in crude oil at a concentration of 1 g per 100 ml of oil. C-4 was able to grow on all crudes supplemented with carbazole except a heavy Monterey crude. Interestingly, C-4 was unable to grow at the expense of any crude oils tested unless supplemented with carbazole.

DISCUSSION

The selection of microorganisms with the metabolic capacity to oxidize specific sulfur- or nitrogen-containing components of crude oil, derived fraction of oil, coal liquids, or shale oil appears feasible and of scientific merit as a potentially useful, applied, energy-related technology. Microorganisms have been isolated that oxidize specific sulfur- and nitrogen-containing aromatic compounds present in fossil fuels. The bacterial-mediated oxidations are strictly aerobic reactions, effecting the conversion of such organic molecules to water-soluble products. Studies with these bacterial isolates demonstrate the following biological characteristics:

1. A specificity to sulfur- or nitrogen-containing aromatic compounds
2. The partial oxidation of dibenzothiophene to water-soluble organic products
3. The use of dibenzothiophene-oxidizing bacteria as strict biocatalysts
4. The inability of these bacterial isolates to oxidize or grow at the expense of a variety of hydrocarbons, including aliphatic alkanes (C_{10}–C_{20}), cycloalkanes, and mononuclear and polynuclear hydrocarbons.

Table 12. Physiological characteristics of dibenzothiophene-oxidizing bacteria.

	Dibenzothiophene Bacteria		
Characteristics	*DBT-2*	*DBT-3*	*DBT-4*
Gram reaction	Negative	Negative	Negative
Plasmids	55 Mdal	None	58 Mdal
Rate of oxidation	275 μmoles/hr	630 μmoles/hr	310 μmoles/hr
Paraffin oxidation	-0-	-0-	-0-
Aromatic hydrocarbon oxidation	-0-	-0-	-0-
Inducible	Yes	Yes	Yes
Oxygen requirement	O_2 dependent	O_2 dependent	O_2 dependent
Growth on dibenzothiophene	Negative	Negative	Negative
Carbazole oxidation	Negative	Negative	Negative
Maximum oxidation temperature	55°C	55°C	55°C

5. A 95% conversion of the organic sulfur present as dibenzothiophene in crude oil to water-soluble products within 5 hr
6. The efficient utilization of nitrogen-containing aromatic compounds as sole sources of carbon, energy, and nitrogen
7. The amenability of these bacteria to mass-culture technology with simple, inexpensive nutritive media
8. The use of these bacteria as specific biocatalysts to perform highly specialized reactions at ambient temperatures

The major physiological characteristics of the dibenzothiophene-oxidizing isolates are summarized in Table 12. Pure cultures have been developed that catalyze the partial oxidation of dibenzothiophene to water-soluble products. The organisms are pregrown on nonhydrocarbon media and are used strictly as biocatalysts at ambient temperatures. The biocatalysts appear specific for sulfur-containing aromatic molecules, with present studies indicating no significant alteration or modification of crude oils. The major physiological characteristics of the carbazole-utilizing isolates are summarized in Table 13. These pure cultures grow at the expense of carbazole as the sole sources of carbon, energy and nitrogen. The organisms appear highly specific for nitrogen-containing aromatic molecules.

This work remains in a basic research phase of development. A future research concern is the evaluation of this biological process against a wider spectrum of crude oils, coal liquids, and coal. Particularly important is to establish the enzymatic competency of these organisms to oxidize more complex sulfur- and nitrogen-containing heterocyclics present in fossil fuels. Willey et al. have determined the structure of a number of complex 3-, 4-, and 5-member ring systems containing sulfur atoms present in coal liquids and shale oil.[20] Further research and development is being directed to the immobilization of the bacteria to inert matrices for studying this reaction process in the design of bioreactors. The application of such biological processes must become adaptable to large-volume process technology to become

Table 13. Physiological properties of microbial carbazole isolates.

	Microbial Carbazole Isolates			
Characteristics	*C-1*	*C-2*	*C-3*	*C-4*
Gram reaction	Negative	Negative	Negative	Negative
Plasmids	150 Mdal; 40 Mdal	60 Mdal	55 Mdal	103 Mdal
Rate of oxidation	15 μmoles/hr	38 μmoles/hr	82 μmoles/hr	523 μmoles/hr
Paraffin oxidation	-0-	-0-	-0-	-0-
Aromatic hydrocarbon oxidation	-0-	-0-	-0-	-0-
Growth on carbazole	Slow (weeks)	Moderate (days)	Moderate (days)	Fast (hr)
Dibenzothiophene oxidation	Negative	Negative	Negative	Negative
Oxygen requirement	O_2 dependent	O_2 dependent	O_2 dependent	O_2 dependent

cost effective. The interphase of biological systems to a medium such as oil poses certain problems and restrictions not normally encountered with microorganisms. The development of a thorough basic data base will allow for the proper assessment and evaluation of this potential biological technology and its application to the desulfurization and denitrogenation of fossil fuels.

ACKNOWLEDGMENTS

This research was supported by the Department of Energy contract #DE-AS09-80ER-10683.

REFERENCES

1 FINNERTY, W.R. 1977. The biochemistry of microbial alkane oxidation: new insights and perspectives. Trends in Biochem. Sci. *2*:73–75.

2 PERRY, J.J. 1979. Microbial cooxidation involving hydrocarbons. Microbiol. Rev. *43*:59–72.

3 GIBSON, D.T. 1972. The microbial oxidation of aromatic hydrocarbons. Crit. Rev. Microbiol. *1*:199–223.

4 KODAMA, K., S. NAKATANI, K. UMEHARA, K. SHIMIZUM, Y. MINOVA, and K. YAMADA. 1970. Microbial conversion of petro-sulfur compounds. Part II. Isolation and identification of products from dibenzothiophene. Agricul. Biol. Chem. *34*:1320–1324.

5 KODAMA, K., K. UMEHARA, K. SHIMIZU, S. NAKATANI, Y. MINODA, and K. YAMADA. 1973. Identification of microbial products from dibenzothiophene and its proposed oxidation pathway. Agricul. Biol. Chem. *37*:45–50.

6 NAKATANI, S., T. AKASAKI, K. KODAMA, Y. MINODA, and K. YAMADA. 1968. Microbial conversion of petro-sulfur compounds. Part II. Culture conditions of dibenzothiophene-utilizing bacteria. Agricul. Biol. Chem. *32*:1205–1211.

7 YAMADA, K., Y. MINODA, K. KODAMA, S. NAKATANI, and T. AKASAKI. 1968. Microbial conversion of petro-sulfur. Part I. Isolation and identification of dibenzothiophene utilizing bacteria. Agricul. Biol. Chem. *32*:840–845.

8 MALIK, K.A., and D. CLAUS. 1976. Abstracts of the 5th International Fermentation Symposium, p. 421.

9 HOU, C.T., and A.I. LASKIN. 1976. Microbial conversion of dibenzothiophene. Dev. Ind. Microbiol. *17*:351–362.

10 LABORDE, A.L., and D.T. GIBSON. 1977. Metabolism of dibenzothiophene by a *Beijerinckia* species. Appl. Environ. Microbiol. *34*:783–790.

11 SAGARDIA, F., J. RIGAU, A. MARTINEA-LAHOZ, F. FUENTES, C. LOPEZ, and W. FLORES. 1975. Degradation of dibenzothiophene and related compounds by a soil *Pseudomonas* in an oil-aqueous environment. Appl. Microbiol. *29*:722–725.

12 BARTHA, R., and D. PRAMER. 1970. Metabolism of acrylamide herbicides. Adv. Appl. Microbiol. *13*:317–341.

13 BURGE, W.D. 1972. Microbial populations hydrolyzing propanil and accumulation of 3,4-dichloroaniline and 3,3′,4,4′-tetrachloroazobenzene in soils. Soil Biol. Biochem. *4*:379–386.

14 CHISAKA, H., and P. KEARNY. 1970. Metabolism of propanil in soils. J. Agricul. Food Chem. *18*:854–860.

15 ENGLEHARDT, G., P.R. WALLNOFER, and R. PLAPP. 1973. Purification and properties of an aryl acifamidase of *Bacillus sphaericus*, catalyzing the hydrolysis of various phenylamide herbicides and fungicides. Appl. Microbiol. *26*:709–718.

16 HELM, V., and H. REVER. 1979. Investigation on the regulation of aniline utilization in *Pseudomonas multivorans* strain Anl. Eur. Appl. Microbiol. Biotechnol. *7*:191–199.

17 REBER, H., V. HELM, and N. KARANTH. 1979. Comparative studies on the metabolism of aniline and chloroaniline by *Pseudomonas multivorans* strain Anl. Eur. J. Appl. Microbiol. Biotechnol. *7*:181–189.

18 WEINBERGER, M., and P. KOLLENBRANDER. 1979. Plasmid-determined 2-hydroxypyridine utilization by *Arthrobacter crystallopoietes*. Can. J. Microbiol. *25*:329–334.

19 KADO, C., and S. LIU. 1981. Rapid procedure for detection and isolation of large and small plasmids. J. Bacteriol. *145*:1365–1373.

20 WILLEY, C., M. IWAO, R.N. CASTLE, and M.L. LEE. 1981. Determination of sulfur heterocycles in coal liquids and shale oil. Anal. Chem. *53*:400–407.

14

BEHAVIOR OF A MICROBIAL POLYSACCHARIDE (EMULSAN) AT OIL-WATER INTERFACES

Z. Zosim, S. Goldman, D.L. Gutnick, and E. Rosenberg

Tel Aviv University

Surfactants and polysaccharides are beneficial to enhanced oil recovery. How these materials orient themselves in emulsions is a basic property that relates to their function. The growth of microorganisms on hydrocarbons is often accompanied by the emulsification of the insoluble carbon source in the culture medium.[1,2,3] In many cases this has been attributed to the production of an extracellular bioemulsifier. Recently the authors have reported the purification, partial chemical characterization, and substrate specificity of the emulsifier of *Acinetobacter* RAG-1.[4,5,6,7] The polymeric substance (referred to as *emulsan*) had an average molecular weight of 9.9×10^5 and intrinsic viscosity of 505 cm^3 per g. Emulsan contains (*i*) a polysaccharide backbone composed of N-acetyl D-galactosamine, N-acetylamino uronic acid, an unidentified amino sugar, (*ii*) esterified fatty acids (0.5 μ moles per mg) consisting primarily of α- and β-hydroxydodecanoic acid, and (*iii*) proteins that can be removed by phenol treatment without destruction of emulsifying activity or loss of viscosity. The experiments presented herein were performed to investigate further the manner by which emulsan interacts with hydrocarbons and the properties of the emulsan-hydrocarbon complex.

MATERIALS AND METHODS

Materials. Olefin-free hexadecane (99% purity) was obtained from Fluka Chemical Co., Switzerland. Other paraffins and aromatic hydrocarbons were reagent-grade products of either Merck or Aldrich Chemicals. Kerosene and gas-oil were obtained from the Haifa Refinery, Haifa, Israel. Agha Jari crude oil was obtained from the Ashkelon-Eilat Pipeline Co., Israel. Rhodamine B ($C_{28}H_{31}ClN_2O_3$), recrystallized from ethanol, had a molar extinction coefficient of 0.9×10^5 at 560 nm, in close agreement with reported values.[8] TM buffer consisted of 0.02 M Tris (hydroxymethyl) aminomethane hydrochloride buffer, pH 7.2; and 10 mM $MgSO_4$. Glycerol (anhydrous) was a highly purified, redistilled product of Merck. Tritiated water (0.5 mCi/ml) was obtained from the Nuclear Research Center, Negev, Israel.

Emulsan, the extracellular emulsifying agent of *Acinetobacter calcoaceticus* RAG-1 (ATCC 31012), was purified from a cell-free supernatant obtained from an ethanol-grown culture both by ammonium sulfate precipitation[9] and cetyltrimethyl ammonium bromide fractionation (I. Belsky, unpublished). The emulsan used in these studies contained 17% protein. Apoemulsan, prepared by a modification of the hot phenol method, contained less than 5% protein.[10] Methyl cellulose (degree of substitution of 1.6) was a product of BDH (England). Xanthan (XFL-14630) was a product of XANCO Oil Field Products Inc.[11] The viscosities of 0.5 mg per ml solutions of emulsan, apoemulsan, methyl cellulose, and xanthan at 30°C were 0.98, 1.40, 1.20 and 2.47 centipoises, respectively.

Emulsion stability measurements. Hydrocarbon-in-water emulsions were prepared by ultrasound treatment, using the Braun Labsonic 1510 instrument. Ten milliliters of the hydrocarbon-water mixture in a cellulose nitrate tube (Beckman) were exposed to sonic oscillation for 15 sec at a constant power value of 30W. Unless otherwise stated, the emulsifiers were added prior to ultrasound treatment. The emulsions that formed were transferred immediately to calibrated Klett tubes, and turbidity was measured using a Klett-Summerson photoelectric colorimeter fitted with a green filter. Reported turbidity values were corrected using a standard calibration curve prepared by serially diluting a hexadecane-2-methylnaphthalene emulsion. Stabilities of emulsions were determined by measuring the turbidity after they had been standing undisturbed at 25 ± 2°C for 1 to 230 hr. The percent stability is defined as the Klett units after 24 hours, divided by the turbidity immediately after sonication, multiplied by 100.

Dye binding studies. A spectrophotometric method was developed for quantitative estimation of Rhodamine B binding to emulsan-stabilized hexadecane droplets. Emulsions were prepared in the presence of varying concentrations of the dye as described above. The emulsions were transferred to glass test tubes and were allowed to stand undisturbed for 21 hr (equilibrium was reached during the first 15 hr). Samples of the emulsions were then centrifuged at 8,000 × g for 20 min at 25°C. Under these conditions the droplets rose to the surface and did not interfere with absorbance readings of the aqueous phase. The residual dye (equilibrium concentration of the dye in the aqueous phase) was determined from trichromic readings ($2A_{560} - A_{540} - A_{580}$) by a comparison with a standard calibration curve of known concentrations of the dye in TM buffer. When Rhodamine B concentration exceeded 0.01 mg/ml, dilutions were made in the buffer before determining absorbance.

General analytical methods. Protein concentrations were determined by the method of Lowry et al. using bovine serum albumin as a standard.[12] Viscosity was measured in an Ostwald-Fenske microviscometer (water value, 55.1 sec) at 30°C, calibrated with 20% and 50% glycerol. Emulsan concentration was determined by a standard emulsification assay and by the microplate modification of the enzyme-linked immunosorbent assay (ELISA test), in which anti-emulsan IgG prepared against purified emulsan was first covalently linked to alkaline phosphatase with gluteraldehyde.[13,14] This material was then bound to varying quantities of emulsan previously immobilized on the walls of an IgG-coated microplate. The ELISA test yielded the same quantitative results regardless of whether the antigen was emulsan or apoemulsan (S. Goldman, in preparation). Size distribution of hexadecane droplets stabilized by emulsan was determined both by microscopy (see "Results") and by using a Coulter Counter Model ZB.

RESULTS

RAG-1 emulsan enhanced both the formation and stability of hexadecane-in-water emulsions prepared by ultrasonic treatment (Figure 1). Using a volume fraction (V_f) of hexadecane of 0.01, the initial turbidities varied from 520 Klett units (no emulsan) to 2,600 Klett units (0.1 mg emulsan per ml). On standing, the hexadecane droplets in the control coalesced and rose rapidly to the surface; by 10 hr the control emulsion was completely broken. In the presence of either 0.05 or 0.1 mg/ml emulsan (hexadecane-emulsan weight ratios of 155 and 77, respectively) , there was a small decrease in turbidity for 2 hr, after which the emulsions remained relatively stable for the next 40 hr. With a higher ratio of hexadecane to emulsan of about 400 (0.02 mg emulsan per ml), the turbidity dropped sharply during the initial 17 hr of standing and then remained relatively stable for the duration of the experiment. Slightly more stable emulsions were obtained when emulsan was added immediately after, rather than prior to, ultrasonic treatment. This might be attributed to a 10% decrease in viscosity of the emulsan during the sonic treatment.

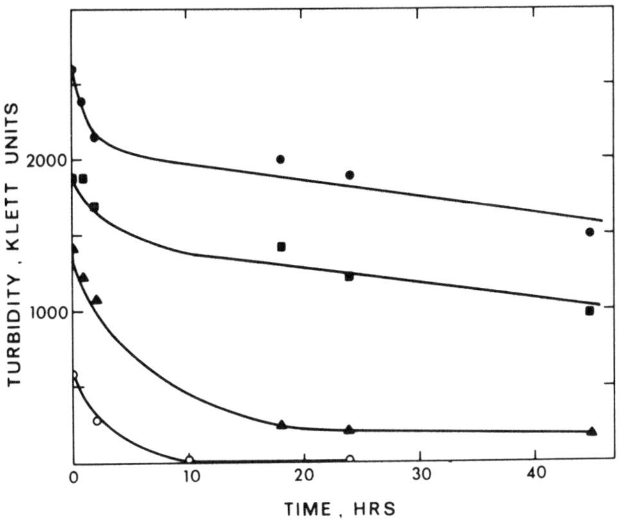

Fig. 1. Stability of hexadecane-in-water emulsions as a function of emulsan concentration. Hexadecane (0.1 ml) in 10 ml TM buffer was emulsified by exposure to sonic oscillation for 15 sec in the presence of zero (○), 0.2 mg (▲), 0.5 mg (■) or 1.0 mg (●) emulsan

The effect of emulsan on the formation and stability of different hydrocarbon-in-water emulsions is summarized in Table 1. Although the stability of all hydrocarbon emulsions tested was enhanced in the presence of 0.05–0.1 mg emulsan per ml, the effect varied significantly with the particular hydrocarbon examined. In general the higher the molecular weight of the liquid hydrocarbon, the more effective emulsan was in inducing and stabilizing the emulsion. This was true both for aliphatic and aromatic hydrocarbons, even though the initial turbidities of aromatic hydrocarbon emulsions were much higher than aliphatic hydrocarbon emulsions. Mixtures containing both aliphatics and aromatics were invariably better substrates for emulsan than were the individual hydrocarbons by themselves. For example, initial

Table 1. Various hydrocarbon-in-water emulsions stabilized by emulsan.[a]

Hydrocarbon	*Emulsan (mg/ml)*	*Turbidity (Klett units)* $t = 0$ *hr*	*Turbidity (Klett units)* $t = 24$ *hr*	*Apparent Stability[b] (%) (24 hr)*
Aliphatics				
Octane	0.1	1,200	112	9.3
Dodecane	0.1	1,700	850	50
Tetradecane	0.1	2,720	1,500	55
Hexadecane	0.1	3,150	1,900	60
Aromatics				
Toluene	0.05	3,600	165	4.6
p-Xylene	0.1	4,400	475	11
	0.05	3,080	260	8.4
2-Methylnaphthalene	0.05	7,600	4,000	53[c]
Tetrahydronaphthalene	0.05	7,600	3,800	50
Mixtures				
Hexadecane: 2-methylnaphthalene	0.05	17,000	12,000	71
Kerosene	0.05	6,700	3,400	51
Gas-oil	0.05	3,250	2,650	85
Crude oil	0.05	6,200	4,400	71

[a]Emulsions were prepared as described in Figure 1 with V_f of hydrocarbon of 0.01.
[b]Apparent stability is defined as the turbidity of the emulsion at 24 hr divided by turbidity at t = O times 100. Controls for each hydrocarbons without emulsan yielded 24-hr turbidity values that were less than 3% of the corresponding values for the emulsan experiment.
[c]This experiment was performed at 35°C to avoid solidification of the hydrocarbon.

turbidities of 2-methylnaphthalene and hexadecane emulsions (formed in the presence of 0.05 mg emulsan per ml) were 7,600 and 3,150, respectively; for the 1:1 mixture of the two hydrocarbons, the initial turbidity was 17,000. Similarly, the stabilities of the individual hydrocarbons were 60% and 53% after 24 hr, while the stability of the emulsified mixture was 71%. Emulsions prepared from each of the complex hydrocarbon mixtures — crude oil, gas-oil, and kerosene — were stabilized by 0.05 mg/ml emulsan.

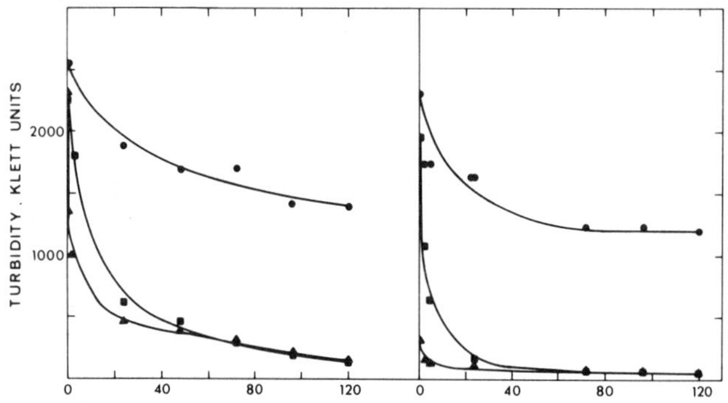

Fig. 2. Stability and reversibility of hexadecane-in-water emulsions. The emulsions were prepared as described in Figure 1 using 1 mg emulsan (●), 1 mg methylcellulose (■) or 1 mg xanthan (▲). After measuring the decrease in turbidities of the emulsions for 120 hr, the samples were mixed gently by hand and turbidities were determined for an additional 120 hr (second period)

Table 2. Stabilization of hexadecane-in-water emulsions by different emulsifiers.[a]

Emulsifier	*Apparent Stability (%), (24 hr)* (0.02 mg/ml)	(0.05 mg/ml)	(0.1 mg/ml)
Emulsan	13	63	74
Apoemulsan	14	62	75
Methylcellulose	4	13	27
Xanthan	10	19	21

[a]Experiments were performed as described in Figure 1 with a V_f of hexadecane of 0.01. Apparent stabilities of emulsions prepared without emulsifier were less than 2%.

Table 2 and Figure 2 compare two RAG-1 emulsifier preparations with two polymeric emulsifiers, methylcellulose and xanthan. Both emulsan and apoemulsan (deproteinized emulsan) were more effective than the two commercial emulsifiers in stabilizing hexadecane-in-water emulsions over the entire concentration range studied. Native emulsan (containing approximately 20% protein) is more effective than apoemulsan in forming emulsions.[15] However, if the hydrocarbon-in-water emulsion is preformed by ultrasound treatment, the protein does not play a crucial role in emulsion stabilization. At a V_f of hexadecane of 0.01 and an emulsifier concentration of 0.1 mg/ml, the turbidities of emulsions (after standing for 120 hr) stabilized by emulsan, xanthan, and methylcellulose were 1,420, 160, and 155 Klett units, respectively (Figure 2, initial period).

The lowering of turbidity may be the result of (*i*) coalescence of the hydrocarbon droplets, giving rise to a complete phase separation, or (*ii*) creaming of stabilized hydrocarbon droplets that float to the surface. In the case of creaming, the hydrocarbon droplets would be expected to reform a homogeneous emulsion with similar stability characteristics upon gentle agitation. The results summarized in Figure 2 (second period) illustrate the turbidities of partially broken emulsan mixtures that were agitated by hand and subsequently allowed to stand for an additional 120 hr. Breakage of xanthan-stabilized emulsions was primarily brought about by coalescence of oil droplets, since breakage could not be reversed. Upon gentle mixing the turbidity rose only to 305 Klett units and then decreased rapidly to 150 Klett units. Microscopic examination provided further evidence that breakage of xanthan-stabilized emulsions was due to coalescence rather than by flocculation. In contrast, breakage of emulsan-stabilized emulsions was due to creaming, since gentle mixing totally reversed the process (2,250 Klett units initially compared to 2,300 Klett units after mixing at 120 hr). It can be seen that breakage during the second period was similar to that which was observed in the initial 120 hr. Furthermore, cream obtained by centrifugation of emulsan-stabilized emulsions readily became suspended in aqueous solution to reform emulsions that did not differ significantly from those obtained initially by ultrasonic treatment. The methylcellulose-stabilized emulsion was an intermediate case; breakage was due to both creaming and coalescence. The hydrocarbon upper layer reformed an emulsion when mixed with water to yield a turbidity similar to that produced by ultrasonic treatment. However, emulsion breakage was much more rapid during the second period than during the initial 120 hr.

A photomicrograph of an emulsan-stabilized hexadecane-in-water emulsion is shown in Figure 3 (lower field). Analysis of a number of such micrographs yielded an average droplet diameter of 4.0 μ when the weight ratio of hexadecane to emulsan was 155 (Figure 4). Emulsions produced under similar conditions yielded an average droplet diameter of 3.6 μ when measured with a Coulter counter. Ninety percent of the droplets had a diameter of less than 6.1 μ. Upon prolonged standing the droplets rose and concentrated at the surface, forming a cream (Figure 3, upper field). After standing for one week, the cream was separated from the aqueous phase. Less than 10% of the emulsan was found in the aqueous phase, as determined both immunologically (see "Materials and Methods") and by the standard emulsifier assay.[16]

The water content of the different samples of cream, calculated from the density of the cream at 20°C compared to the known densities of water and hexadecane, was found to be 40–60%. Following centrifugation at 8,000 × g for 40

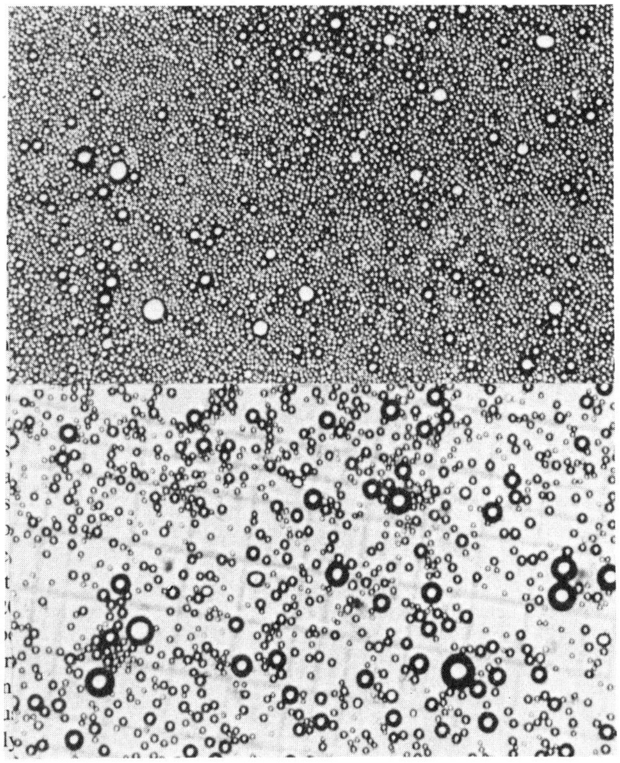

Fig. 3. Light microscopy of hexadecane-in-water emulsions stabilized by emulsan. Hexadecane (0.2 ml) in 10 ml TM buffer was emulsified by sonic oscillation for 15 sec in the presence of 0.2 mg emulsan. The lower field shows the emulsion immediately after sonic treatment; the upper field shows the cream formed after several days of standing. Magnification in both cases is 389 X.

min, more viscous creams were obtained that consisted of about 65% hexadecane and 35% water. Even after the centrifugation procedure, there was no coalescence of oil droplets. The water content of the cream was also determined using tritiated water as a tracer. With a V_f of hexadecane of 0.1 and an emulsan concentration of 0.5 mg/ml, the resulting cream contained 43% water after standing 3 days and 30% water after standing 10 days, followed by centrifuging at 8,000 × g for 30 min.

A qualitative experiment demonstrating binding of the water-soluble dye Rhodamine B to emulsan-stabilized hexadecane droplets is shown in Table 3. When the hexadecane emulsion was prepared in the presence of both emulsan and Rhodamine B, the dye was rendered nondializable for at least 24 hr. However, the dye was released during the first hour of dialysis if either emulsan or hexadecane were omitted or if the emulsion were stabilized by methylcellulose instead of emulsan. Even when much higher concentrations of methylcellulose were used to stabilize the emulsions, the dye was not retained in the dialysis bag. Thus, Rhodamine

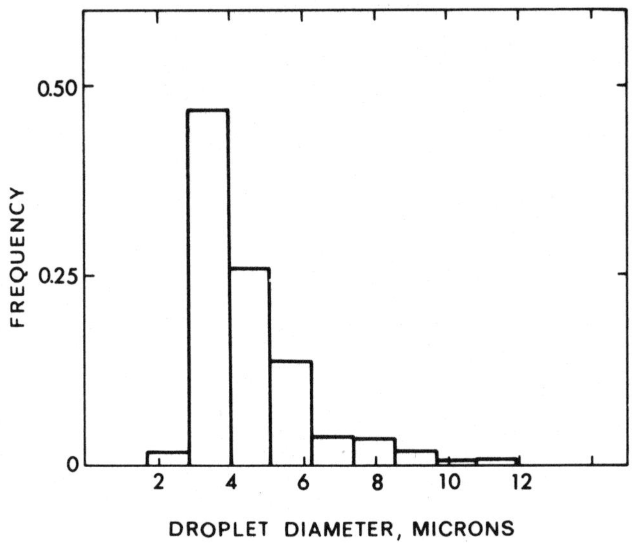

Fig. 4. Size distribution of hexadecane-in-water droplets stabilized by emulsan. Emulsions were prepared in 10 ml TM buffer containing 0.1 ml hexadecane and 0.5 mg emulsan. Diameters of 1,724 droplets were determined from enlargements of micrographs

binding was a property of the emulsan-hydrocarbon complex and not due simply to the greater hydrocarbon-water interface.

Adsorption of Rhodamine B to hexadecane-in-water droplets stabilized by emulsan was further investigated by determining spectrophotometrically the residual Rhodamine following separation of the cream from the aqueous phase (Figure 5). At low concentrations of Rhodamine, more than 80% of the dye was adsorbed. Since the cream phase represented less than 5% of the total volume of liquid, the effective concentration of dye into the cream was greater than 75-fold. At saturating concentrations of dye, about 2.2 μ moles of dye were bound per 1.0 mg emulsan.

The data of Figure 5 were replotted according to the Freundlich empirical adsorption equation in logarithmic coordinates.[17] With initial dye concentrations from 0.2 to 6 μmoles, a straight line was obtained, indicating a single type of binding. The apparent dissociation constant for the Rhodamine-emulsan complex was 3.4×10^{-6}.

Table 3. Binding of Rhodamine B to emulsified hexadecane stabilized by RAG-1 Emulsan.

Reaction Mixture[a]	*Observation after 24 hr of Dialysis*[b] *Inside (10 ml)*	*Dialysate (800 ml)*
1. Complete	Intense pink, turbid	Colorless
2. Complete minus hexadecane	Colorless, clear	Faint pink
3. Complete minus emulsan	Colorless, two phases	Faint pink
4. Complete minus hexadecane and emulsan	Colorless, clear	Faint pink
5. Complete minus emulsan plus 1.0 mg methylcellulose	Colorless, turbid	Faint pink

[a]The complete reaction mixture consisted of 0.1 ml hexadecane, 1.0 mg emulsan, and 0.1 mg Rhodamine B in a final volume of 10 ml TM buffer.

[b]After exposing the mixture to sonic oscillation as described in "Materials and Methods," the entire volume was placed in a cellophane bag and dialyzed for 24 hr against 800 ml distilled water.

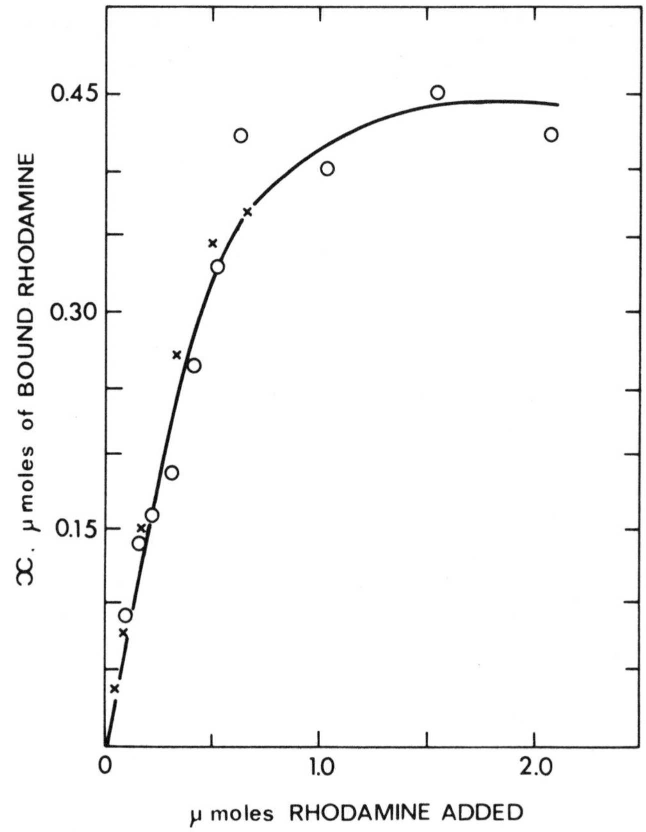

Fig. 5. Adsorption of Rhodamine B to hexadecane-in-water droplets stabilized by emulsan. Experiments were performed with (*i*) V_f of hexadecane of 0.01 and emulsan concentration of 0.1 mg per ml (X) or (*ii*) V_f of hexadecane of 0.02 and emulsan concentration of 0.02 mg per ml (○). Rhodamine binding was measured spectrophotometrically following separation of the cream from the aqueous phase (see "Materials and Methods")

The influence of pH on the binding of Rhodamine B to hexadecane-in-water droplets stabilized by emulsan is shown in Figure 6. Below pH 4, the dye was released from the interface. Release of the dye was not a result of emulsan coming off the oil-water interface (Figure 7). Emulsan

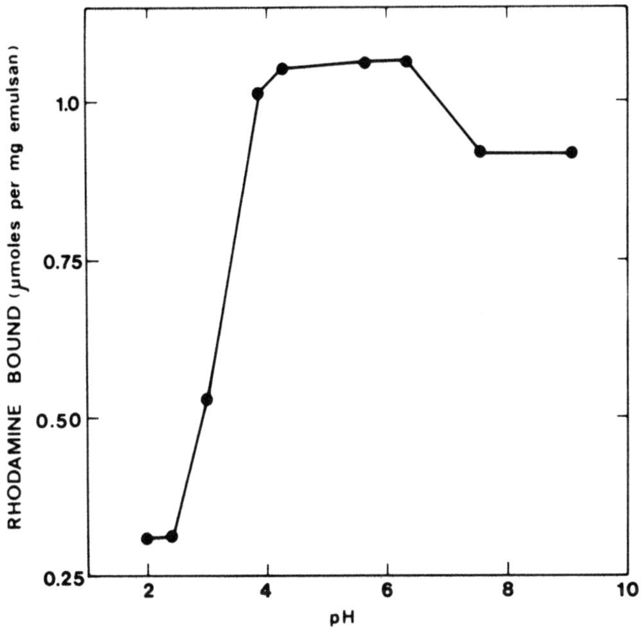

Fig. 6. Effect of pH on the adsorption of Rhodamine B to hexadecane-in-water droplets stabilized by emulsan. Experiments were performed with emulsions containing V_f of hexadecane of 0.01, emulsan concentration of 0.05 mg per ml, and 0.1 mM Rhodamine. The pH of the mixtures was adjusted before the emulsion was formed by sonic oscillation. Rhodamine binding was measured spectrophotometrically.

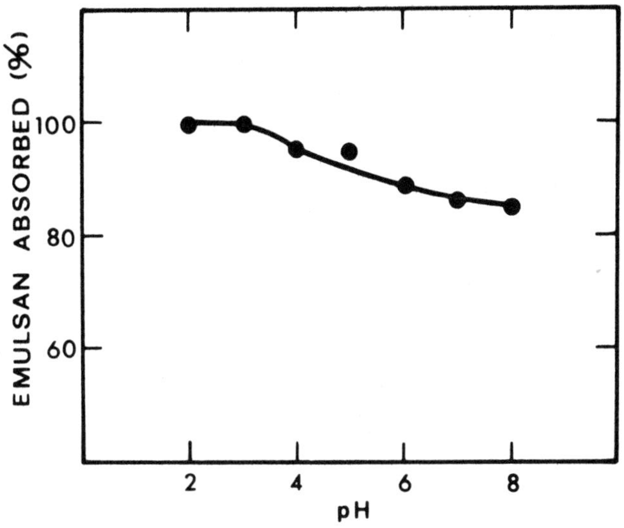

Fig. 7. Effect of pH on adsorption of emulsan onto hexadecane-water interfaces. Experiments were performed with emulsions of hexadecane V_f of 0.01 and emulsan initial concentration of 0.05 mg per ml. Residual emulsan in the aqueous phase was determined immunologically (see "Materials and Methods")

remained bound to the oil droplets over the entire pH range studied (pH 2–9).

DISCUSSION

The data presented here demonstrate that emulsan, the water-soluble extracellular bioemulsifier of *Acinetobacter calcoaceticus* RAG-1, stabilizes a wide variety of hydrocarbon-in-water emulsions. The emulsan binds tightly to the surface of hydrocarbon droplets, presumably forming a strong polymeric film on the droplet surface that prevents coalescence. This provides a rationale for the previously reported finding that the average size of hydrocarbon droplets formed in water by mechanical agitation depended upon the weight ratio of hydrocarbon to emulsan rather than simply on emulsan concentration.[18] Furthermore, the tight binding of emulsan to hexadecane-water interfaces explains the failure to find emulsan in clarified culture broths of hexadecane-grown *Acinetobacter* RAG-1.

The stability-reversibility of phenomenon, as summarized in Figure 2, may be considered to reflect the affinity of the various emulsifiers for hexadecane droplets; xanthan < methylcellulose < emulsan. This relationship may be due to the presence of hydrophobic groups (or sites) on the structure of the three polysaccharides. Emulsan is the most hydrophobic of the gel polymers, containing C_{12} fatty acid side chains in addition to N-acyl groups.

From the weight ratio of hydrocarbon to bound emulsan and the average droplet diameter (as determined by two independent methods — microscopy and Coulter-counter techniques), it is possible to estimate the thickness of the emulsan film around the oil droplet. For a weight ratio of 155, the mean droplet diameter was 3.8 μ. The volume of hexadecane per droplet, V_H, is $4/3r_H^3$, where r_H is the radius of the inner hexadecane sphere; the volume of emulsan per droplet, V_E, is $4/3r_T^3 - 4/3r_H^3 = 4/3\,(r_T^3 - r_H^3)$, where $r_T = 1.9\ \mu$ was the radius of the total droplet (hexadecane sphere plus emulsan shell).

dividing:

$$\frac{V_H}{V_E} = \frac{r_H^3}{6.86 - r_H^3}$$

also:

$$\frac{V_H}{V_E} = \frac{m_H\, d_E}{m_E\, d_H}$$

where the ratio of masses $m_H/m_E = 193$ (corrected for fraction of emulsan bound) and the densities of emulsan and hexadecane are 1.404 and 0.773 g/cm^3, respectively.[19]

$$\frac{V_H}{V_E} = 350.5 = \frac{r_H^3}{6.86 - r_H^3}$$

or $r_H = 1.898\ \mu$ and the thickness of the emulsan film,

$r_T - r_H$, equals 0.002 μ or 20Å. In the same way for a weight ratio of hexadecane to emulsan of 39, which gave a mean droplet diameter of 2.0 μ, it was found that the film thickness was 42Å. These values are minimum estimates since they assume the tightest possible packing of the emulsan molecules. The data are consistent with early results, showing that polysaccharides tend to lie flat on the oil-water interface.[20] In general, coalescence is inversely related to the thickness of the film formed on the droplet; strong polymeric films with a thickness greater than 5Å have been shown to prevent coalescence.[21]

Hexadecane droplets stabilized with emulsan can be separated from the bulk water phase by centrifugation. This cream phase, small hydrocarbon droplets coated with emulsan and water, is referred to as *emulsanasol*. With a weight ratio of hexadecane to emulsan of 155, the resulting emulsanasol contained 30–50% water. The emulsanasols have properties differing from the pure hydrocarbon and aqueous solutions of emulsan. One such interesting property is the ability to bind and thereby concentrate the water-soluble dye Rhodamine B. At saturating concentrations of the dye, 2.2 μ moles were bound per milligram emulsan in the emulsanasol. Since emulsan contains only 1.5 microequivalents of carboxyl residues per mg, simple ionic binding between the tertiary amine groups of Rhodamine and the carboxyl groups could not explain the phenomenon.[22] Rather, a particular conformation of emulsan on the interfacial film in contact with the hydrocarbon may better explain the binding properties of emulsanasols.

The fact that the anionic nature of emulsan also plays a crucial role in binding Rhodamine can be inferred from the observation that protonation of the carboxyl group of emulsan (pk = 3.05) blocked dye binding.

SUMMARY

Emulsan is a polymeric extracellular emulsifying agent produced by *Acinetobacter* RAG-1. Hydrocarbon-in-water emulsions (V_f of hydrocarbon of 0.01–0.10) were stabilized by small quantities of emulsan (0.02–0.2 mg/ml). Although both aliphatic and aromatic hydrocarbon emulsions were stabilized by emulsan, mixtures containing both aliphatics and aromatics were better substrates for emulsan than were the individual hydrocarbons by themselves. The emulsan remained tightly bound to the hydrocarbon even after centrifugation as determined by (*i*) residual emulsan in the aqueous phase and (*ii*) the fact that the resulting cream readily dispersed in water to reform stable emulsions. With hexadecane-emulsan weight ratios of 39 and 155, the noncoalescing oil droplets had average droplet diameters of 2.0 and 4.0 μ, respectively. Dialysis studies showed that the water-soluble dye Rhodamine B adsorbed tightly to the interface of hexadecane-emulsan droplets, although the dye did not bind to either hexadecane or emulsan alone. At saturating concentrations of Rhodamine B, 2.2 μ moles of the dye were bound per milligram emulsan. The dye was released from the interface at pH's below 4, whereas emulsan remained adsorbed to droplets from pH 2–9.

REFERENCES

1 ERICKSON, L.E., and T. NAKAHARA. 1975. Growth in cultures with two liquid phases: hydrocarbon uptake and transport. Proceed. Biochem. *10*:9–13.

2 GUTNICK, D.L. and E. ROSENBERG. 1977. Oil tankers and pollution: a microbiological approach. Ann. Rev. Microbiol. *31*:379–396.

3 ZAJIC, J.E., and C.J. PANCHAL. 1976. Bioemulsifiers. CRC Crit. Rev. Microbiol. *5*:39–66.

4 ROSENBERG, E., A. ZUCKERBERG, H. RUBINOWITZ, and D.L. GUTNICK. 1979. Emulsifier of *Arthrobacter* RAG-1: isolation and emulsifying properties. Appl. Environ. Microbiol. *37*:402–408.

5 BELSKY, I., D.L. GUTNICK, and E. ROSENBERG. 1979. Emulsifier of *Arthrobacter* RAG-1: determination of emulsifier-bound fatty acids. FEBS Letters *101*:175–178.

6 ZUCKERBERG, A., A. DIVER, Z. PEERI, D.L. GUTNICK and E. ROSENBERG. 1979. Emulsifier of *Arthrobacter* RAG-1: chemical and physical properties. Appl. Environ. Microbiol. *37*:414–420.

7 ROSENBERG, E., A. PERRY, D.T. GIBSON, and D.L. GUTNICK. 1979. Emulsification of *Arthrobacter* RAG-1: specificity of hydrocarbon substrate. Appl. Environ. Microbiol. *37*:409–413.

8 SINGLETERRY, C.R., and L.A. WEINBERGER. 1951. The size of soap micelles in benzene from osmotic pressure and from the depolarization of fluorescence. J. Amer. Chem. Soc. *73*:4574–4579.

9 ROSENBERG, ZUCKERBERG, RUBINOWITZ, and GUTNICK, op cit.

10 ZUCKERBERG, DIVER, PEERI, GUTNICK, and ROSENBERG, op cit.

11 MORRIS, E.R. 1977. Molecular origin of xanthan solution properties, p. 81–89. *In* P.A. Sandford and A. Laskin (ed.), Extracellular microbial polysaccharides. American Chem. Soc. Symp. 45, Washington, D.C.

12 LOWRY, O.H., N.J. ROSEBROUGH, A.L. FARR, and R.J. RANDALL. 1951. Protein measurement with the folin phenol reagent. J. Biol. Chem. *193*:265–275.

13 ROSENBERG, ZUCKERBERG, RUBINOWITZ, and GUTNICK, op cit.

14 CLARK, M.F., and A.N. ADAMS. 1977. Characteristics of the microplate method of enzyme-linked immunosorbent assay for the detection of plant viruses. J. Gen. Virology *34*:475–483.

15 ZUCKERBERG, DIVER, PEERI, GUTNICK, and ROSENBERG, op cit.

16 ROSENBERG, ZUCKERBERG, RUBINOWITZ, and GUTNICK, op cit.

17 SHAW, D.J. 1970. p. 120–132. *In*: Introduction to colloid and surface chemistry, 2nd ed. Butterworths, London.

18 ROSENBERG, ZUCKERBERG, RUBINOWITZ, and GUTNICK, op cit.

19 ZUCKERBERG, DIVER, PEERI, GUTNICK, and ROSENBERG, op cit.

20 ADAM, N.K. 1933. Surface films of cellulose derivatives on aqueous solutions. Trans. Faraday Soc. *29*:90–106.

21 BISWAS, B., and D.A. HAYDON. 1962. Coalescence of droplets stabilized by viscoelastic adsorbed films. Koll. Z. und Zeitschr. fur Polymere 1262. *185*:31–38.

22 ZUCKERBERG, DIVER, PEERI, GUTNICK, and ROSENBERG, op cit.

15 THE INFLUENCE OF THE SURFACTANT HLB ON THE GROWTH OF *ACINETOBACTER CALCOACETICUS* ON HEXADECANE

R.J. Neufeld

McGill University, Quebec

J.E. Zajic

University of Texas at El Paso

The growth of microbes on liquid hydrocarbons has been the subject of considerable interest for such wide-ranging applications as degradation of oil spills, bitumen separation from tar sands, and enhanced in situ oil recovery.[1,2,3] Utilization of hydrocarbons by microbes often is accompanied by the production of extracellular surface-active materials that stimulate growth due to increased emulsification of the hydrocarbon phase.[4,5] At the same time, it has been recognized that synthetic surfactants can promote improved microbial growth on hydrocarbons.[6,7]

The selection of an appropriate surfactant is at times difficult due to the large number of surfactant types available on the market with wide-ranging properties and characteristics. A numerical rating scheme known as the hydrophile-lipophile balance (HLB) number was first introduced in 1947 by Griffin and has come into wide use in the industry. An emulsifier consists of a molecule that combines both hydrophilic and hydrophobic groups. The balance of the size and strength of these two opposing groups determines the HLB. It is useful to know the emulsifier's relative affinity for oil and water in selecting the appropriate compound that will form the desired emulsion. An emulsifier that contains predominantly lipophilic groups will be oil soluble and will have an HLB number of less than 6. Because these emulsifiers are more soluble in the oil phase, w/o emulsions will be favored. HLB values ranging from 10 to 18 are conducive to the formation of o/w emulsions. Wetting agents, detergents, and solubilizers have HLB ranges of 8 to 10, 12 to 14, and 16 to 18, respectively. The HLB of a blend of emulsifiers is obtained by averaging the HLB values of the individual components.[8]

Lupton and Marshall examined the effects of surfactants on the degradation of a model oil by *Acinetobacter* SL1.[9] Of interest were the roles that some physical properties of the surfactants, such as ionic class, micelle formation, and hydrophilic-lipophilic balance (HLB), played in enhancing the ability of the microbe to metabolize oil. Six surfactants were studied with HLB values ranging from 8 to 17. A close correlation was observed between growth and the HLB of the surfactant. Tween 85, with an HLB value of 10, resulted in optimal emulsification and growth, while surfactants with HLB values greater than or less than 10 were not as effective.

To examine the effects in more detail, a homologous series of nonionic surfactants was chosen for the present investigation. A nonionic series was preferred due to possible toxic effects of the ionic surfactants.[9] The pluronic polyol (BASF Wyandotte Corp.) surfactants have hydrophilic groups at both ends of a changeable hydrophobic base. Varying the sizes of the hydrophilic and hydrophobic portions permits the altering of surfactant properties such as HLB. Thus the surfactant is formed by adding water-soluble polyoxyethylene groups to both ends of a water-insoluble polyoxypropylene chain. The hydrophilic groups may be controlled in length to constitute from 10% to 80% (by weight) of the final molecule. Choosing surfactants within this homologous series offers a wide range of HLB values while eliminating variations in chemical species.

MATERIALS AND METHODS

Culture maintenance and testing. *Acinetobacter calcoaceticus* was used in the present investigation. Isolation, maintenance, and identification of the culture was described previously.[10] Growth and testing were performed in the basic mineral salts medium previously described, supplemented with 2% (v/v) n-hexadecane (Humphrey Chemical Co.). The cultures were agitated in 500-ml flasks containing 100 ml of medium at 200 rpm on a temperature-controlled (30°C) gyrotory shaker (New Brunswick Scientific).

Biomass determination. The cell dry weight was determined using a technique that provided an estimate between the bound cells, those adhering to the hydrocarbon phase, and free cells in the aqueous phase. For a description of the technique, refer to Neufeld et al.[10]

Surfactants tested. The 23 surfactants tested were pluronic polyols, provided by BASF Wyandotte Corp. The HLB values of the surfactants ranged from 0.5 to 30.5. Surfactants were added to the test flasks at a concentration of 1% (w/v). Biomass determinations were complicated by the insolubility of five surfactants with the lowest HLB values. This problem was solved by filtering the culture broth through cheesecloth prior to analyzing for biomass.

RESULTS AND DISCUSSION

The general structure of the pluronic polyol is:

$$HO\,(CH_2CH_2O)_a\,(\underset{\displaystyle CH_3}{\underset{|}{C}}HCH_2O)_b\,(CH_2CH_2O)_c\,H$$

The hydrophilic polyoxyethylene groups (a and c), when varied in length, permit the modification of the surfactant properties such as HLB. The physical properties of the surfactants were provided by the manufacturer and are presented in Table 1. Included with the properties are liquid surface tension (γ_{wv}) and interfacial tension (γ_{ow})

The effects of the surfactant HLB on the growth of *Acinetobacter* were examined using these 23 pluronic surfactants. The flasks were sampled after 24 and 48 hr growth. The results are presented in Figure 1.

The results appear quite erratic; however, the trends are real, as they were confirmed in two identical subsequent experiments. Control flasks did not contain any surfactant.

Table 1. Pluronic surfactants and their physical properties

Pluronic*	HLB	γ_{wv} (mN/m) 0.1% (w/v)	γ_{ow} (mN/m) 0.1% (w/v)	Solubility** Water	Hydrocarbon
L121	0.5	33.0	10.7	ins	ins
L101	1	ins	ins	ins	ins
L81	2	ins	ins	ins	>1
L61	3	ins	ins	ins	>1
L122	4	33.0	7.5	ins	ins
L72	6.5	39.0	4.5	>10	ins
L62	7	42.8	10.8	>10	ins
L42	8	46.5	15.0	>10	>1
P103	9	34.4	1.8	>10	ins
L63	11	43.3	10.0	>10	ins
L43	12	47.3	14.5	>10	>1
P94	13.5	39.6	6.9	>10	ins
P84	14	42.0	10.0	>10	ins
L64	15	43.2	12.3	>10	ins
P85	16	42.5	10.5	>10	ins
P65	17	46.3	13.5	>10	ins
L35	18.5	48.8	19.6	>10	ins
F127	22	40.6	7.0	>10	ins
F87	24	44.0	17.4	>10	ins
F108	27	41.2	10.0	>10	ins
F88	28	48.5	20.5	>10	ins
F68	29	50.3	19.8	>10	ins
F38	30.5	52.6	24.7	>10	ins

*surfactants: L(liquid), P(paste), F(flake)

** ins:solubility <1g/100 ml
<1:solubility >1g/100 ml, <10g/100ml
>10:solubility >10g/100 ml

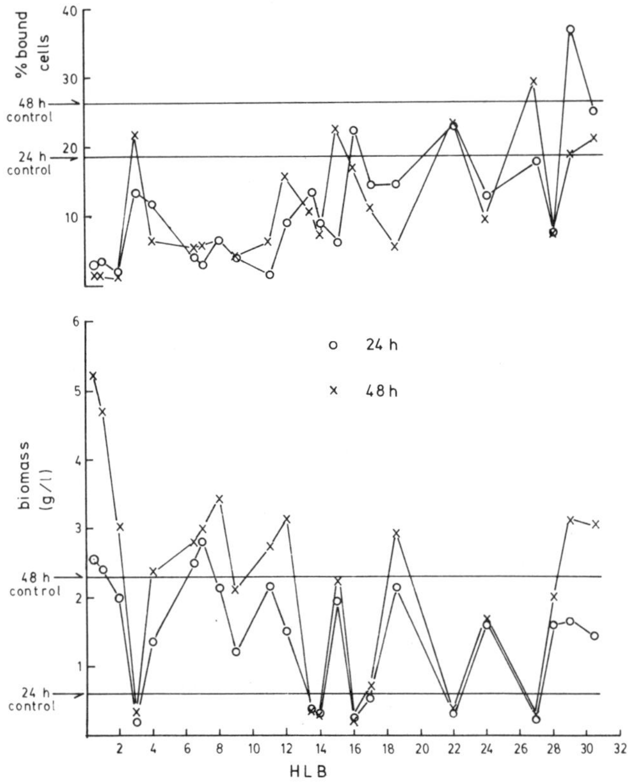

Fig. 1. The effects of HLB on the growth of *Acinetobacter calcoaceticus* on hexadecane and the partition characteristics of the cells with the hydrocarbon phase

It is clear that many of the surfactants tested enhanced growth on hexadecane. Those that were particularly noticeable had HLB values of 0.5, 1, 2, 6.5, 7, 8, 11, 12, 18.5, 19, and 30.5. Surfactants that clearly suppressed growth had HLB values of 3, 13.5, 14, 16, 17, 22, and 27. Thus, the role that HLB plays in growth is not clear. At an HLB of 24, considerably improved growth was observed after 24 hr; however, little change occurred between 24 and 48 hr.

Normally, pluronic surfactants with HLB values below 6 are hydrophobic, while those with values of 8 to 30 identify surfactants that are hydrophilic in character. An HLB value within the range 4–6 will stimulate the formation of water in oil (w/o) emulsions, which may explain the improved growth observed within this HLB range. It is possible that the microbe could be stabilized within the hydrocarbon phase, thus promoting intimate contact between the cell and substrate; however, an examination of Figure 1 will indicate that there tends to be a lower proportion of cells bound to the hydrocarbon phase at the lower HLB levels than there are at the higher HLB values. In the HLB range 4–6, less than 7% of the cell population was observed bound to the hydrocarbon phase after 48 hr growth.

With the exception of F108 (HLB27), the percentage of the cell population growing bound to the hydrocarbon phase was lower than in the control condition with no surfactant present after 48 hr growth. The situation was similar after 24 hr growth with only four exceptions, those involving surfactants with HLB values of 16, 22, 29, and 30.5. Thus it appears that the addition of synthetic surfactant to the culture broth lowers the affinity of the cells to the hydrocarbon phase, while it enhances their growth in many cases. These observed trends do not appear to be solely a function of the surface or interfacial properties of the microbial broth, as a correlation does not exist between growth and these parameters.

As may be seen in Figure 2, the partitioning of the cells between the aqueous and hydrocarbon phases may be a function of growth, as the trends seem to suggest that at the higher biomass levels fewer of the cells are bound to the hydrocarbon phase. Whether this trend is due to a physical change in the cell surface or to a modification of interfacial properties by the cells or cellular products is at this point unclear.

To eliminate the possibility of the microbe metabolizing some of the surfactants studied, causing the observed

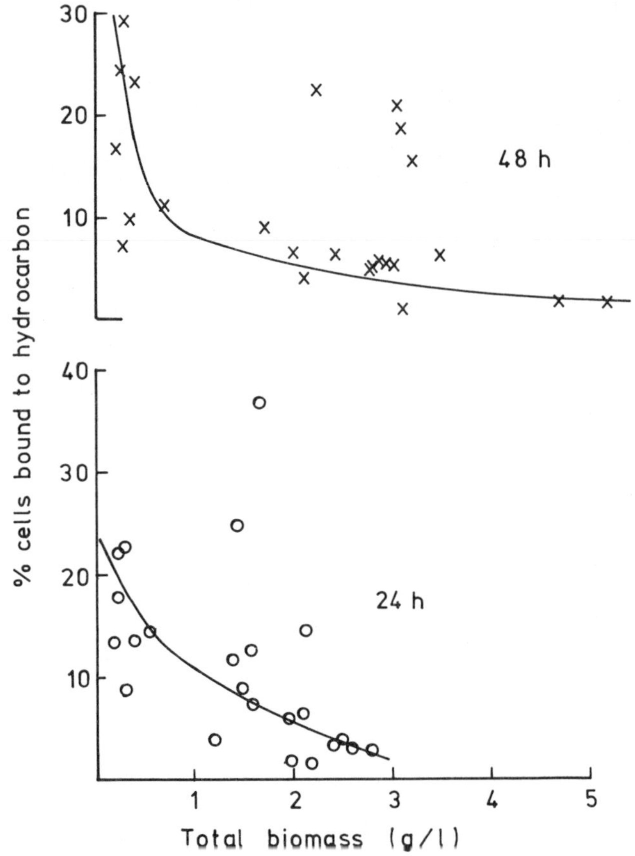

Fig. 2. The relationship between growth and partitioning of the cells with the hydrocarbon phase in the presence of synthetic surfactants

Table 2. Growth by *A. calcoaceticus* on selected pluronic surfactants.

Surfactant	*Biomass (g/l)*	*HLB*	*Terminal pH*
No surfactant (control)	0.9	—	7.1
L121	0.9	0.5	7.1
L101	1.0	1	7.1
L42	1.1	8	7.0
L43	0.8	12	7.0
L35	0.9	18.5	7.0
F68	1.0	29	7.0

effects, six surfactants that were demonstrated to enhance growth were added to mineral salts medium without hexadecane. The biomass levels were determined after 48 hr incubation; the results are presented in Table 2. Growth did not occur on any of the surfactants tested.

Figure 3 consists of a grid with the molecular weight of the poly(oxypropylene)hydrophobe on one axis and the weight percent of the poly(oxyethylene)hydrophilic unit in the total molecule for the pluronic series on the other axis. The growth results are presented on the grid to illustrate areas in which enhanced growth, no change in growth, and repressed growth occurred. With the exception of L61, the repressed growth regions occur in the center of the grid

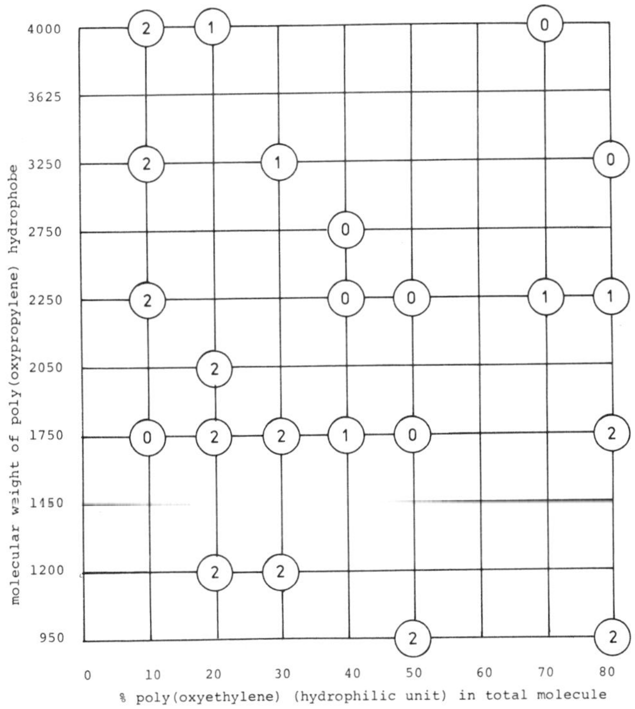

Fig. 3. Pluronic grid: Growth characteristics on hexadecane in the presence of pluronic surfactants; 0, no growth; 1, similar growth to control (no surfactant); 2, enhanced growth compared to control

extending to the upper right-hand corner. Areas in which enhanced growth occurred include the left portion of the grid, bottom, and lower right side. Figure 4 provides a summary of surfactant characteristics such as molecular weight, water solubility, wetting, detergency, and toxicity (information provided by the manufacturer). Toxicity results were based on target organ studies, acute and chronic oral toxicity

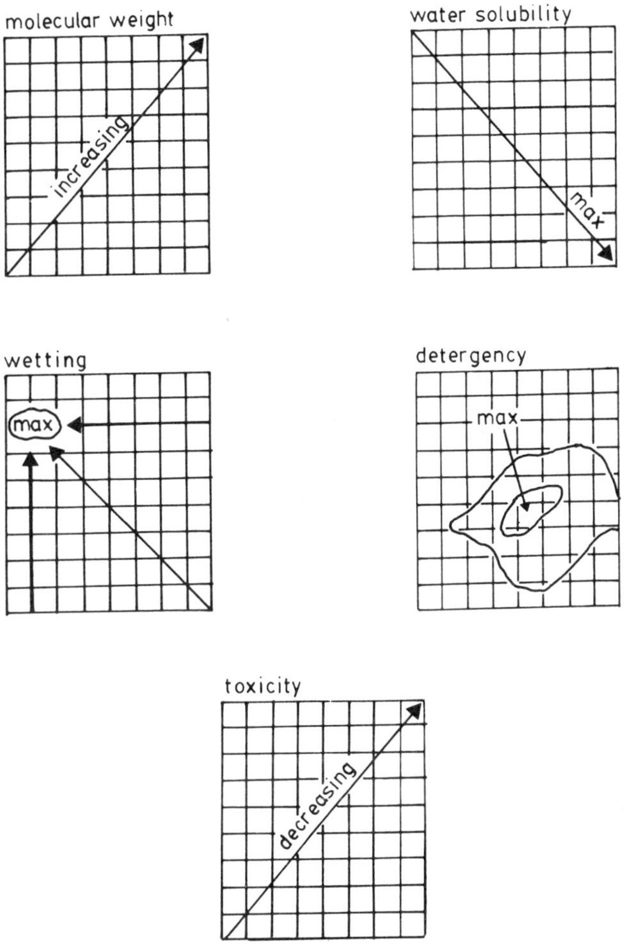

Fig. 4. Properties of pluronic surfactants illustrated on the pluronic grid (see Figure 3)

investigations, and eye and skin sensitivity tests. The results indicate that toxicity decreased with increasing molecular weight, the opposite situation to that shown in Figure 3, where it appears that the inhibitory effects were most pronounced with an increasing molecular weight of the surfactant. While maximum wetting by the surfactant may account for enhanced growth in the presence of some surfactants such as L121, L101, L81, and L72, it would do little to account for enhanced growth with L42, L43, L35, and L38, which are at the bottom of the grid. It could be expected that a better wetting of the cell surface by the hydrocarbon, mediated with surfactant, would have a positive influence on growth through direct contact between the cell and substrate.

Maximum detergency may be expected by surfactants in the center of the grid, possibly explaining the repressed growth observed with P94, P84, P85, and P65. This may be explained in terms of damage to cell wall and membrane components of the cell due to the washing and solubilizing action of the surfactant.

The effect of surfactant HLB on microbial growth was studied in further detail using surfactant mixtures in which the HLB values could be carefully controlled. Since HLB is a relative measure of the hydrophobic and hydrophilic groups in a surfactant molecule, the HLB of a mixture of surfactants may be calculated by averaging the HLB values based on the relative proportion of the surfactants.[11] Thus a 50-50 mixture of two surfactants with HLB values of 8 and 10 will yield an HLB of 9 for the combination.

Two sets of conditions were established. L101 (HLB 1) and F68 (HLB 29), two surfactants that enhanced growth on hexadecane, were mixed in proportions yielding mixtures with HLB values of 3, 13.5, 16, 22, and 27, values that were demonstrated previously to repress growth. The two surfactants, L61 (HLB 3) and F108 (HLB 27), which individually repressed growth, were mixed in proportions yielding HLB values of 5, 8, 15, 18.5, and 24, values that previously had been shown to enhance growth on hexadecane. The amount of surfactant mixture added to the growth flasks was the same as in the previous experiment (1% w/v). Results are presented in Figure 5.

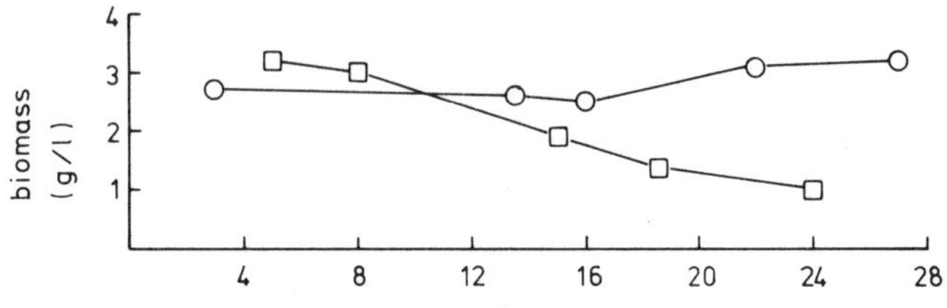

Fig. 5. Growth on hexadecane in the presence of surfactant mixtures. L101 and F68 (circles) were mixed in proportions providing HLB values which had been shown to repress growth. L101 and F68 individually enhance growth on hexadecane. L61 and F108 (squares) were mixed in proportions providing HLB values that had been shown to enhance growth. L61 and F108 individually repressed growth on hexadecane

The first set of surfactant mixtures, L101 and F68, resulted in slightly enhanced growth (compared to controls with no surfactant present) even though the HLB values of the mixtures had been previously demonstrated to repress growth. The levels of biomass were quite similar, with an upward trend at the higher HLB values (22 and 27). The second set of surfactant combinations (L62 and F108) that individually had been demonstrated to repress growth showed enhanced growth when in combination at HLB 5 and 8 with a downward trend toward HLB values of 15, 18.5, and 24. Nevertheless, growth under all combinations of this second surfactant set was better than when the surfactants were tested individually.

While these data are difficult to interpret, they do lend themselves to a few observations. HLB may not be the most appropriate criterion in establishing the ability of a particular surfactant to enhance or repress growth of a microbe on a hydrocarbon substrate. Other factors such as detergency, wetting, and emulsification, while not always lending themselves easily to quantification, may be more useful parameters to consider. It is clear from these results that the properties of a mixture of surfactants may be considerably different from those of the individual surfactants. Thus two surfactants that individually repress growth may enhance growth on a particular substrate when used in combination.

A number of these results appear to contradict those of Lupton and Marshall, who demonstrated a close correlation between the growth of *Acinetobacter* SL1 and the degree of emulsification of the substrate (model oil) by a range of homologous nonionic surfactants.[12] The most effective surfactant tested for promoting growth and degradation of model oil was Tween 85 with an HLB value of 10, while surfactants with markedly higher or lower HLB values were not as effective. Even the above workers examined only six nonionic surfactants (Tween and Span series) compared to 23 in this study, the trend which they observed, peaking at an HLB of 10, could not be demonstrated with the pluronic series in the present study.

It has been clearly demonstrated in this and other studies that careful selection of an appropriate surfactant will markedly improve microbial growth on hydrocarbon substrates. While surfactant HLB may provide a useful measure of a surfactant's characteristics and properties, it does not seem to be a reliable indication in terms of effectiveness of enhancing microbial growth on hydrocarbons.

SUMMARY

An attempt was made to establish a relationship between the hydrophilic-lipophilic balance (HLB) of a surfactant and the ability of the surfactant to enhance microbial growth on a liquid hydrocarbon. A series of 23 chemically homologous surfactants with HLB values ranging from 0.5 to 30.5 were tested. Surfactant mixtures were also prepared to modify the surfactant HLB. While some of the surfactants greatly enhanced growth, others inhibited growth with no clear relationship to the surfactant HLB. The surfactants tended to lower the affinity of the microbial cells for the hydrocarbon phase while enhancing growth in many cases. HLB does not appear to be the most appropriate criterion in establishing the ability of a particular surfactant to enhance growth of microbes on hydrocarbon substrates.

ACKNOWLEDGMENTS

This work was supported by the Natural Sciences and Engineering Research Council of Canada. The authors would like to thank Dr. D.G. Cooper for helpful discussions.

REFERENCES

1 OLIVIERI, A., P. BACCHIN, A. ROBERTIELLO, N. ODDO, L. DEGEN, and A. TONOLO. 1976. Microbial degradation of oil spills enhanced by a slow release fertilizer. Appl. Environ. Microbiol. *31*:629–634.

2 ZAJIC, J.E., and D.F. GERSON. 1978. Microbial extraction of bitumen from Athabasca oil sand. *In:* O. Strausz (ed.), Oil sands and oil shale. Springer-Verlag.

3 BUBELA, B. 1978. Role of geomicrobiology in enhanced oil recovery: status quo. The APEA Journal. *18*:161–166.

4 ERICKSON, L.E., and T. NAKAHARA. 1974. Growth in cultures with two liquid phases: hydrocarbon uptake and transport. Process Biochem. *20*:9–13.

5 NEUFELD, R.J., J.E. ZAJIC, and D.F. GERSON. 1980. Cell surface measurements in hydrocarbon and carbohydrate fermentations. Appl. Environ. Microbiol. *39*:511–517.

6 LUPTON, F.S., and K.C. MARSHALL. 1979. Effectiveness of surfactant in the microbial degradation of oil. Geomicrobiol. Journal *1*:235–247.

7 BREVIL, C., and D.J. KUSHNER. 1980. Effects of lipids, fatty acids, and other detergents on bacterial utilization of hexadecane. Can. J. Microbiol. *26*:223–231.

8 ZAJIC, J.E., and C.J. PANCHAL. 1976. Bioemulsifiers. CRC Critical Rev. Microbiol. *5*:39–66.

9 LUPTON and MARSHALL, op cit.

10 NEUFELD, ZAJIC, and GERSON, op cit.

11 ZAJIC and PANCHAL, op cit.

12 LUPTON and MARSHALL, op cit.

16

BACTERIA-INDUCED DEMULSIFICATION

W.L. Cairns, D.G. Cooper, and N. Kosaric

University of Western Ontario

Destabilization of emulsions formed during recovery and processing of petroleum crudes is a perpetual and costly problem for which better solutions are continuously sought. The complexity of the demulsification problem centers around the variability in the composition of the organic and aqueous components of field and process emulsions. The use of surfactants in enhanced oil recovery might be expected only to compound the traditional problem. Demulsifying agents must therefore be flexible, either in being effective over a broad range of emulsion properties or in being readily varied to match the different emulsion properties.

Previous studies have shown that kerosene-water emulsions that have been stabilized with synthetic surfactants can be effectively demulsified by the addition of small volumes (1–5% v/v) of whole bacterial cultures.[1,2,3] Table 1 shows typical results, indicating that both oil-in-water and water-in-oil types of emulsions and emulsions stabilized with neutral, cationic, and anionic surfactants are all susceptible to bacteria-induced demulsification. Table 1 also indicates that demulsifying ability can be found in cultures of more than one genus.

Some of the variables that determine the activity of bacterial demulsifiers within the whole culture are the bacterial species, the culture age, and the growth medium composition.

The component of the whole culture responsible for the demulsifying activity is the bacterial cell itself. Figure 1 summarizes the fractionation scheme for a culture of *Nocardia* sp. grown on a hydrocarbon-containing medium. The half-lives of comparable aliquots from each fraction indicate that the demulsifying activity resides with the cell and that the activity is not readily extracted from the surface by hydrocarbon solvents.

A PROPOSED MECHANISM FOR BACTERIA-INDUCED DEMULSIFICATION

A mechanism is suggested in Figure 2 how a bacterial particle could stabilize an emulsion. If the bacterial surface were appreciably but not totally wet by the discontinuous (droplet) phase of an emulsion, the bacterium would eventually find an equilibrium position at the interface between the continuous and discontinuous phase, such that the particle would be more than half submerged into the discontinuous phase. If two elements of the discontinuous phase (two emulsion droplets or an emulsion droplet and the already-separated discontinuous phase) contact, wet, and spread on the same bacterial surface, the two elements will coalesce on the surface before equilibrium is achieved (Figure 2a). On the other hand, if the bacterial cell surface

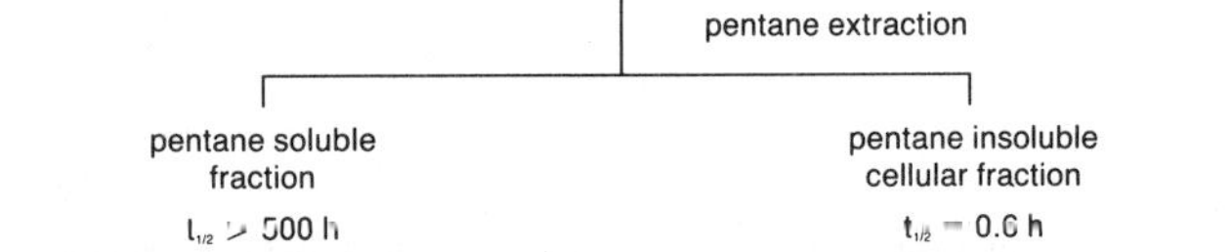

Fig. 1. Fractionation scheme of whole cultures of *Nocardia* sp. showing half-lives of test emulsion with comparable quantities of material from each fraction. Details in ref. 1

Table 1. Demulsification of kerosene-water emulsions by whole 7-day-old cultures of bacteria grown on hydrocarbon-containing media

Organism	*Half-Life (hr) of Test Emulsions Stabilized by Different Surfactants (Details in Ref. 1,2,3)*			
	Neutral, O/W	*Anionic, O/W*	*Cationic, O/W*	*Neutral, W/O*
Control (culture medium)	>200	>200	>200	>200
Nocardia sp.	<1	<1	<1	2
Rhodococcus sp.	<1	160	14	<1
Corynebacterium sp.	<1	<1	<1	14

is not appreciably wet by the discontinuous phase, two elements of the discontinuous phase can reach their equilibrium position of wetting before contact and before subsequent coalescence occurs (Figure 2b). In this latter case the bacterial particle behaves as an emulsion stabilizer rather than as a demulsifier. The magnitude of the equilibrium contact angles (α and β of Figure 2) between the continuous phase and the bacterial cell surface would indicate the wettability of the bacterial surface by the discontinuous phase and hence the suitability of the bacterial cell as either an emulsion breaker or an emulsion stabilizer.

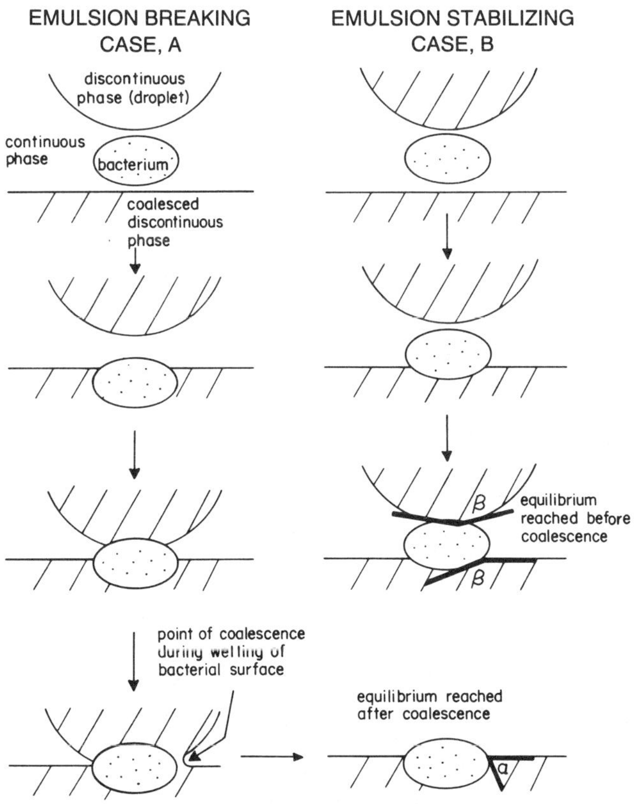

Fig. 2. An illustration of how bacteria might either promote (A) or inhibit (B) coalescence of emulsions. A or B will occur depending on how effectively the bacterial cell surface is wet by the discontinuous emulsion phase. Wettabilities are reflected in the equilibrium contact angles (α and β) shown here for a water-in-oil emulsion

The concept of using solid surfaces as demulsifiers is not original. Small particles of nonbiological origin have previously been noted as demulsifiers,[4] and solid surfaces interacting with emulsion droplets are the basis of demulsifying techniques using fibers[5,6] and films.[7]

TESTING THE PROPOSED MECHANISM

A test of the proposed mechanism would be to demonstrate a correlation between the half-lives of emulsions exposed to bacterial surfaces and the degree of wettability of those surfaces by the discontinuous phase of the emulsion. Contact angles would be a suitable measure of that wettability.[8]

The organism selected for the demulsification-wettability study was *Corynebacterium petrophilum.* Cells were grown in 500-ml flasks on 200 ml of a medium containing 2% dextrose and 0.8% nutrient broth. The flasks were maintained at 20°C ± 2°C on a shaker rotating at 250 rpm. The cultures were stabilized at different ages with 0.8% formaldehyde, which from previous studies had been shown to have no appreciable effect on contact angle measurements or on demulsification by the bacterial cell surfaces. The stabilized cultures were refrigerated at 2°C until used. Prior to use, the cultures were centrifuged to pellet the cells. The cells were resuspended in tap water and again centrifuged to wash them free of medium. The washing procedure was repeated and the cells were finally resuspended in tap water at a concentration of 5 g/l.

Contact angles were measured using a procedure modified from that of van Oss et al.[9] Five milliliters of washed, resuspended cells were filtered onto a 0.8-micron cellulosic microporous disc (Amicon Canada Ltd., Oakville, Ontario) and were washed with 5 ml of distilled water. The filtered cells were then positioned in the optical system diagrammed in Figure 3 and the indicated contact angles (θ) were measured. All values reported are those angles measured through the aqueous phase. Drops were applied to the filtered cells either by allowing 5 μl of aqueous phase to fall

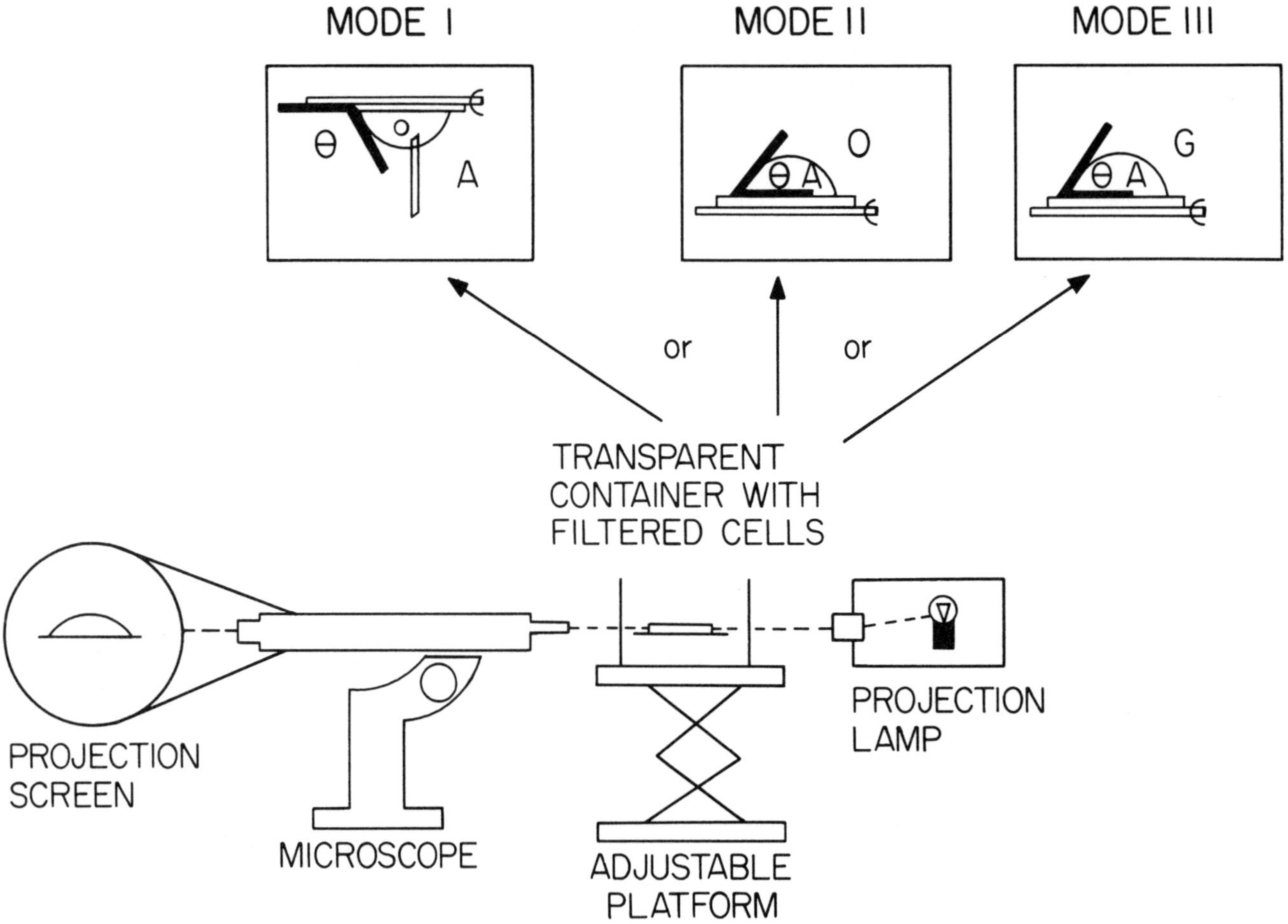

Fig. 3. Diagrams of the optical system used to measure contact angles and of the three modes used to measure wettability of filtered bacterial cells (C)
Mode I Organic drop (O) in bulk aqueous phase (A) (analagous to an O/W emulsion)
Mode II Aqueous drop (A) in buik organic phase (O) (analagous to a W/O emulsion)
Mode III Aqueous drop (A) in air (G)

through the organic phase onto the cells (Mode II), or by extruding upward onto the cell surface a small organic phase drop from a syringe tip submerged in the aqueous phase (Mode I). In this latter method, the drop remained adherent to the syringe, and contact angle was found to be independent of drop size over the range used as long as angles were measured after equilibrium had been reached following the extrusion of organic phase from the syringe.

Demulsification ability of the washed bacterial cells was tested using the emulsions shown in Table 2. All emulsions were oil-in-water (O/W) emulsions. The structure and properties of the surfactants are shown in Table 3. The emulsions were prepared and half-lives were determined as elsewhere described.

Table 4 shows the half-lives ($t_{1/2}$) of the various emulsions when 0.1 ml of differently aged washed cells (resuspended at a concentration of 5 g/l) were added to 10 ml of the various emulsions. The values from Table 4 are plotted in Figure 4, from which it can be seen that: (*i*) as the cells got older, they generally had improved demulsification ability, and (*ii*) all emulsions showing a differential response to various cell ages were similarly influenced.

Figure 5 shows the contact angles of distilled water drops on kerosene-wet cells (Mode II, figure 3) as a function of culture age. There is a noticeable similarity in the shapes of the curves in Figure 4 and the curve in Figure 5. The peaks centering around day 4 in both the half-life and contact angle curves were produced by a temporary malfunction of the shaker between days 3 and 4. This probably allowed the cultures to become oxygen-deficient for a short time. Figure 6 is a plot of the logarithm of emulsion half-life vs. contact angle of the bacterial surface producing the demulsification. The correlation strongly suggests that wettability of the cell surface was a contributing factor to the demulsifying ability

Table 2. Emulsions used to test bacteria-induced demulsification

*Emulsion Code**	*Emulsion pH*	*Surfactant Molarity*	*Interfacial Tension (mN/m)*
A200 (0.5)	9	1.78×10^{-3}	22
A200 (1.5)	9	5.34×10^{-3}	11
A200 (2.0)	9	7.12×10^{-3}	8
RW150 (6.25)	5.5	6.82×10^{-3}	19
T/S (0.3)	7	2.58×10^{-4} **	13
T/S (1.0)	7	8.62×10^{-4}	10
T/X (3.16)	7	27.2×10^{-4}	8
L92 (0.01)	7	2.74×10^{-6}	15
L92 (0.1)	7	2.74×10^{-5}	9
L92 (0.68)	7	1.86×10^{-4}	3

*The values in brackets indicate the total surfactant concentration in g/l.

**An apparent molarity is reported that is the sum of the molarities of the two component surfactants: 27 mole % Tween 60 and 73 mole % Span 60.

Table 4. Half-lives of emulsions exposed to equal quantities of washed *C. petrophilum* cells of different ages

	*Half-Lives (hr)**					
	Age of Culture (days)					
Emulsion	*2*	*3*	*4*	*7*	*14*	*21*
A200 (0.5)	5.6	0.77	0.19	0.20	<0.01	0.01
A200 (1.5)	—	—	—	—	—	—
A200 (2.0)	—	—	—	—	—	—
RW150 (6.25)	0.01	<0.01	<0.01	<0.01	<0.01	<0.01
T/S (0.3)	492	36	102	68	4	16
T/S (1.0)	—	1,111	1,921	210	57.6	
T/S (3.16)	—	—	—	1,871	118	
L92 (0.01)	<0.01	<0.01	<0.01	<0.01	<0.01	<0.01
L92 (0.1)	0.15	0.09	0.04	0.02	<0.01	<0.01
L92 (0.68)	9.7	3.5	4.1	3.6	0.01	0.04

*A dash for $t_{1/2}$ indicates that there was no detectable demulsification up to 1,000 hr. A blank space indicates that no tests were done with that age of cell.

Table 3. Surfactants used in emulsion preparation

Emulsion Code	*Surfactant Name and/or Formulation**	*Approximate Molecular Weight*
L92	Pluronic L92 $HO(CH_2-CH_2-O)_a(CH(CH_3)-CH_2-O)_b(CH_2-CH_2-O)_cH$	3,640
T/S	Tween 60 (polyoxyethylene 20 sorbitan monostearate)	3,087
	Span 60 (sorbitan monostearate)	446
A200	Avirol 200 (ammonium lauryl sulfate)	281
RW150	$RNH(OCH_2CH_2)_{15}OH$ R taken as $CH_3-(CH_2)_{15}-$	917

*Surfactants were obtained from the following sources:
L92 — BASF Wyandotte Corp., Wyandotte, Michigan
Tween 60, Span 60 — J.T. Baker Chemical Co., Phillipsburg, N.J.
Avirol 200 — Henkel Inc., Hoboken, N.J.
RW150 — Rohm and Haas, Philadelphia, PA

of whole cells, and that demulsification of at least O/W emulsions was enhanced with increased hydrophobicity of the cell surface.

THE ROLE OF SURFACTANTS IN BACTERIA-INDUCED DEMULSIFICATION

Figure 6 shows that although the log half-life vs. contact angle plots of all emulsions resemble each other, the absolute values of the half-lives produced by a given cell surface varied with both the surfactant type and its concentration used to stabilize the emulsions. It is therefore important to question how surfactants modulate the demulsifying ability of cell surfaces.

Surfactants might, for example, change the physical parameters (e.g., drop size and therefore total interfacial area) of the emulsions or perhaps change the nature of the interaction between cell surface and emulsion drop by

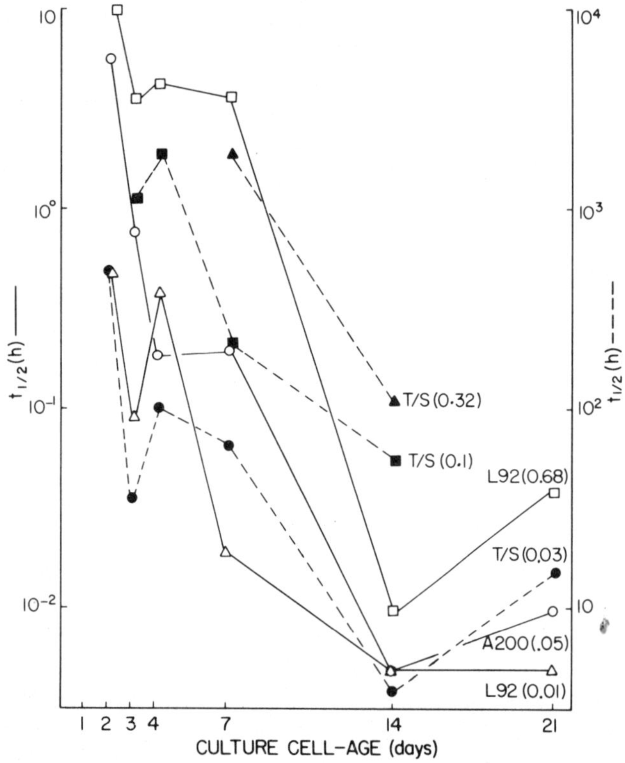

Fig. 4. Half-lives of emulsions exposed to equal quantities of washed *C. petrophilum* cells from culture of different ages. Emulsion codes correspond to those in Table 2

altering the surfaces of the droplet and/or the bacteria. Changes in the surfaces of the droplet or bacteria could influence adhesion of the droplet on that surface.

Figure 7 shows that a correlation exists between the half-lives of the neutral surfactant-stabilized emulsions (L92, T/S) and the molar concentration of surfactant. The charged-surfactant-stabilized emulsions (anionic A200 and cationic RW 150) do not have half-life values that follow the same pattern. The influence of surfactant bulk phase concentration, molecular size, configuration, and charge on the number of surfactant molecules per unit area of oil-water interface and the influence of surfactant chemistry and charge on the interactions of surfactant with oil-water interface and bacterial cell surface are a few of the factors which would necessarily be considered in explaining Figure 7. It is beyond the scope of this paper to undertake such an explanation; however, it is possible to state that the demulsifying ability of any given cell surface appears to depend on the structure and concentration of the surfactant present.

The surfactant-concentration dependence of demulsification by a given cell surface is also reflected in a similar surfactant-concentration dependence of contact angles. Figure 8 shows that when the contact angles of both young and old cells were measured as a function of surfactant concentration, the angles generally increased with increasing surfactant concentration below the critical micelle concentration (CMC) but decreased with further increases in surfactant concentration above the CMC.

The different influences of surfactant concentrations above and below the CMC on contact angles were probably due to interaction of surfactant with the liquid-liquid interface and with the solid-liquid interfaces at certain concentrations. The influence of surfactant concentration on the wetting of high-energy (i.e., hydrophilic) surfaces such as metals and minerals has previously been examined.[10] Wettability of such surfaces depended on the preference of the surfactant for the three interfaces involved (solid-air, solid-wetting liquid, and air-wetting liquid). At low surfactant concentrations the preferential binding of surfactant to the solid-air interface resulted in increased hydrophobicity of the solid surface and decreased wetting by water. At concentrations somewhat below the CMC, the solid-air interface approached saturation, and the surfactant began

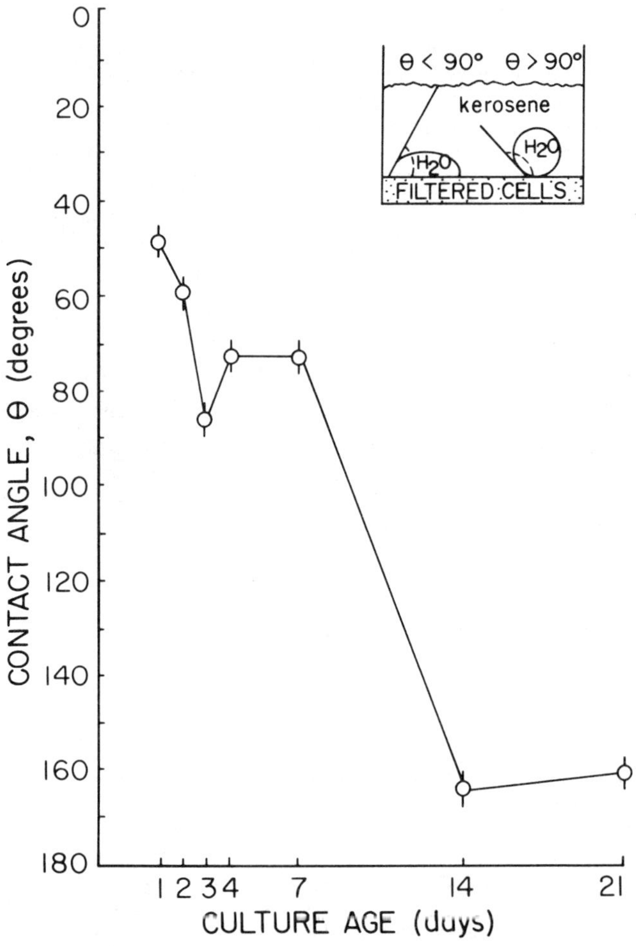

Fig. 5. Contact angles (Mode II) of distilled water on filtered, washed cells of *C. petrophilum* from cultures of different ages

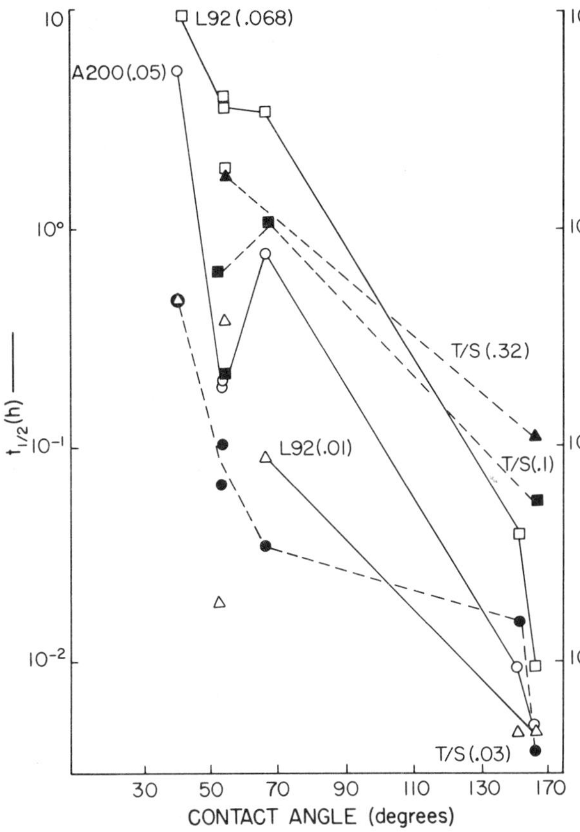

Fig. 6. Half-lives of emulsions (Figure 4) as a function of contact angles (Figure 5) of water with the demulsifying cells

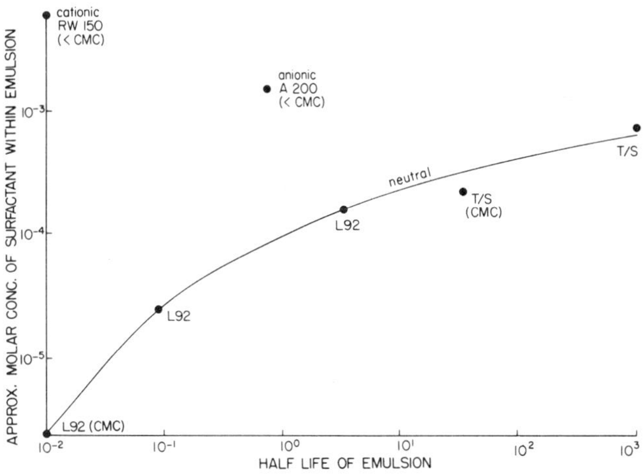

Fig. 7. Half-lives of the studied emulsions stabilized by different surfactants and at different concentrations in the presence of cells having a given contact angle

concentrating at the other interfaces. The water drop surface therefore changed from hydrophilic to hydrophobic. There was then an increased wetting of the mineral surface at concentrations greater than the CMC, whereas increasing surfactant concentrations below the CMC resulted in decreased wetting of the surface.

It is suggested that similar phenomena could account for the reversal of wetting as seen in Figure 8. To the extent that wetting with water of bacterial surfaces and demulsification of oil-in-water emulsions by those surfaces are inversely related, the surfactant-concentration dependence of contact angle and of emulsion half-life would be as diagrammed in Figure 9. Figure 10 presents a schematic illustration of how surfactant might modulate the interaction between a kerosene drop and bacterial surfaces. At low concentrations the surfactant could preferentially adsorb to either the cell surface or the drop surface with resulting improvement in adhesional interaction and consequential drop spreading. As the concentration would be increased, the opposite surface would also become saturated with surfactant and, as it did, adhesional interactions and drop spreading would decrease. Again, to the extent that adhesion and spreading of the oil

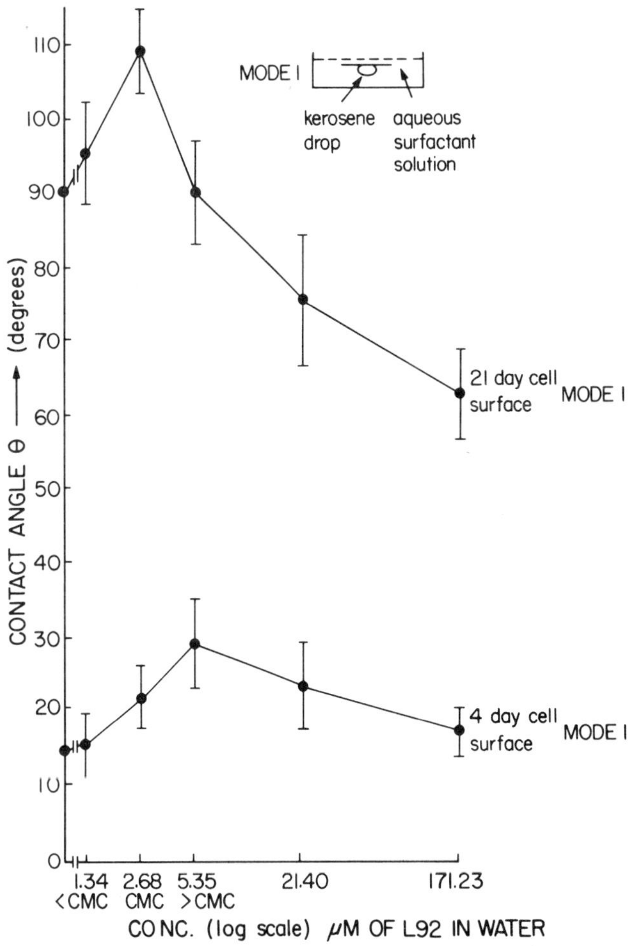

Fig. 8. Contact angles (Mode I) of aqueous L92 surfactant solutions of differing concentration on young hydrophilic and older, more hydrophobic filtered cells. The critical micelle concentration is indicated.

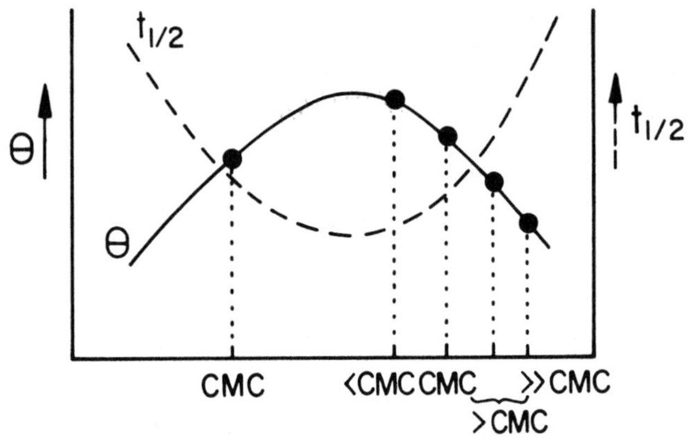

Fig. 9. An illustration of how contact angle (θ) and emulsion half-life ($t_{1/2}$) might be related to the surfactant (emulsifier) concentration. Surfactant concentration is indicated in the figure as being less than, equal to, or more than the critical micelle concentration (CMC)

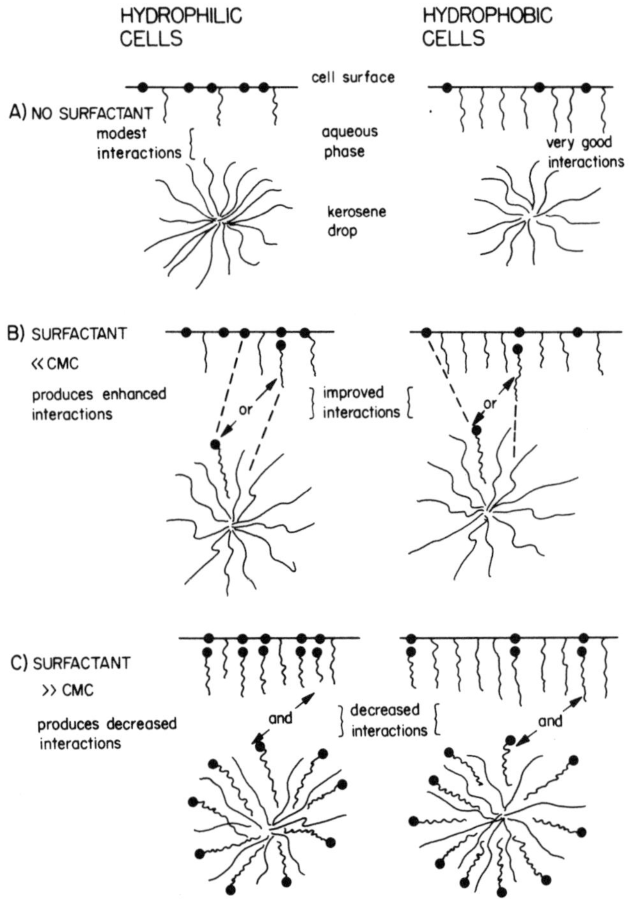

Fig. 10. An illustration of how surfactants at concentrations below and above the CMC might differently influence the interaction of an oil drop in an O/W emulsion with hydrophilic and hydrophobic cell surfaces

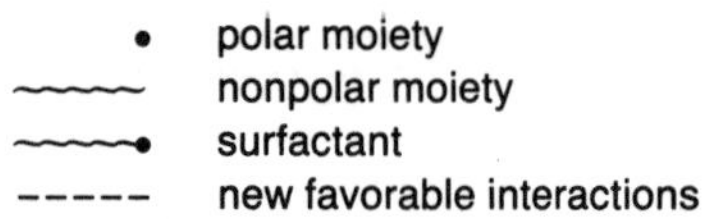

drop on the bacterial surfaces determine the demulsification rate of O/W emulsions, it is expected that progressive increases in surfactant concentration above the CMC should result in increasingly longer half-lives as experimentally observed (Figure 7).

SUMMARY

A mechanism is proposed for bacteria-induced demulsification. The mechanism invokes first the contact of an emulsion drop with the bacterial cell outer surface, followed by spreading of the drop on the surface, and finally coalescence of the drop with another drop or the already-separated discontinuous phase that has similarly contacted the cell surface and has spread on it.

Consistent with the proposed mechanism, a correlation was observed between the stability of oil-in-water emulsions in the presence of a given cell surface and the contact angle of that cell surface. The more hydrophobic the cell surface, the more readily it broke oil-in-water emulsions.

The chemical properties and concentration of surfactants in an emulsion were shown to influence the stability of the emulsion in the presence of a given bacterial surface. Surfactant concentration was also shown to influence contact angles of the cell surface in a manner which depended upon the concentration of the surfactant relative to its critical micelle concentration. A qualitative explanation was offered for how surfactants might modulate the demulsifying activity of bacterial cell surfaces.

ACKNOWLEDGMENTS

This work was sponsored by both the National Research Council of Canada and the Department of Energy, Mines and Resources of the Government of Canada.

REFERENCES

1 CAIRNS, W.L., D.G. COOPER, J.E. ZAJIC, J.M. WOOD, and N. KOSARIC. 1982. Characterization of *Nocardia amarae* as a potent biological coalescing agent of water-oil emulsions. Appl. Environ. Microbiol. *43*:362–366.

2 COOPER, D.G., J.E. ZAJIC, W.L. CAIRNS, and N. KOSARIC, 1980. Demulsification and microbes. Part 1. Biochemical engineering research reports, vol. VI. The University of Western Ontario, London.

3 COOPER, D.G., W.L. CAIRNS, and N. KOSARIC. 1981. Demulsification and microbes. Part 2. Biochemical engineering research reports, vol. VIII. The University of Western Ontario, London.

4 MIZRAHI, J., and E. BARNEA. 1970. The effects of solid additive on the formation and separation of emulsions. Brit. Chem. Engin. *15*:497.

5 CHIEU, N.J., E.F. GLOYNA, and R.S. SCHECHTER. 1977. Coalescence of emulsified oily wastewater by fibrous beds. Proc. Industrial Waste Conference *30*:611.

6 SHERONY, D.F., R.C. KINTNER, and D.T. WASAN. 1978. Coalescence of secondary emulsions in fibrous beds. Surface and Colloid Science *10*:99.

7 SPARKS, R.E., and I.D. STAFFORD. 1970. Continuous breaking of emulsions by coalescence on supported liquid films. Engin. Bull. *137*:745.

8 ZISMAN, W. A. 1964. Relation of the equilibrium contact angle to liquid and solid constituents, p. 1-51. *In* R.F. Gould (ed.), Contact angle, wettability and adhesion. Advances in chemical series, vol 43. Amer. Chem. Soc., Washington, D.C.

9 VAN OSS, C.J., C.F. GILLMAN, and A.W. NEUMANN. 1975. Phagocytic engulfment and cell adhesiveness as cellular surface phenomena, p. 7-19. Marcel Dekker Inc., New York.

10 JOHNSON, R.D., AND R.H. DETTRE. 1969. Wettability and contact angles. *In* E. Matijevic (ed.), Surface and colloid science. Wiley-Interscience, New York.

17 BACTERIAL ADHERENCE TO HYDROCARBONS

M. Rosenberg, D.L. Gutnick, and E. Rosenberg
Tel Aviv University

The importance of cell adherence in growth physiology of many bacterial species is well documented. For example, *Caulobacter* cells divide only after they attach to solid surfaces.[1] The initial step in the life cycle of *Bdellovibrio* is attachment to Gram-negative bacteria.[2,3] Specific cell-to-cell interactions pervade the entire life cycle of the myxobacteria.[4,5] Adhesion is a prerequisite to the successful colonization by microorganisms of animals and plants.[6,7] Microbial adhesion to surfaces has been the subject of several recent reviews.[6,7,8,9,10]

In cases where the carbon or energy source is a water-insoluble material, such as cellulose or chitin, bacterial adhesion can facilitate growth.[10,11,12] But cell contact is not an absolute requirement because extracellular enzymes can degrade these polymers into water-soluble substrates. However, growth of microorganisms on hydrocarbons presents a special problem since hydrocarbons are immiscible with water and their breakdown cannot occur extracellularly. The first step in aromatic or aliphatic hydrocarbon degradation is introduction of molecular oxygen into the molecules by cell-associated enzymes.[13,14,15]

Two general types of hydrocarbon-cell interactions have been postulated, depending upon the state and size of the oil droplets relative to the size of the microbial cells: (*i*) adherence of cells to large oil droplets and (*ii*) pseudosolubilization involving the cellular assimilation of emulsified small hydrocarbon droplets.[16] The relative contribution of these two types of interactions to the growth of bacteria on hydrocarbons has been difficult to ascertain experimentally. The frequently observed relationship between hydrocarbon-aqueous interfacial area and growth rate can be used to support either hypothesis.[17]

Adherence and growth of microorganisms on hydrocarbons is one example of a wide spectrum of phenomena in which bacterial hydrophobicity and adherence to hydrophobic surfaces are involved. The hydrophobic nature of the outermost bacterial surface has been cited as a factor in partitioning microorganisms at the air-water interface;[18,19] in the adherence of bacteria to nonwettable plastic surfaces,[20,21,22,23] and in the attachment of bacteria to phagocytes and other mammalian cells.[24,25,26,27,28]

A rapid and quantitative method for measuring bacterial adherence to hydrocarbons was described previously.[29,30] The present report deals with the use of this technique in (*i*) examining the cell surface hydrophobicity of various bacterial species, some of which do not metabolize hydrocarbons, and (*ii*) selecting and isolating nonadherent mutants. The isolation of a nonadherent mutant of the hydrocarbon-degrading bacterium *Acinetobacter calcoaceticus* RAG-1 has enabled us to examine the role of adherence in bacterial growth on hexadecane.

MATERIALS AND METHODS

Bacterial Strains. Tables 1 and 2 contain a list of bacteria used in this study. *Acinetobacter calcoaceticus* MR-481 is a nonadhering derivative of strain RAG-1.

Media and Growth Conditions. For measuring adherence to test hydrocarbons or preparing inocula for growth experiments, 0.1 ml of bacteria from an overnight culture was inoculated into 50 ml of growth medium and was incubated in a New Brunswick model G-53 gyrotory shaker at 150 rpm at 30°C. Cells were harvested at the end of the

Table 1. Adherence of bacteria to hexadecane.

Strain	*Source*	*Growth Medium*[a]	*Adherence to Hexadecane (%)*[b]
Acinetobacter calcoaceticus RAG-1	this laboratory	BHI	97
Bacillus cereus	local isolate	NB	1
Bacillus subtilis 168	E.Z. Ron	NB	0
Enterobacter aerogenes CDC 659/66	E.Z. Ron	NB	9
Escherichia coli B	E.Z. Ron	NB	0
Escherichia coli K12	E.Z. Ron	NB	2
Escherichia coli J5 (rough)	E.Z. Ron	NB	11
Micrococcus lysodeikticus	I. Friedberg (ATCC 4698)	NB	0
Myxococcus xanthus	this laboratory	CT	35
Proteus mirabilis S1959	S. Rottem	NB	0
Pseudomonas aeruginosa PAS279	J. Shapiro	NB	0
Serratia marcescens	R. Zack	NB	96
Staphlococcus albus	R. Zack	NB	0
Staphlococcus aureus	R. Zack	NB	91
Streptococcus pyogenes M5	I. Ofek	BHI	84

[a]Cells were harvested at the end of the exponential phase
[b]The standard adherence assay was performed with 0.2 ml hexadecane.

exponential phase, washed twice, and resuspended in PUM buffer (22.2 g $K_2HPO_4 \cdot 3H_2O$; 7.26 g KH_2PO_4; 0.8 g urea; 0.2 g $MgSO_4 \cdot 7H_2O$; and distilled water to 1,000 ml, pH 7.1). Growth media were BHI, brain heart infusion broth (Difco); hexadecane medium, PUM buffer supplemented with 0.2% (v/v) hexadecane (olefin free, 99% purity, Fluka, Switzerland); P-hexadecane medium, 5 g NH_4NO_3, 2.5 g K_2HPO_4, 1 g $MgSO_4 \cdot 7H_2O$, 2 ml hexadecane, 50 mg threonine, and tap water to 1,000 ml; and NB, nutrient broth (Difco) supplemented with 0.5% NaCl. CT medium contained 1% casitone (Difco) and 0.2% $MgSO_4 \cdot 7H_2O$[31].

In the growth experiments, washed cells of RAG-1 and MR-481 were inoculated into acid-washed Klett tubes (14-mm diameter) containing 4.2 ml of an aqueous medium to give an initial turbidity of 5 Klett units. The acetate growth medium consisted of PUM buffer supplemented with 0.2% sodium acetate. Hexadecane growth medium consisted of 0.5 ml hexadecane (olefin free, 99% purity, Fluka, Switzerland) layered onto 4.2 ml of PUM buffer. The test tubes were incubated upright in a New Brunswick gyrotory water bath shaker (Model G-76) at 330 rpm at 30°C. Turbidity was monitored directly in the growth tubes by means of a Klett-Summerson colorimeter fitted with a green filter.

Adherence to Hydrocarbons. The technique for measuring adherence of bacteria to hydrocarbons has been

Table 2. Adherence of *Acinetobacter* strains to hexadecane.

Strain Designation	*Source*	*Growth on Hexadecane*[a]	*Adherence to Hexadecane (%)*[b]
RAG-1 (ATCC 13012)	this laboratory	+	82
HO1-N	W.R. Finnerty	+	99
BD4	E. Junl	+	0
BD413	E. Juni	+	4
B5W str 71 (ATCC 15149)	K. Bryn	+	84
CDC 7827 (ATCC 17959)	K. Bryn	+	86
B5W (ATCC 17903)	K. Bryn	−	0
ATCC 17906	K. Bryn	−	52
ATCC 17977	K. Bryn	−	7
ATCC 17985	K. Bryn	−	41

[a]Growth was examined on PUM agar in the presence of hexadecane vapors.
[b]Adherence was measured on cells grown to stationary phase on BHI broth; the standard adherence assay was performed with 0.2 ml hexadecane.

described previously.[32] Cells were washed twice and were resuspended in PUM buffer to an initial A_{400} of approximately 1.5. Various volumes of the test hydrocarbons (n-hexadecane, n-octane, or p-xylene) were added to round-bottom, acid-washed test tubes (10-mm diameter) containing 1.2 ml of washed-cell suspension, and the phases were vortexed uniformly for 120 sec. After allowing 15 min for the phases to separate, the absorbance of the lower aqueous phase of each tube was measured at 400 nm (Gilford model 240 spectrophotometer, 1-cm light path). Adherence was calculated as the percentage loss in absorbance relative to that of the initial cell suspension.

Isolation of the Nonadherent Mutant MR-481. A spontaneous mutant of RAG-1 that is deficient in its ability to adhere to hydrocarbons was isolated by the following technique: RAG-1 cells were inoculated into 50 ml of a nutrient broth medium (NB) and were incubated in a New Brunswick Model G-53 gyrotory shaker at 150 rpm at 30°C. Cells were harvested in stationary phase following 18 hr growth, washed twice, and resuspended in PUM buffer to an A_{400} of approximately 1.5. Two ml of n-octane were added to 5 ml of the cell suspension, and the phases were vortexed for 5 min. Following phase separation the lower aqueous phase was tranferred to a second test tube. Octane was again added, and the phases were mixed as described previously. This extraction process was continued until cells in the lower aqueous phase were too few to be visible under microscopic examination. The final aqueous phase served as the inoculum for a second nutrient broth culture that was grown to stationary phase as previously described. Cells were again harvested, washed, and extracted with octane, and the final aqueous phase was then inoculated into a third nutrient broth culture. After four such growth and extraction cycles, considerable turbidity remained in the aqueous phase, even following several extraction attempts with octane. After four such growth and extraction cycles, considerable turbidity remained in the aqueous phase, even following several extraction attempts with octane. The final aqueous phase was plated on nutrient agar. Three individual colony types were isolated and tested for adherence to hydrocarbons. One clone of cells, termed MR-481, exhibited no significant affinity towards the test hydrocarbons and was chosen for further study.

Emulsan Preparation. The general method for preparing the extracellular emulsifying agent of *A. calcoaceticus* RAG-1, referred to as emulsan, was described previously.[33] The particular preparation used in this study was further purified by precipitating an aqueous solution of emulsan with cetyltrimethylammonium bromide (BDH Chem. Ltd.), redissolving the precipitate in 0.5 M Na_2SO_4, and dialyzing extensively against distilled water. The material was then deproteinized by the phenol method and dialyzed.[34] The final product had a reduced viscosity of 490 cc per g, an emulsifying activity of 150 units per mg, an ester content of 0.5 μ moles per mg, and a protein content of less than 0.5%. This emulsan preparation was dissolved in PUM buffer (2 mg/ml) and was sterilized by autoclave prior to its addition in the growth experiments.

RESULTS

Bacterial Adherence to Hydrocarbons. The basic assay for bacterial adherence to hydrocarbons is illustrated in

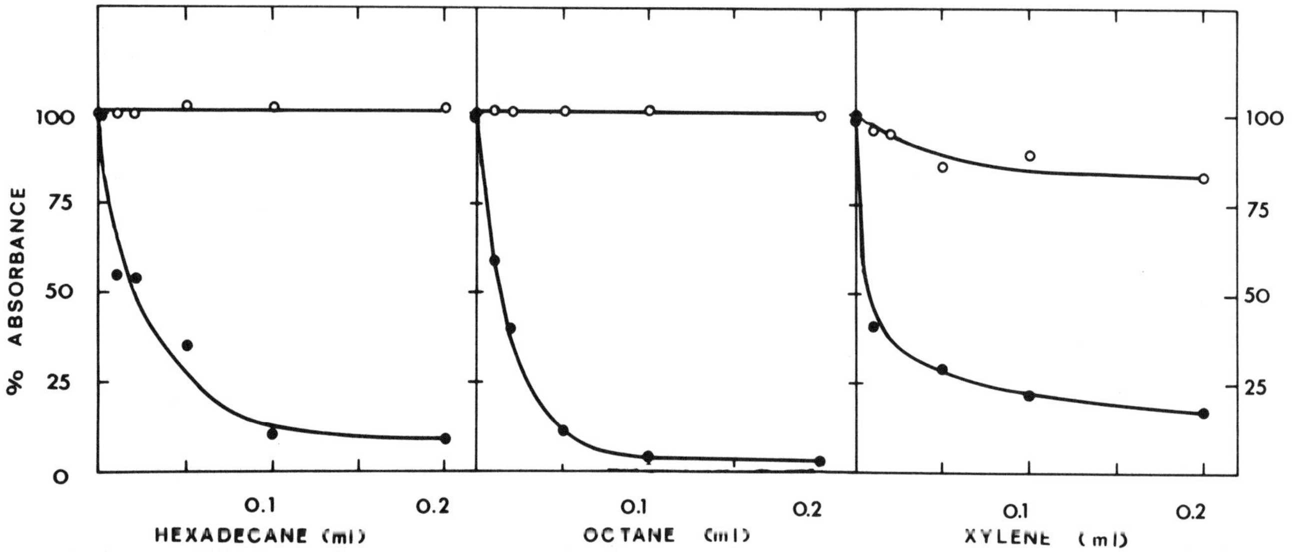

Fig. 1. Adherence of *Eschericia coli* B (○) and *Acinetobacter calcoaceticus* RAG-1 (●) to hydrocarbons. Cells were harvested at the end of the exponential growth phase, washed twice in PUM buffer, and then resuspended to an initial A_{400} of 1.2–1.4. The adherence assay was performed as described in "Materials and Methods." The ordinate is the percentage of initial absorbance of the cell suspension

Figure 1. The assay was performed with varying volumes of three test hydrocarbons: n-hexadecane, n-octane, and p-xylene. Following phase separation the turbidity of the lower aqueous phase was compared to that of the initial suspension. When the assay was performed with cells having no affinity for the test hydrocarbons (e.g., *E. coli*), the hydrocarbon droplets rose and coalesced soon after mixing and no significant changes were observed in the turbidity of the lower aqueous phase. However, when adherent cells were similarly tested (e.g., *A. calcoaceticus* RAG-1), a creamy upper layer was obtained following mixing. Microscopic observation of this layer revealed an oil-in-water emulsion consisting of hydrocarbon droplets, each covered by a layer of bacteria. The extraction of cells into this emulsion layer was accompanied by a corresponding decrease in turbidity of the lower aqueous phase. The fraction of adherent cells was determined by comparing the optical densities of the lower aqueous phase before and after mixing. Bacterial adherence was a function of volume of hydrocarbon added. The cell-coated hydrocarbon droplets in the upper layer remained stable for a period of days. Addition of 5% isopropanol resulted in immediate droplet coalescence and the concommitant release of extracted cells into the aqueous phase, thereby demonstrating that the bacteria had partitioned to the hydrocarbon-water interface.

Tables 1 and 2 summarize a number of experiments in which cell surface hydrophobicity of a variety of bacteria was measured by adherence to hexadecane. The majority of laboratory strains examined exhibited low affinity toward the test hydrocarbon. Among the bacterial strains exhibiting high affinity for hexadecane were hydrocarbon-degrading *Acinetobacter calcoaceticus* strains RAG-1, HO1-N, B5W str 71, and CDC 7828; and bacteria that do not metabolize hydrocarbons: *Serratia marcescens, Staphylococcus aureus*, and *Streptococcus pyogenes*.

Pseudomonas aeruginosa (which can utilize hexadecane as the sole carbon and energy source) exhibited no significant affinity toward any of the test hydrocarbons when grown to late exponential phase in nutrient broth (Figure 2). Since adherence in this organism might have required the presence of hydrocarbons during growth, *P. aeruginosa* cells were also assayed after growth on hexadecane (Figure 2). The affinity of alkane-grown *P. aeruginosa* toward hexadecane increased somewhat, e.g., 32% of such cells were removed from the aqueous phase in the presence of 0.2 ml hexadecane. In contrast, *A. calcoaceticus* RAG-1 exhibited much higher affinity when grown on hexadecane; 93% of the cells could be removed by as little as 0.05 ml hexadecane.

Role of Adherence in the Growth of A. calcoaceticus *RAG-1 on Hexadecane. Acinetobacter calcoaceticus* RAG-1 adheres avidly to hydrocarbons.[35,36,37] A spontaneous mutant of RAG-1, deficient in its ability to adhere to hydrocarbons, was selected by enriching for cells that did not adhere to octane. This mutant, referred to as MR-481, was unable to adhere to any of the three test hydrocarbons, even when grown under conditions for which RAG-1 adherence was optimal (Figure 3).

Mutant MR-481 resembled the wild type in the following characteristics: (*i*) colonial and cell morphology,

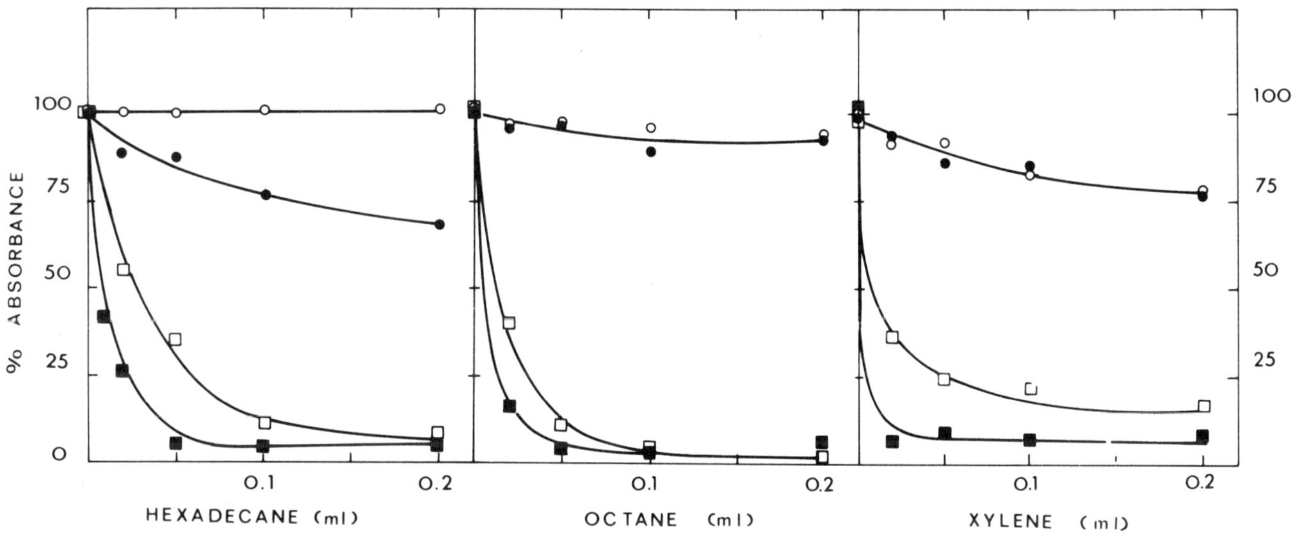

Fig. 2. Adherence of *Pseudomonas aeruginosa* PAS 279 and *Acinetobacter calcoaceticus* RAG-1 to hydrocarbons. The experiment was performed as described in Figure 1 with cells harvested at the end of the exponential growth phase: *P. aeruginosa*, NB medium (○), P-hexadecane medium (●); *A. calcoaceticus*, NB medium (□), hexadecane medium (■)

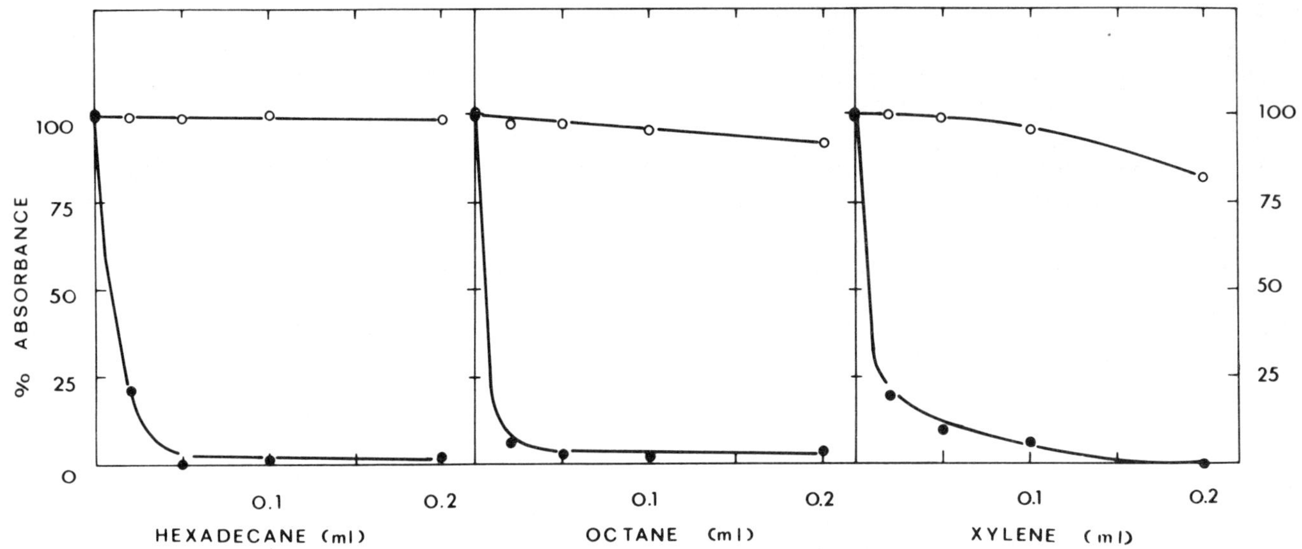

Fig. 3. Adherence of RAG-1 and MR-481 to hydrocarbons. Stationary phase cells were harvested from brain heart infusion broth, washed twice, and resuspended in PUM buffer to an A_{400} = 1.45 to 1.50. To 1.2 ml of suspended RAG-1 (●) or MR-481 (○) cells were added various volumes of test hydrocarbon, and the mixtures were vortexed for 120 sec. Following phase separation the turbidity of the lower aqueous phase was measured. The ordinate is the percent of initial absorbance of the cell suspension

(*ii*) sensitivity to two bacteriophages, ap2 and ap3, that are specific for *A. calcoaceticus* RAG-1 strains,[38] (*iii*) production of extracellular emulsifying activity during growth on ethanol as the sole carbon and energy source,[39] and (*iv*) doubling time on brain heart infusion broth (38 min at 30°C with aeration).

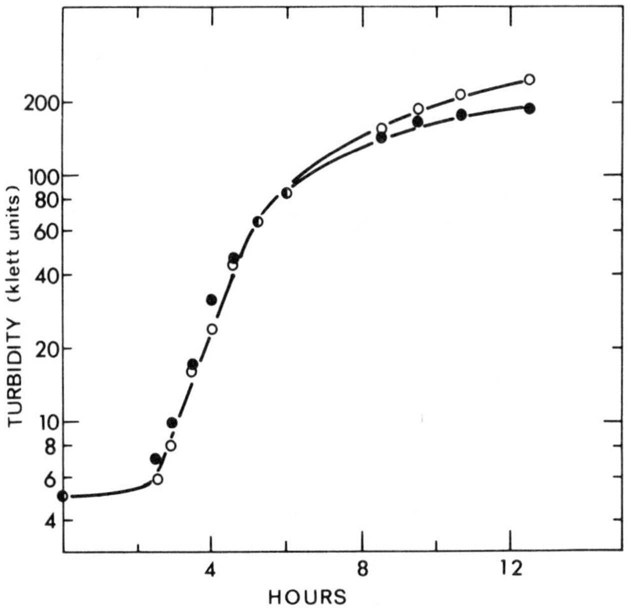

Fig. 4. Growth of RAG-1 and MR-481 in an acetate medium. Stationary phase cells were harvested from brain heart infusion broth, washed twice, and resuspended in PUM buffer supplemented with 0.2% sodium acetate. RAG-1 (●) and MR-481 (○) cells were inoculated into a 4.2-ml acetate medium in Klett tubes to an initial turbidity of 5 Klett units

In all growth experiments, inocula were prepared in the same manner as was described for measuring adherence in Figure 3. In a defined medium with sodium acetate as the soluble carbon and energy source, RAG-1 and MR-481 showed similar growth kinetics (Figure 4). Following a 2-hr lag, both strains grew with a doubling time of 48 min.

Using hexadecane as the sole source of carbon and energy (and under limited dispersion conditions), RAG-1 grew, whereas the nonadherent mutant MR-481 failed to grow for at least 54 hr (Figure 5). Growth of RAG-1 was accompanied by breaking the upper hydrocarbon layer into droplets. Microscopic examination of the upper phase indicated that these droplets were covered by patches of adhering cells. Hydrocarbon droplets were not observed in the aqueous phase under these conditions; thus, the observed turbidity of the lower aqueous phase was due to unbound cells. The values presented for growth of RAG-1 on hexadecane should be considered minimal since attached bacteria were not measured. In the corresponding tubes inoculated with mutant MR-481, neither growth nor substantial breakage of the hexadecane layer into droplets was observed (Figure 6).

Effect of Emulsan on the Growth of MR-481 on Hexadecane. *A. calcoaceticus* RAG-1 produces a potent extracellular emulsifying agent, referred to as emulsan.[40,41,42,43] It was of interest to examine the effect of a highly purified preparation of emulsan (see "Materials and

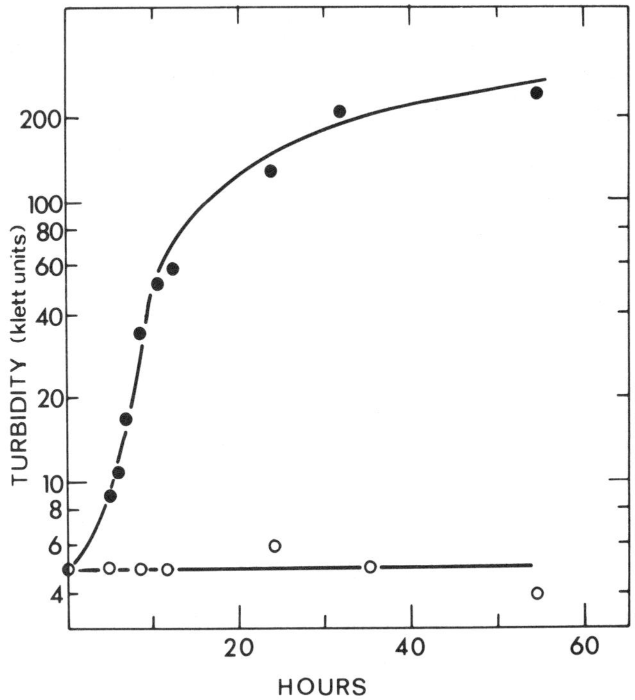

Fig. 5. Growth of RAG-1 and MR-481 in a hexadecane medium. Klett tubes containing 4.2 ml PUM buffer were inoculated with either RAG-1 (●) or MR-481 (○) to an initial turbidity of 5 Klett units. Tubes were overlayed with 0.5 ml hexadecane and were incubated as described in "Materials and Methods"

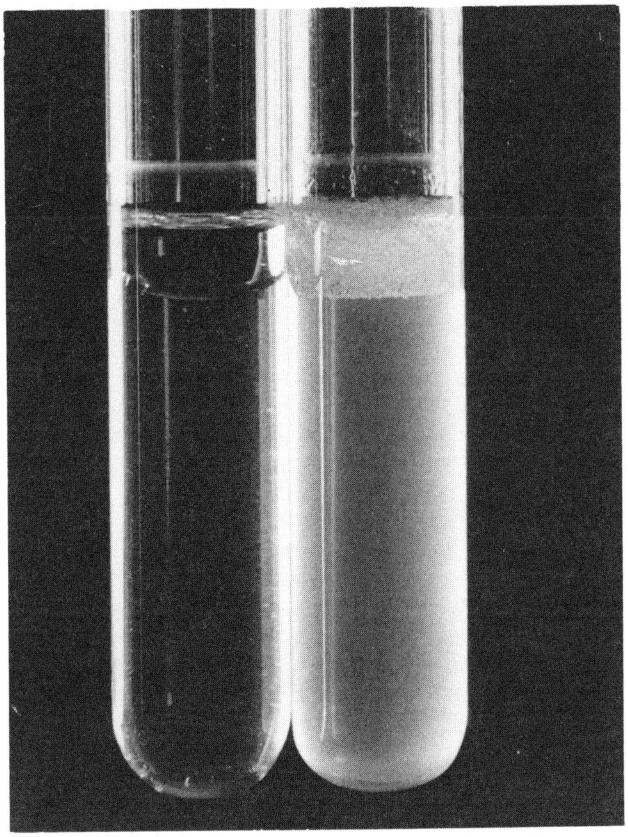

Fig. 6. Growth of MR-481 (left) and RAG-1 (right) on hexadecane medium. Tubes were photographed following 48 hr of incubation under the conditions described in Figure 3

Methods") on the growth on hexadecane of RAG-1 and the nonadherent mutant MR-481 (Figure 7). In the absence of hexadecane, emulsan did not serve as a carbon and energy source for either RAG-1 or MR-481. However, emulsan did enable growth of MR-481 on hexadecane. Growth of MR-481 in the presence of both emulsan and hexadecane began after about 6 hr, whereas neither component alone supported growth for 54 hr. At the weight ratio of hexadecane-emulsan used (1,925:1), the hexadecane droplets formed were too large to disperse into the aqueous phase and thus did not interfere with turbidity measurements. Emulsan had no significant effect on the growth of RAG-1 on hexadecane under these conditions.

Growth of MR-481 on hexadecane was also induced if the emulsan was added after several hours of incubation (Figure 8). In each case of emulsan addition, the resultant breakage of the hexadecane layer into droplets was followed by subsequent growth of the mutant cells. The time required for the commencement of growth following emulsan addition appeared to increase with the length of incubation prior to adding the bioemulsifier.

Growth of MR-481 on hexadecane in the presence of emulsan was not the result of genotypic nor phenotypic reversion of MR-481 from nonadherent to adherent cells. As shown in Table 3, there was no significant increase in the adherence of MR-481 grown on hexadecane in the presence of emulsan compared to the initial inoculum grown in a rich medium (Figure 3).

RAG-1 cells that were harvested and washed following growth on hexadecane both in the presence and absence of emulsan exhibited high affinity toward hexadecane (Table 3). Thus, the appearance of free RAG-1 cells in the lower phase during growth on hexadecane did not result from a loss in their adherent properties.

DISCUSSION

Mudd and Mudd reported over fifty years ago that certain microorganisms tend to concentrate at oil-water interfaces.[44] Although this observation has been corroborated frequently during studies on the growth of microorganisms on liquid hydrocarbons, there has been no direct evidence that adherence is an essential aspect of microbial growth on hydrocarbons.[45,46,47,48,49,50] Certain investigators have argued

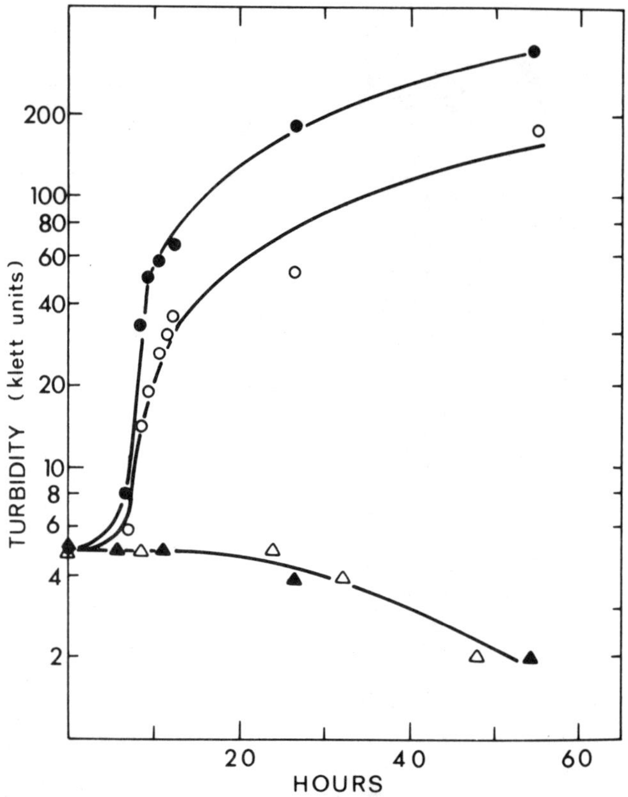

Fig. 7. Effect of emulsan on the growth of MR-481 on hexadecane. Inocula preparation and incubation conditions were as described in Figure 3. RAG-1 (●) and MR-481 (○) were grown on a hexadecane medium in the presence of 50 μg/ml emulsan. RAG-1 (▲) and MR-481 (△) controls were grown in PUM buffer supplemented with 50 μg/ml emulsan in the absence of hexadecane

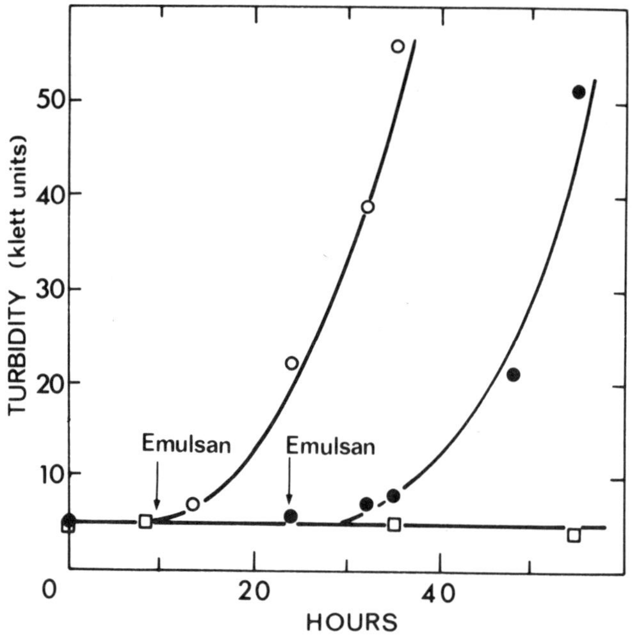

Fig. 8. Induction of growth of MR-481 on hexadecane by emulsan addition. Inocula preparation and incubation conditions were as described in Figure 3. Emulsan was added to a final concentration of 50 μg/ml at 10 hr (○) or 24 hr (●). Tubes containing no emulsan served as controls (□)

Table 3. Adherence of RAG-1 and MR-481 following growth on hexadecane.

Strain	Turbidity (A_{400})[a] Initial	Final	% Adherence
RAG-1[b]	1.42	0.09	94
RAG-1[c]	1.40	0.05	96
MR-481[c]	1.43	1.36	4.9

[a]Cells were washed twice and resuspended in PUM buffer to initial turbidities as indicated. Hexadecane (0.2 ml) was added to 1.2 ml of cell suspension and the mixture was vortexed for 120 sec. Following phase separation, the turbidity of the aqueous phase was measured.

[b]RAG-1 cells were harvested following 54 hr growth on a hexadecane medium, washed twice, and resuspended in PUM buffer.

[c]RAG-1 and MR-481 were harvested following 54 hr growth on a hexadecane medium in the presence of 50 μg/ml emulsan, washed twice, and resuspended in PUM buffer.

that microbial growth on hydrocarbons is due to emulsification of the water-insoluble substrate rather than by direct contact between cells and hydrocarbon droplets.[51,52,53] *A. calcoaceticus* RAG-1 is an interesting microorganism in this regard because it adheres avidly to hydrocarbons and produces an extracellular emulsifying agent, emulsan, that has been extensively characterized.[54,55,56,57,58,59,60]

Isolation of a mutant strain of *A. calcoaceticus* RAG-1 deficient in its ability to adhere to hydrocarbon enabled an examination of the role of adherence in the growth of *A. calcoaceticus* RAG-1 on hexadecane. The nonadherent mutant strain, MR-481, failed to grow on hexadecane under the moderate agitation conditions that supported growth of wild-type cells. Two trivial explanations for the inability of MR-481 to grow on hexadecane — insufficient oxygen and a faulty hydrocarbon-metabolizing enzyme system — were excluded by the following data: (*i*) MR-481 and RAG-1 cells grew with similar kinetics on an acetate medium under conditions identical to those employed for growth on hexadecane; (*ii*) MR-481 was able to grow on a hexadecane medium following addition of an emulsifier; and (*iii*) MR-481 cells were able to grow on a solid medium supplemented with hexadecane vapors as the sole carbon and energy source. We conclude from these data that adherence is a prerequisite for growth of RAG-1 on liquid hexadecane in the absence of strong mechanical agitation or emulsification of substrate.

Both RAG-1 and MR-481 produce a potent extracellular emulsifying agent that enables growth of nonadhering MR-481 cells on hexadecane. The observation that MR-481 does not grow on hexadecane even though it produces normal amounts of emulsan can readily be explained by considering emulsification as a cell-density dependent phenomenon.[61,62,63] At the low inoculum employed (5×10^7 cells/ml), the concentration of emulsan that the cells produced was too low to emulsify the hydrocarbon and permit growth.

These studies suggest that adherence is a prerequisite for growth on liquid hydrocarbon under two conditions: low cell density and limited emulsification. The ability of MR-481 to grow on hexadecane following emulsan addition demonstrates that the presence of extracellular emulsifying agents can enable growth of nonadherent bacteria on hydrocarbons. The poor adherence of *P. aeruginosa* PAS 279 to hydrocarbon (Figure 2) suggests that its growth, similar to that of MR-481, depends on dispersion of the hydrocarbon phase by extracellular emulsification or vigorous mechanical agitation. In most natural environments conditions of low cell density and limited emulsification prevail; thus, the growth of poorly adhering Pseudomonas strains in shake flasks and fermentors carried out in the presence of extraneous emulsifiers or vigorous agitation probably does not reflect growth in the open environment, where emulsifiers and dispersed hydrocarbons would tend to diffuse away from the cell.[64,65,66,67]

The adherence to hydrocarbon of bacteria that do not metabolize hydrocarbons (Tables 1 and 2) can be explained in terms of general cell surface hydrophobicity.[68,69] *S. pyogenes*, for example, has been reported to adhere to buccal epithelial cells through hydrophobic interactions.[70] Its adherence to hydrocarbons may thus be mediated by such interactions. Similarly, hydrophobic cell surface components that enable *Serratia* to partition at the air-water interface may also promote their adherence to hydrocarbons as observed in this study.[71]

Bacterial hydrophobicity has been cited as an important factor in several biological phenomena, including bacterial pathogenicity,[72] and in adherence and growth of bacteria on various hydrophobic surfaces.[73,74,75,76] Adherence to hydrocarbons provides a simple quantitative technique for measuring cell surface hydrophobicity and can be used for the selection and isolation of nonadhering mutants. Adherence of RAG-1 to hydrocarbons may reflect a general phenomenon in microbiology: a biological advantage is conferred on microbial cells that can escape by means of hydrophobic surface components from the bulk aqueous phase to an interface that provides nutrition and enables growth.

SUMMARY

A simple, quantitative assay method was used to measure the adherence of a variety of bacteria to liquid hydrocarbons. Bacterial species differed greatly in their ability to adhere to hydrocarbons; even within the same species, *Acinetobacter calcoaceticus*, different strains varied greatly in their cell surface hydrophobicity. No direct correlation was found between adherence to hydrocarbons and the ability to metabolize hydrocarbons. For example, *Staphylococcus aureus*, a bacterium incapable of degrading hydrocarbons, adhered strongly to all test hydrocarbons, whereas *Pseudomonas aeruginosa* adhered poorly to hydrocarbons, despite its ability to grow on hydrocarbons under laboratory conditions.

The high affinity of *Acinetobacter calcoaceticus* RAG-1 towards liquid hydrocarbons enabled the isolation of a spontaneous nonadherent mutant, MR-481. Strain MR-481 exhibited no significant affinity toward the three test hydrocarbons yet resembled the wild type in many properties, including production of the extracellular emulsifying agent, emulsan. In order to study more closely the role of adherence in growth on hydrocarbons, RAG-1 and MR-481 were compared for growth on hexadecane under conditions of limited agitation and at low initial cell density. Adherent RAG-1 cells were able to grow rapidly under these conditions, whereas nonadherent MR-481 cells failed to grow for at least 54 hr. However, addition of emulsan either initially or at various times after inoculation enabled the nonadherent MR-481 cells to grow on hexadecane. Growth was not the result of reversion of MR-481 from nonadherent to adherent cells.

The data demonstrate that adherence is a crucial factor in the growth of *A. calcoaceticus* RAG-1 on hexadecane in the absence of extracellular emulsification of the substrate.

ACKNOWLEDGMENTS

We would like to thank R. Martinez for valuable advice during the performance of these experiments and N. Kaplan for critical evaluation of the manuscript.

REFERENCES

1 SHAPIRO, L. 1976. Differentiation in the caulobacter cell cycle. Ann. Rev. Microbiol. *30*:377–407.

2 STARR, M.P., and J.C.C. HUANG. 1972. Physiology of the Bdellovibrios. *In* A.H. Rose and D.W. Tempest (ed.), Advances in microbial physiology. *8*:215–261. Academic Press, London.

3 VARON, M., and M. SHILO. 1968. Interaction of *Bdellovibrio bacteriovorus* and host bacteria. I. Kinetic studies of attachment and invasion of *Escherichia coli* B by *B. bacteriovorus*. J. Bacteriol. *95*:744–753.

4 DWORKIN, M. 1973. Cell-to-cell interactions in the myxobacteria. Symp. Soc. Gen. Microbiol. *23*:125–142.
5 STANIER, R.Y. 1942. The cytophaga group: a contribution to the biology of myxobacteria. Bacteriol. Rev. *6*:143-196.
6 OFEK, I., and E.H. BEACHEY. 1980. General concepts and principles of bacterial adherence in animals and man, p. 1–31. *In* E.H. Beachey (ed.), Bacterial adherence. Chapman and Hall, London and New York.
7 LIPPINCOTT, J.A., and B. LIPPINCOTT. 1980. Microbial adherence in plants, p. 375–398. *In* E.H. Beachey, (ed.), Bacterial adherence. Chapman and Hall, New York.
8 DANIELS, S.L. 1980. Mechanisms involved in sorption of microorganisms to solid surfaces, p. 7–58. *In* G. Bitton and K.C. Marshall (ed.), Adsorption of microorganisms to surfaces. John Wiley & Sons, New York.
9 MARSHALL, K.C. 1976. Interfaces in microbial ecology. Harvard University Press, Cambridge, Mass.
10 LATHAM, M.J. 1980. Adhesion of rumen bacteria to plant cell walls, p. 339–350. *In* R.C.W. Berkeley, J.M. Lynch, J. Melling, P.R. Rutter, and B. Vincent (ed.), Microbial adhesion to surfaces. Society of Chemical Industry, London.
11 STANIER, op cit.
12 SUNDARRJ, C.O., and J.V. BHAT. 1972. Breakdown of chitin by *Cytophaga johnsoni*. Arch. fur Mikrobiologie *85*:159–167.
13 GIBSON, D.T. 1971. The microbial oxidation of aromatic hydrocarbons. Crit. Rev. Microbiol. *1*:199–223.
14 RATLEDGE, C. 1978. Degradation of aliphatic hydrocarbons, p. 1–46. *In* R.J. Watkinson (ed.), Developments in biodegradation of hydrocarbons. I. Applied Science Publishers, London.
15 FOSTER, J.W. 1962. Bacterial oxidation of hydrocarbons, p. 241–261. *In* O. Hayaishi (ed.), Oxygenases. Academic Press, New York.
16 KIRSCHNER ZILBER, I., E. ROSENBERG, and D. GUTNICK. 1980. Incorporation of ^{32}P and growth of pseudomonad UP-2 on n-tetracosane. Appl. Environ. Microbiol. *40*:1086–1093.
17 NAKAHARA, T., L.E. ERICKSON, and J.R. GUTIERREZ. 1977. Characteristics of hydrocarbon uptake in cultures with two liquid phases. Biotech. Bioeng. *19*:9–25.
18 BLANCHARD, D.C., and L.D. SYZDEK. 1978. Seven problems in bubble and jet drop researches. Limnol. Oceanogr. *23*:389–400.
19 MARSHALL, op cit.
20 DEXTER, S.C., J.D. SULLIVAN JR., J. WILLIAMS III, and S.W. WATSON. 1975. Influence of substrate wettability on the attachment of marine bacteria to various surfaces. Appl. Microbiol. *30*:298–308.
21 FLETCHER, M., and G.I. LOEB. 1979. The influence of substratum characteristics on the attachment of a marine pseudomonad to solid surfaces. Appl. Environ. Microbiol. *37*:67–72.
22 GERSON, D.F., and D. SCHEER. 1980. Cell surface energy, contact angles and phase partition. III. Adhesion of bacterial cells to hydrophobic surfaces. Biochem. Biophys. Acta *602*:506–510.
23 ROSENBERG, M. 1981. Bacterial adherence to polystryene: a replica method of screening for bacterial hydrophobicity. Appl. Environ. Microbiol. *42*:375–377.
24 BEACHEY, E.H., W.A. SIMPSON, and I. OFEK. 1980. Interaction of surface polymers of *Streptococcus pyogenes* with animal cells, p. 389–405. *In* R.C.W. Berkeley, J.M. Lynch, J. Melling, P.R. Rutter, and B. Vincent (ed.), Microbial adhesion to surfaces. Society of Chemical Industry, London.
25 PERERS, L., L. ANDÄKER, L. EDEBO, O. STENDAHL, and C. TAGESSON. 1977. Association of some enterobacteria with the intestinal mucosa of mouse in relation to their partition in aqueous polymer two-phase systems. Acta. Path. Microbiol. Scand. Sect. B *85*:308–316.
26 ROSENBERG, M., A. PERRY, E.A. BAYER, D.L. GUTNICK, E. ROSENBERG, and I. OFEK, 1981. Adherence of *Acinetobacter calcoaceticus* RAG-1 to human epithelial cells and to hexadecane. Infect. Immun. *33*:29–33.
27 SMYTH, C.J., P. JONSSON, E. OLSSON, O. SÖDERLIND, J. ROSENGREN, S. HJERTÉN, and T. WADSTRÖM. 1978. Differences in hydrophobic surface characteristics of procine enteropathogenic *Escherichia coli* with or without K88 antigen as revealed by hydrophobic interaction chromatography. Infect. Immun. *22*:462–472.
28 VAN OSS, C.J. 1978. Phagocytosis as a surface phenomenon. Ann. Rev. Microbiol. *32*:19–39.
29 ROSENBERG, M., D. GUTNICK, and E. ROSENBERG. 1980. Adherence of bacteria to hydrocarbons: a simple method for measuring cell-surface hydrophobicity. FEMS Microbiol. Lett. *9*:29–33.
30 ROSENBERG, M., E. ROSENBERG, and D. GUTNICK. 1980. Bacterial adherence to hydrocarbons, p. 541–542. *In* R.C.W. Berkeley, J.M. Lynch, J. Melling, P.R. Rutter, and B. Vincent (ed.), Microbial adhesion to surfaces. Society of Chemical Industry, London.
31 ROSENBERG, E., K.H. KELLER, and M. DWORKIN 1977. Cell density-dependent growth of *Myxococcus xanthus* on casein. J. Bacteriol. *129*:770–777.
32 OFEK and BEACHEY, op cit.
33 ROSENBERG, E., A. ZUCKERBERG, C. RUBINOWITZ, and D.L. GUTNICK 1979. Emulsifier of *Arthrobacter* RAG-1: isolation and emulsifying properties. Appl. Environ. Microbiol. *37*:402–408.
34 ZUCKERBERG, A., A. DIVER, Z. PEERI, D.L. GUTNICK, and E. ROSENBERG. 1979. Emulsifier of *Arthrobacter* RAG-1: chemical and physical properties. Appl. Environ. Microbiol. *37*:414–420.
35 ROSENBERG, GUTNICK, and ROSENBERG, op cit.
36 ROSENBERG, PERRY, BAYER, GUTNICK, ROSENBERG, and OFEK, op cit.
37 ROSENBERG, ROSENBERG, and GUTNICK, op cit.
38 PINES, O., and D.L. GUTNICK. 1981. Relationship between phase resistance–emulsan production and interaction of phages with the cell surface of *Acinetobacter calcoaceticus* RAG-1. Arch. Microbiol.
39 NAKAHARA, ERICKSON, and GUTIERREZ, op cit.

40 BELSKY, I., D.L. GUTNICK, and E. ROSENBERG. 1979. Emulsifier of *Arthrobacter* RAG-1: determination of emulsifier-bound fatty acids. FEBS Lett. *101*:175–178.

41 REISFELD, A., E. ROSENBERG, and D. GUTNICK 1972. Microbial degradation of crude oil: factors affecting the dispersion in sea water by mixed and pure cultures. Appl. Environ. Microbiol. *24*:363–368.

42 ROSENBERG, ZACKERBERG, RUBINOWITZ, and GUTNICK, op cit.

43 ZUCKERBERG, DIVER, PEERI, GUTNICK, and ROSENBERG, op cit.

44 MUDD, S., and E.B.H. MUDD. 1924. The penetration of bacteria through capillary spaces. IV. A kinetic mechanism in interfaces. J. Exp. Med. *40*:633–645.

45 GUTNICK, D.L., and E. ROSENBERG. 1977. Oil tankers and pollution: a microbiological approach. Ann. Rev. Microbiol. *31*:379–396.

46 KIRSCHNER ZILBER, ROSENBERG, and GUTNICK, op cit.

47 MARSHALL, op cit.

48 McLEE, A.G., and S.L. DAVIES. 1972. Linear growth of a *Torulopsis* sp. on n-alkanes. Can. J. Microbiol. *18*:315–319.

49 NAKAHARA, ERICKSON, and GUTIERREZ, op cit.

50 NEUFELD, R.J., J.E. ZAJIC, and D.F. GERSON. 1980. Cell surface measurements in hydrocarbon and carbohydrate fermentations. Appl. Environ. Microbiol. *39*:511–517.

51 CHAKRAVARTY, M., H.D. SINGH, and J.N. BARUAH. 1975. A kinetic model for microbial growth on liquid hydrocarbons. Biotech. Bioeng. *17*:399–412.

52 VELANKAR, S.K., S.M. BARNETT, C.W. HOUSTON, and A.R. THOMPSON. 1975. Microbial growth on hydrocarbons-some experimental results. Biotech. Bioeng. *17*:241–251.

53 WANG, D.I.C., and A. OCHOA. 1972. Measurements on the interfacial areas of hydrocarbon in yeast fermentation and relationships to specific growth rates. Biotech. Bioeng. *14*:345–360.

54 ROSENBERG, GUTNICK, and ROSENBERG, op cit.

55 ROSENBERG, PERRY, BAYER, GUTNICK, ROSENBERG, and OFEK, op cit.

56 ROSENBERG, ROSENBERG, and GUTNICK, op cit.

57 BELSKY, GUTNICK, and ROSENBERG, op cit.

58 REISFELD, ROSENBERG, and GUTNICK, op cit.

59 ROSENBERG, ZUCKERBERG, RUBINOWITZ, and GUTNICK, op cit.

60 ZUCKERBERG, DIVER, PEERI, GUTNICK, and ROSENBERG, op cit.

61 LANKFORD, C.E., J.R. WALKER, J.B. REEVES, N.H. NABBUT, B.R. BYERS, and R.J. JONES. 1966. Inoculum-dependent division lag of *Bacillus* cultures and its relation to an endogenous factor(s). (''Schizokinen''). J. Bacteriol. *91*:1070–1079.

62 ROSENBERG, KELLER, and DWORKEN, op cit.

63 ROSENBERG, M., and E. ROSENBERG. 1981. Role of adherence in growth of *Acinetobacter calcoaceticus* RAG-1 on hexadecane. J. Bacteriol. *148*: 51–57.

64 CHAKRAVARTY, SINGH, and BARUAH, op cit.

65 HISATSUKA, K., T. NAKAHARA, Y. MINODA, and K. YAMADA. 1975. Capacity to oxidize n-alkane in EDTA-treated cells of *Pseudomonas aeruginosa* S_7B_1. Agr. Biol. Chem. *39*:999–1005.

66 VALENKAR, BARNETT, HOUSTON, and THOMPSON, op cit.

67 WANG and OCHOA, op cit.

68 MUDD and MUDD, op cit.

69 ROSENBERG, GUTNICK, and ROSENBERG, op cit.

70 BEACHEY, SIMPSON, and OFEK, op cit.

71 BLANCHARD and SYZDEK, op cit.

72 VAN OSS, op cit.

73 BEACHEY, SIMPSON, and OFEK, op cit.

74 MARSHALL, op cit.

75 OFEK and BEACHEY, op cit.

76 ROSENBERG and ROSENBERG, op cit.

ADDITIONAL REFERENCES

BAYER, E.A., E. ROSENBERG, and D. GUTNICK. 1981. The isolation of cell surface mutants of *Acinetobacter calcoaceticus* RAG-1. J. Gen Microbiol.

18 GLYCOPROTEIN SURFACTANTS STABILIZE LONG-LIVED GAS MICROBUBBLES IN THE ENVIRONMENT

Joseph S. D'Arrigo

University of Hawaii

The existence of long-lived gas microcavities or microbubbles in ordinary water and other aqueous liquids (e.g., body fluids) has been postulated or demonstrated by numerous investigators over the last four decades.[1-24] However, there is far less agreement in the literature as to the predominant physicochemical mechanism by which such microbubbles are stabilized. A detailed knowledge of this physicochemical stabilization mechanism in aqueous media is of practical importance to fields as varied as offshore drilling, naval hydrodynamics, hydraulic and ocean engineering, meteorology, chemical oceanography, underwater acoustics, surface chemistry, marine biology, the continual medical problem of decompression sickness, and enhanced oil recovery. Recent experimental work using carbohydrate (agarose) gels derived from marine algae has provided much evidence that largely nonionic or hydrophobic surfactants surround and stabilize the long-lived gas microbubbles present in such aqueous gels.[25-30] The concept of surfactant-stabilized microbubbles is in complete accord with the fact that surface-active substances in aqueous liquids, usually detected as trace organic contaminants, have long been known to affect both the cavitation threshold and the stability of bubbles formed in these liquids.[31-43] Even in the open sea, which is known to contain dissolved and nonliving particulate organic matter and specifically surface-active substances, an extensive microbubble population has been detected by various workers.[44-68] Consequently, it appeared likely that the naturally occurring, largely hydrophobic surfactants stabilizing long-lived gas microbubbles in agarose gels were present not only in marine algae (from which the gels were derived) but probably also in sea water and other environmental sources as well.[69-72] Accordingly, a systematic screening of a wide variety of natural substances revealed that (filtered) aqueous extracts of ordinary forest soil contained unusually high concentrations of gas microbubbles. This finding was reassuring and consistent with the fact that humic substances, similar (although not identical) to those occurring in soils, have been found to be widespread in sea water by various investigators.[73-77] An ongoing experimental program was thereafter expanded to characterize further and compare chemically the surfactants stabilizing long-lived microbubbles in both agarose gels and forest soil extracts.

INITIAL EXPERIMENTAL WORK WITH AGAROSE GELS

Over the last decade, bubble formation in gelatin has provided a useful model for studying the etiology of decompression sickness in humans.[78,79] From these and other studies, it has become clear that bubbles originate in water and probably also in humans as long-lived gas microbubbles, which must be held intact by surface-active films that are gas permeable.[80] In order to study the physical properties of the surfactant monolayer films surrounding gas microbubbles, a different gel method was developed that utilizes agarose as the gelling substance.[81,82] Unlike gelatin, which is composed of a complex mixture of charged peptide chains, agarose is an uncharged, relatively inert, homogeneous polysaccharide that is commercially available in highly purified form.[83,84] Consequently, agarose gels make well-controlled, surface-chemical studies of bubble formation possible.

The pressure vessel, glass counting chambers, and their Plexiglas holder used in the experiments reviewed below are all similar to those described elsewhere.[85,86] As viewed through the Plexiglas wall of the pressure vessel, the inside dimensions of each rectangular counting chamber are 15 mm wide and 6 mm along the line of sight.

Each chamber was filled with agarose solution to a depth of 7 mm. After gelation, the agarose samples were exposed to a fixed pressure schedule [i.e., 0 psig (for 22 hr at $21°C) \xrightarrow{1\ \mathrm{psig/sec}}$ (for 22 hr at $21°C \xrightarrow{\leq 15\ \mathrm{sec}}$ 0 psig)] using pure nitrogen gas. In control experiments the 22-hr saturation period was found to be more than adequate to saturate the aqueous gel medium completely; however, only bubbles formed between the depths of 1 and 4 mm in the gel were counted.[86] This procedure avoided any extraneous effects on bubble number that might arise either from the observed deposition of precipitates at the bottoms of some counting chambers after chemical modification of the sol prior to gelation or from the mere presence of a gas-gel interface at the top of the sample. Therefore, the total volume of agarose gel examined in each sample amounted to 0.27 ml. In this specified volume all bubbles observed appeared randomly distributed, and none of them were in contact with the walls of the glass counting chambers. The one exception to the above methodological description involved all of the CAV-CON gel samples (Lot. No. 1005; see Table 4), which were supplied by the manufacturer in rectangular thermoplastic cuvettes with optical windows on two opposite sides. Each cuvette had a 10 × 10-mm interior bottom and contained 1.0 ml of gel. Furthermore, all bubbles (below a horizontal plane tangent to the bottom of the curved gas-gel interface) were counted since, despite provision of the same spectrum of electrolyte additions described above, each of the specially purified CAV-CON agarose gel samples (0.2% w/w) were without detectable turbidity or sediment. Finally, the above-mentioned 22-hr saturation period was found to be entirely adequate to saturate the approximately 10-mm-deep CAV-CON gels.

Experimental work was accordingly carried out using this newly developed agarose gel method to determine some of the physical properties of the surfactant(s) believed to stabilize the long-lived gas microbubbles in these aqueous gels.[86,87] It was found that addition of up to 1mM UO_2^{2+}, La^{3+}, or SO_4^{2-} to the gel medium had only slight effects, if any, on the number of bubbles produced by a given decompression, regardless of the gas used.[88] Similarly, lowering the pH of the gel medium to 4 had little or no effect on bubble formation. While the above negative results could readily be explained by hypothesizing that the surfactant monolayers surrounding long-lived gas microbubbles do not contain a significant number of ionized chemical groups, the suspected nonionic surfactants stabilizing the microbubbles should nevertheless be salted out from the aqueous gels if exposed to ion concentrations reaching into the decimolar range.

With this in mind, the next study involved a systematic examination of the effects of a wide variety of salts at a fixed concentration of 0.4 M on bubble production within agarose gels.[89] It was found that the relative effectiveness of the different ions tested in decreasing the degree of bubble formation in the aqueous gels (Table 1) exhibited many similarities with published data in the physiocochemical literature for salting out of identified nonionic surfactants.[90–115] Specifically, the pronounced and very similar anion sequences obtained (using sodium and potassium salts), combined with earlier data, suggested that the polar portions of these nonionic surfactants represent either weak bases and/or amide groups. This view was in accordance with the relative position of Mg^{2+} in the two cation sequences obtained (using chloride and nitrate salts), which indicated strong salting out in both cases and hence rendered it quite unlikely that ether linkages contribute to the hydrophilicity of the nonionic surfactants stabilizing gas microbubbles.

In a subsequent study three additional commercial preparations of ultrapure agarose were compared (Tables 2–4) for each of the 42 different electrolytes tested in the earlier investigation (D'Arrigo, 1979) and for one selected organic compound (Table 5) in order to identify better reproducible and significant trends in the salting-out data on natural, long-lived gas microbubbles.[116] The data in Tables 2–4 demonstrate, as was also seen in the earlier study (Table 1), that the valence of the ions is not a factor of extreme and overriding importance as concerns the electrolyte effects on bubble production within the different commercial agarose gels. This finding confirms the belief that the observed electrolyte effects are not the result of flocculation of charged colloids, which occurs through a nonspecific screening mechanism.[116–118] Instead, the anion and cation sequences obtained with regard to the extent of bubble suppression contain many similarities with published data in the physicochemical literature for salting out of identified nonionic surfactants.[90–115] Moreover, the precise pattern of electrolyte effects on bubble formation can yield specific clues as to the structural characteristics of the nonionic surfactants stabilizing gas microbubbles present in agarose gels.

Inspection of Tables 2–4 reveals that the anion sequences obtained with the different commercial agarose gels were more pronounced and far more similar using sodium and potassium salts than were the corresponding cation sequences obtained with chloride and nitrate salts. This result, which closely resembles the pattern of data obtained in the earlier study (Table 1), suggests that the polar portions of these nonionic surfactants represent either weak bases and/or amide groups. Three additional results, however, weigh in favor of the amide group in particular.

First, the anion sequences obtained in this (and the previous) study, along with the quantitative data presented in Tables 1–4, clearly suggest that polyvalent anions

Table 1. Bubble production in marine colloids agarose gels (Lot No. 62867; 0.2% w/w; 0.27 ml) with different salts (0.4 M).

Salt	*No. of Trials*	*Number of Bubbles (Mean + SEM)*	*Salt*	*No. of Trials*	*Number of Bubbles (Mean + SEM)*
Na_3 citrate	12	3.00 ± 0.48	K_3 citrate	12	4.00 ± 1.17
Na_2SO_4	16	5.38 ± 0.76	K_2SO_4	16	9.75 ± 1.15
Na_2 tartrate	16	5.94 ± 0.76	KF	16	12.00 ± 1.44
NaF	16	8.63 ± 0.85	K tartrate	12	12.67 ± 1.25
Na acetate	12	16.08 ± 1.67	K acetate	16	15.63 ± 1.79
NaSCN	12	18.75 ± 1.92	KSCN	16	16.06 ± 1.07
NaBr	12	20.25 ± 0.96	KBr	16	18.25 ± 1.27
$NaNO_3$	16	20.44 ± 1.33	KI	16	20.19 ± 2.03
NaI	12	20.50 ± 1.22	KNO_3	20	20.65 ± 1.28
NaCl	16	23.56 ± 2.52	KCl	20	25.65 ± 1.24
(agarose control)	140	23.89 ± 0.49	(agarose control)	164	23.32 ± 0.43
$CdCl_2$	16	18.31 ± 1.24	$Mg(NO_3)_2$	16	17.19 ± 1.55
$MgCl_2$	16	21.25 ± 1.31	$Sr(NO_3)_2$	16	17.81 ± 1.27
LiCl	16	23.00 ± 1.23	$NaNO_3$	16	20.44 ± 1.33
NaCl	16	23.56 ± 2.52	KNO_3	20	20.65 ± 1.28
$CaCl_2$	12	24.08 ± 1.39	$LiNO_3$	16	20.88 ± 1.20
NH_4Cl	12	25.08 ± 1.23	$AgNO_3$	16	22.81 ± 1.84
KCl	20	25.65 ± 1.24	NH_4NO_3	12	23.17 ± 1.36
$BaCl_2$	12	32.25 ± 1.64	$Ca(NO_3)_2$	12	23.33 ± 1.53
$NiCl_2$	16	40.44 ± 2.32	$Cd(NO_3)_2$	16	26.69 ± 1.40
$SrCl_2$	16	49.69 ± 6.06	$Zn(NO_3)_2$	16	30.25 ± 2.41
$CoCl_2$	16	59.06 ± 13.81	$Ni(NO_3)_2$	16	67.25 ± 11.23
(agarose control)	168	25.17 ± 0.49	(agarose control)	172	24.69 ± 0.34

Source: J.S. D'Arrigo, "Physical Properties of the Nonionic Surfactants Surrounding Gas Cavitation Nuclei." *Journal of Chemical Physics*, 71, 1979, pp. 1809–1813.

Table 2. Bubble production in Miles agarose gels (Lot No. 1477; 0.2% w/w; 0.27 ml) with different salts (0.4 M).

Salt	*No. of Trials*	*Number of Bubbles (Mean ± SEM)*	*Salt*	*No. of Trials*	*Number of Bubbles (Mean ± SEM)*
Na_3 citrate	12	2.00 ± 0.51	K_3 citrate	12	3.08 ± 0.62
NaF	12	3.58 ± 0.51	K_2SO_4	12	7.33 ± 1.04
Na_2SO_4	12	4.50 ± 0.54	K_2 tartrate	12	8.83 ± 0.81
Na_2 tartrate	12	7.08 ± 0.58	KF	12	9.17 ± 0.60
Na acetate	12	8.42 ± 1.25	K acetate	12	10.00 ± 1.33
NaCl	12	13.08 ± 1.92	KSCN	12	16.67 ± 1.21
$NaNO_3$	12	17.83 ± 1.65	KNO_3	12	17.42 ± 1.07
NaSCN	12	19.08 ± 1.84	KBr	12	19.33 ± 1.20
NaI	16	19.69 ± 1.53	KI	12	21.75 ± 1.92
NaBr	16	21.75 ± 1.26	KCl	12	22.25 ± 1.32
(agarose control)	128	23.75 ± 0.51	(agarose control)	120	25.44 ± 0.43
NaCl	12	13.08 ± 1.92	$Zn\ (NO_3)_2$	12	16.33 ± 2.72
LiCl	12	15.33 ± 0.82	KNO_3	12	17.42 ± 1.07
$Mg\ Cl_2$	12	15.58 ± 1.33	$NaNO_3$	12	17.83 ± 1.65
NH_4Cl	12	15.75 ± 1.34	$Mg\ (NO_3)_2$	16	18.00 ± 1.14
$Cd\ Cl_2$	12	15.83 ± 0.95	$Li\ NO_3$	12	18.25 ± 1.67
$BaCl_2$	12	20.42 ± 1.20	NH_4NO_3	12	20.75 ± 1.59
$CoCl_2$	12	21.08 ± 1.61	$AgNO_3$	12	23.67 ± 2.10
$CaCl_2$	12	21.17 ± 1.08	$Ca\ (NO_3)_2$	12	25.00 ± 1.96
KCl	12	22.25 ± 1.32	$Cd\ (NO_3)_2$	12	28.67 ± 2.35
$NiCl_2$	12	29.17 ± 1.20	$Ni\ (NO_3)_2$	12	73.67 ± 3.29
$SrCl_2$	12	36.58 ± 2.03	$Sr\ (NO_3)_2$	12	107.17 ± 8.54
(agarose control)	132	27.24 ± 0.49	(agarose control)	136	26.84 ± 0.54

Source: J.S. D'Arrigo, "Structural Features of the Nonionic Surfactants Stabilizing Long-Lived Bubble Nuclei," *Journal of Chemical Physics*, 72, 1980, pp. 5133–5138.

Table 3. Bubble production in FMC Corp. agarose gels (Lot No. 60808; 0.2% w/w; 0.27 ml) with different salts (0.4 M).

Salt	*No. of Trials*	*Number of Bubbles (Mean ± SEM)*	*Salt*	*No. of Trials*	*Number of Bubbles (Mean ± SEM)*
Na_2 tartrate	12	9.67 ± 0.90	K_2SO_4	12	9.08 ± 0.82
Na_3 citrate	12	11.83 ± 1.18	K_3 citrate	12	11.17 ± 1.28
Na_2SO_4	12	14.00 ± 1.31	K_2 tartrate	12	11.92 ± 1.95
NaF	12	24.75 ± 2.48	KF	12	30.83 ± 2.53
Na acetate	12	27.50 ± 1.29	KNO_3	12	38.42 ± 2.13
NaCl	12	32.58 ± 4.20	K acetate	12	39.00 ± 1.44
NaSCN	12	39.92 ± 1.80	KCl	12	40.67 ± 2.79
$NaNO_3$	12	48.92 ± 2.33	KSCN	12	42.33 ± 2.27
NaI	12	49.00 ± 3.10	KI	12	47.58 ± 2.43
NaBr	12	52.75 ± 2.31	KBr	12	49.50 ± 2.39
(agarose control)	120	50.78 ± 0.71	(agarose control)	120	50.64 ± 0.62
NH_4Cl	12	30.08 ± 1.95	NH_4NO_3	12	35.92 ± 2.18
$MgCl_2$	12	31.33 ± 1.51	$Zn(NO_3)_2$	12	38.42 ± 2.07
NaCl	12	32.58 ± 4.20	KNO_3	12	38.42 ± 2.13
$BaCl_2$	12	35.58 ± 3.79	$LiNO_3$	12	44.08 ± 1.85
$CaCl_2$	12	37.25 ± 2.42	$Mg(NO_3)_2$	12	45.17 ± 2.08
KCl	12	40.67 ± 2.79	$Cd(NO_3)_2$	12	47.83 ± 3.43
$NiCl_2$	12	42.25 ± 2.42	$AgNO_3$	8	48.38 ± 2.62
LiCl	12	42.50 ± 2.71	$NaNO_3$	12	48.92 ± 2.33
$CoCl_2$	12	45.67 ± 2.85	$Ca(NO_3)_2$	12	52.00 ± 1.71
$SrCl_2$	12	51.42 ± 2.22	$Ni(NO_3)_2$	12	73.75 ± 3.86
$CdCl_2$	12	51.50 ± 2.47	$Sr(NO_3)_2$	12	88.92 ± 8.36
(agarose control)	132	50.66 ± 0.70	(agarose control)	128	50.55 ± 0.88

Source: J.S. D'Arrigo, "Structural Features of the Nonionic Surfactants Stabilizing Long-Lived Bubble Nuclei," *Journal of Chemical Physics*, 72, 1980, pp. 5133–5138.

(e.g., citrate^{3-}, tartrate^{2-}, and SO_4^{2-}) have a moderately powerful salting-out effect on the nonionic surfactants stabilizing gas microbubbles. This finding is noteworthy since a variety of amides examined earlier by different investigators showed similarly strong (yet not extreme) salting out by all polyvalent anions tested, and amide-containing nonelectrolytes are known to possess surface-active properties often.[119-121]

Second, it can be seen from Table 5 that phenol (1% v/v) reduced bubble production, compared with distilled/deionized water controls, by over 50% in all three of the ultrapure agarose preparations available for this experimental study. The importance of this observation derives from the fact that phenol, in aqueous solution at room temperature, is known to bind strongly to a variety of amide-containing compounds, including naturally occurring fibers (e.g., wool and horse hair).[122] Binding of phenol may be via the formation of hydrogen-bond complexes with alkylamide groups in these protein fibers in water. Refractive index measurements on aqueous solutions of simple amides (e.g., acetamide, dimethylformamide, and diacetamidomethane) and phenol have in fact suggested the existence of a 1:1 complex of the two solutes.[123] The complex formation, however, also can very well involve an aromatic-amide interaction that may be ascribed to an interaction of the polarizable electrons of phenol's aromatic ring with the amide or, possibly, to the formation of a weak molecular complex between the amide and unoccupied orbitals of the aromatic system.[124] Regardless of the precise mechanism(s) of phenol binding, such complex formation with the nonionic surfactants stabilizing gas microbubbles can be expected to destabilize the presumed surfactant monolayer structure surrounding these long-lived microbubbles by altering the hydrophilic-lipophilic balance (HLB) of individual surfactant molecules at the gas-aqueous gel interface.[125-133] The result might well be a sizable effect of phenol on bubble production in these aqueous gels as found in this study (Table 5).

The third finding which suggests that amide groups, and not weak bases, represent the polar portions of the nonionic surfactants stabilizing gas microbubbles concerns the observed effects of certain cations on bubble production in agarose gels. Although there was considerable variability among the cation sequences (using chloride and nitrate salts) obtained for any one of the agarose gel types tested (Tables 1–4), this variability was least with the specially purified CAV-CON agarose gel (Table 4) and was sufficiently low to allow some useful information to be extracted. In particular, it was noted upon testing with the chloride and nitrate salts

Table 4. Bubble production in CAV-CON agarose gels (Lot. No. 1005; 0.2% w/w; 0.95 ml) with different salts (0.4 M).

Salt	No. of Trials	Number of Bubbles (Mean ± SEM)	Salt	No. of Trials	Number of Bubbles (Mean ± SEM)
Na_3 citrate	12	2.17 ± 0.42	K_3 citrate	12	1.00 ± 0.39
Na_2 tartrate	12	2.92 ± 0.53	K_2 tartrate	12	2.58 ± 0.53
NaF	12	4.08 ± 2.14	K_2SO_4	12	4.83 ± 0.82
Na_2SO_4	12	5.08 ± 1.00	KF	12	6.42 ± 0.70
NaCl	12	14.50 ± 2.21	K acetate	12	17.50 ± 1.12
Na acetate	12	26.25 ± 2.97	KBr	12	21.50 ± 1.73
$NaNO_3$	12	31.17 ± 2.00	KCl	12	22.58 ± 2.25
NaBr	12	34.25 ± 3.21	KI	12	33.33 ± 2.14
NaI	12	36.42 ± 1.40	KNO_3	12	43.92 ± 2.45
NaSCN	12	43.33 ± 2.21	KSCN	12	48.33 ± 1.87
(agarose control)	120	43.63 ± 0.60	(agarose control)	120	46.71 ± 1.41
NH_4Cl	12	8.25 ± 1.41	NH_4NO_3	11	26.27 ± 4.10
NaCl	12	14.50 ± 2.21	$NaNO_3$	12	31.17 ± 2.00
KCl	12	22.58 ± 2.25	Mg $(NO_3)_2$	12	42.33 ± 3.86
$CaCl_2$	12	23.00 ± 1.26	KNO_3	12	43.92 ± 2.45
$BaCl_2$	12	24.83 ± 1.32	$LiNO_3$	10	47.40 ± 5.14
LiCl	12	26.83 ± 1.99	Sr $(NO_3)_2$	12	115.7 ± 13.06
Mg Cl_2	12	27.25 ± 2.78	Zn $(NO_3)_2$	12	135.1 ± 8.95
$NiCl_2$	12	49.33 ± 2.22	Ca $(NO_3)_2$	12	162.2 ± 26.98
$CdCl_2$	12	103.92 ± 10.82	$AgNO_3$	12	195.3 ± 16.51
$CoCl_2$	12	112.67 ± 28.87	Cd $(NO_3)_2$	12	212.3 ± 9.76
$SrCl_2$	12	124.58 ± 6.18	Ni $(NO_3)_2$	12	>250
(agarose control)	120	44.92 ± 0.88	(agarose control)	132	44.26 ± 1.35

Source: J.S. D'Arrigo, "Structural Features of the Nonionic Surfactants Stabilizing Long-Lived Bubble Nuclei," *Journal of Chemical Physics,* 72, 1980, pp. 5133–5138.

that this gel type yielded the same, and hence presumably more reliable, bubble-reduction (or salting-out) sequence for three important cations: $Na^+ > K^+ > Li^+$ (Table 4). The importance of this particular finding stems from the following observation of Long and McDevit described in their classic analysis of the relevant early literature concerning salt effects on nonelectrolyte solutes:

> When the specific effects of salts on polar nonelectrolytes [e.g., nonionic surfactants] are considered more closely, it is found that, although in a gross sense the salt order is similar to that for nonpolar molecules, there are significant variations in detail. Furthermore, these variations fall into two distinct groups. For one, which includes nonelectrolytes that have a definite basic character, there is an increased sensitivity to changes in the anion of the salt and also a considerable shift towards increased salting in by lithium salts. In contrast, for the group which includes the weak acids, the salting out by lithium salts is increased, relatively, and is usually larger than for sodium salts. A relatively greater sensitivity for changes in the cation is observed

Table 5. Effect of phenol (1% v/v) on bubble production in different commercial agarose gels.

Agarose Gel	Solution	No. of Trials	Number of Bubbles (Mean ± SEM)
Miles (Lot No. 1477; 0.2% w/w; 0.27 ml)	Distilled water control	12	22.25 ± 1.35
	Phenol (1% v/v in H_2O)	12	7.58 ± 1.20
FMC Corp. (Lot No. 60808; 0.2% w/w; 0.27 ml)	Distilled water control	12	52.75 ± 3.24
	Phenol (1% v/v in H_2O)	12	18.83 ± 1.46
CAV-CON (Lot No. 1005; 0.2% w/w; 0.95 ml)	Distilled water control	12	38.17 ± 6.65
	Phenol (1% v/v in H_2O)	12	10.5 ± 4.18

Source: J.S. D'Arrigo, "Structural Features of the Nonionic Surfactants Stabilizing Long-Lived Bubble Nuclei," *Journal of Chemical Physics,* 72, 1980, pp. 5133–5138.

for this group. Finally, there is an intermediate group where the salt order [i.e., $Na^+ > K^+ > Li^+$ regarding salting out] is similar to that for nonpolar electrolytes.[134]

This salting-out sequence identified with intermediate polar nonelectrolytes is identical to the Na-K-Li sequence obtained with both chloride and nitrate salts using the specially purified CAV-CON agarose gel. Consequently, this finding argues against the notion that the polar portions of the nonionic surfactants stabilizing long-lived gas microbubbles represent weak bases. Instead, amide groups, which are essentially neutral but large dipoles, within the nonionic surfactants are much more likely to give rise to the $Na^+ > K^+ > Li^+$ bubble-reduction (or salting-out) sequence repeatedly observed using the CAV-CON gel.

In view of the three separate findings described above, it is concluded that the polar portions of the nonionic surfactants stabilizing long-lived gas microbubbles represent primarily amide groups. This view is also consistent with the relative position of Mg^{2+} in all the cation sequences. Despite considerable variability among the cation sequences obtained with the different commercial agarose gels, the relative position of Mg^{2+} repeatedly indicated appreciable (and often strong) salting out in each case (Tables 1–4). This last result provides some additional information about the structure of the nonionic surfactants in agarose gels. The magnesium cation is known to form an oxonium compound readily with the lone-pair electrons of ether oxygen atoms of (synthetic) nonionic polyoxyethylated surfactants to cause salting in.[135,136] Therefore, in view of the consistent salting-out effect displayed by Mg^{2+} in agarose gels (Tables 1–4), one must conclude that ether linkages in all likelihood do not contribute to the hydrophilicity of the nonionic surfactants stabilizing microbubbles.

As ether linkages commonly occur in a wide variety of the synthetic nonionic surfactants present in trace amounts in many different water sources throughout our environment, it appears that such artificially produced chemical contaminants do not contribute measurably to the stability of gas microbubbles in a variety of aqueous (agarose) gels and, perhaps, aqueous media in general.[137–145] The nonionic or hydrophobic surfactants stabilizing long-lived gas microbubbles are probably mostly, if not all, of natural origin.

RECENT EXPERIMENTAL WORK COMPARING AQUEOUS SOIL EXTRACTS WITH AGAROSE GELS

Since the above experimental work with agarose gels clearly indicated that the polar portions of the surfactants stabilizing microbubbles represent primarily amide groups (i.e., peptide linkages and/or acid amide groups), a reasonable hypothesis was that the surfactants were proteinaceous in nature. In such a case their considerable hydrophobicity might well be explained in terms of hydrophobic amino acid residues (e.g., aromatic amino acids) comprising a large or significant structural portion of these surfactants that surround microbubbles. To test this hypothesis, three different types of protein-specific chemical tests were performed on long-lived microbubbles derived from aqueous solutions of agarose powder and from filtered aqueous extracts of Hawaiian forest soil.[146]

The pressure vessel, glass counting chambers, and their Plexiglas holder used in these experiments were all similar to those described above. The actual saturation pressure used in all experiments employing agarose gels not containing soil extract was 85 psig. With this saturation pressure, essentially all of the bubbles formed in the gels in control experiments originated from the agarose powder itself and not the distilled water used to make the gels, i.e., bubble number was found to be directly proportional to agarose concentration (in the range 0.2–0.8% w/w). In additional control experiments it was determined that, with the particular high-purity agarose powder used (see below) in the present experiments at a concentration of 0.2% or 0.3% w/w, no bubbles formed in the 0.27-ml agarose gels when saturation pressures below 40 psig were used. Consequently, only saturation pressures of 30 or 40 psig were used in all experiments employing agarose gels made with filtered (see below) aqueous extracts of forest soil (as opposed to distilled water). Owing to the unusually high concentrations of gas microbubbles occurring in the aqueous soil extracts, the 0.27-ml agarose gels formed from such extracts were found to contain enough large (and hence lower-cavitation-threshold) microbubbles, out of a probable spectrum of many sizes, to yield 20–40 bubbles consistently upon decompression from saturation pressures of 30 or 40 psig. Accordingly, chemical modification of the bubble production elicited by these low saturation pressures would then provide structural information about those long-lived microbubbles originating specifically from forest soil as opposed to the agarose powder.[146] The actual presence of surfactant-stabilized gas microbubbles in both damp (i.e., undissolved) soil and dry agarose powder becomes understandable in light of the fact that microbubble dissolution in carefully filtered sea water has always been observed to result in the formation of an organic particle.[147] Similarly, the dissolution of air bubbles even in distilled degassed water has been found to give rise to individual deposits or organic impurity. Moreover, such newly formed, microscopic deposits appear to contain a minuscule air pocket, since it was also found "that a long exposure of the water to a vacuum was sufficient to reduce the volume of the impurity considerably."[148]

The samples of aqueous soil extract, utilized in each of the three protein-specific tests, were all prepared in the same manner. Approximately 500 g of damp surface soil obtained from a forested area on windward Oahu was added to 2 l of distilled water and was refrigerated. The mixture was gently swirled at intervals for one day, and then all undissolved material was allowed to settle during the second day. Thereafter, the supernatant was carefully removed and filtered with suction through a precleaned Gelman Acropor 3 μm (pore diameter) membrane filter used in conjunction with a Gelman thick glass-fiber prefilter. Aliquots of 8 ml each were then pipetted into separate glass vials and were stored frozen for later use over the following 3 months in the soil extract experiments. When thawed for an experiment, none of the aliquots of soil extract contained any sediment. When used with different agarose concentrations to form gels in control measurements, they yielded an essentially constant number of bubbles for a given saturation pressure $\leq$40 psig. The reproducibility of such control measurements, observed with each of several different samples of Hawaiian forest soil collected, provided further support for the belief that surfactant-stabilized gas microbubbles exist in high concentration in such aqueous soil extracts, since all solid particulate debris larger than 3 μm had been removed previously by the filtration process.[149]

The data presented in Table 6 demonstrate that 0.3% (w/v) ninhydrin had a marked effect at pH 5.0 on bubble production in agarose gels containing distilled water. This concentration of ninhydrin reduced bubble production at pH 5.0 to only one-third of that observed in the control buffered at the same pH, whereas no significant effect at pH 8.0 was observed. This pH dependence of the chemical effect is important for the reason that ninhydrin is known to react appreciably with α-amino acids, peptides, and proteins in various buffered aqueous solutions containing citrate buffer (as used in this study) at pH 5.0.[150] While ninhydrin is also known to react with many amino compounds and ammonia, such interfering substances are essentially absent from the highly purified SeaKem HGT(P) agarose powder that was used, which contains a very slight contamination from only anionic residues (i.e., sulfate and pyruvate).[151-154] It therefore appears clear that the surfactants stabilizing long-lived gas microbubbles in agarose powder are proteinaceous; this finding agrees well with the previous experimental results (D'Arrigo, 1980), indicating that the polar portions of these naturally occurring, largely hydrophobic surfactants represent primarily amide groups (i.e., peptide linkages and/or acid amide groups).

In addition, ninhydrin was observed to have a similarly strong, pH-dependent effect on bubble production in agarose gels containing aqueous soil extract (Table 7). In this case, 0.3% (w/v) ninhydrin at pH 5.0 reduced bubble production, as compared with the control buffered at the same pH, by approximately 35%, whereas only a slight (if any) effect occurred at pH 8.0. Hence, the surfactants stabilizing long-lived microbubbles in aqueous soil extract also appear to be proteinaceous in nature. This finding is consistent with the fact that approximately one-quarter to one-half of the nitrogenous organic matter in soils is proteinaceous material that contains a wide variety of α-amino acids that have been identified successfully in both Hawaiian soils and soils from other locations.[155-178] Most of this proteinaceous material appears to be absorbed (via hydrogen bonding) onto humic-acid polymers, thereby forming complexes resistant to decomposition by biological agencies.[179-185] Furthermore, a humoprotein complex has been successfully isolated by Simonart, Batistic, and Mayaudon from podzolic, meadow, and brown forest soils, and the humoprotein content was observed to be highest in the forest soil. The proteinaceous fraction itself was subsequently isolated and its maximum adsorption lay in the spectral region 260–280 μm. These last two findings regarding brown forest soil and the absorption maximum of the protein fraction suggested that the aforementioned hydrophobicity of the evidently proteinaceous surfactants stabilizing long-lived gas microbubbles, present in both Hawaiian forest soil and agarose powder, might well be explained as follows: Hydrophobic amino acid residues, particularly aromatic amino acids that absorb maximally in

Table 6. Effect of 0.3% (w/v) ninhydrin on bubble production in buffered agarose gels (0.3% w/w, 0.27 ml) containing distilled water.

Gel Treatment	*No. of Trials*	*Number of Bubbles (Mean ± SEM)*
pH 5.0 (0.1 M Na_3 citrate) buffer	16	35.69 ± 2.67
pH 5.0 buffer + 0.3% ninhydrin	16	11.63 ± 1.09
pH 8.0 (0.1 M Na_3 citrate) buffer	16	23.19 ± 2.58
pH 8.0 buffer + 0.3% ninhydrin	16	20.50 ± 2.83

Source: J.S. D'Arrigo, "Aromatic Proteinaceous Surfactants Stabilize Long-Lived Gas Microbubbles from Natural Sources." *Journal of Chemical Physics,* 75, 1981, pp. 962–968.

Table 7. Effect of 0.3% (w/v) ninhydrin on bubble production in buffered agarose gels (0.3% w/w, 0.27 ml) containing aqueous soil extract.

Gel Treatment	*No. of Trials*	*Number of Bubbles (Mean ± SEM)*
pH 5.0 (0.1 M Na_3 citrate) buffer	16	32.94 ± 2.01
pH 5.0 buffer + 0.3% ninhydrin	16	21.63 ± 1.88
pH 8.0 (0.1 M Na_3 citrate) buffer	16	36.56 ± 1.41
pH 8.0 buffer + 0.3% ninhydrin	16	33.69 ± 2.10

Source: J.S. D'Arrigo, "Aromatic Proteinaceous Surfactants Stabilize Long-Lived Gas Microbubbles from Natural Sources," *Journal of Chemical Physics,* 75, 1981, pp. 962–968.

Table 8. Effect of 100 μm methylene blue dye on bubble production in agarose gels (0.2% w/w, 0.27 ml) containing distilled water and pH 5.0 (1 mm $NaHCO_3$) buffer.

Gel Treatment	*No. of Trials*	*Number of Bubbles (Mean ± SEM)*
Nonirradiated	20	28.25 ± 1.42
100 μm dye; nonirradiated	20	23.05 ± 0.99
2-hr (intense visible light) irradiation	20	25.15 ± 1.19
100 μm dye + 2-hr irradiation	20	13.95 ± 0.99

Source: J.S. D'Arrigo, "Aromatic Proteinaceous Surfactants Stabilize Long-Lived Gas Microbubbles from Natural Sources," *Journal of Chemical Physics,* 75, 1981, pp. 962–968.

Table 9. Effect of 100 μm methylene blue dye on bubble production in agarose gels (0.2% w/w, 0.27 ml) containing aqueous soil extract and pH 5.0 (10 mm $NaHCO_3$) buffer.

Gel Treatment	*No. of Trials*	*Number of Bubbles (Mean ± SEM)*
Nonirradiated	16	28.00 ± 1.37
100 μm dye; nonirradiated	16	23.81 ± 1.98
2-hr (intense visible light) irradiation	16	28.31 ± 1.76
100 μm dye + 2-hr irradiation	16	20.69 ± 1.83

Source: J.S. D'Arrigo, "Aromatic Proteinaceous Surfactants Stabilize Long-Lived Gas Microbubbles from Natural Sources," *Journal of Chemical Physics,* 75, 1981, pp. 962–968.

the spectral range 275–280 μm, comprise a large or significant structural portion of the surface-active proteinaceous compounds that surround microbubbles.[186]

To test this hypothesis, a series of photochemical experiments was conducted that involved the use of a common dye, methylene blue. This dye is known to be an effective sensitizer of the photo-oxidation of some amino acids; the specific nature of this reaction allows the degradation of certain amino acids in proteins. Much previous work by other investigators has demonstrated clearly that the aromatic amino acids are particularly susceptible to photo-oxidative destruction in the presence of methylene blue.[187–189] It was therefore quite informative to find in this experimental study (Tables 8, 10, and 11) that intense, 2-hr irradiation with visible light did in fact lead to a marked reduction in the bubble-production capability of transparent agarose gels containing distilled water to which a low concentration (100 μM) of methylene blue had been added. Specifically, at pH 5.0 (Table 8), 4.5 (Table 10), and 4.0 (Table 11), the dye along with visible-light irradiation resulted in bubble reductions (as compared to control gels having the same pH and irradiation treatment, but no dye) of 45%, 42%, and 42%, respectively, whereas the dye without an accompanying irradiation treatment resulted in bubble reductions (as compared to dye-free, nonirradiated controls) of only 18%, 26%, and 21%. The corresponding bubble-reduction percentages in agarose gels containing aqueous soil extract at pH 5.0 (Table 9) were 27% irradiated and 15% nonirradiated. (In this case, the smaller percentages observed probably resulted from a reduction of the free methylene blue concentration due to binding of the dye to humic-acid polymers.[190] It should be emphasized that the approximately twofold increase of the bubble-reduction percentage in both types of gels and at each pH when the visible-light irradiation treatment accompanied the use of methylene blue strongly argues that a genuine photo-oxidative destruction of proteinaceous material occurred in the agarose gels.

Table 10. Effect of 100 μm methylene blue dye on bubble production in agarose gels (0.2% w/w, 0.27 ml) containing distilled water and pH 4.5 (1 mm $NaHCO_3$) buffer.

Gel Treatment	*No. of Trials*	*Number of Bubbles (Mean ± SEM)*
Nonirradiated	16	21.00 ± 1.41
100 μm dye; nonirradiated	16	15.44 ± 0.81
2-hr (intense visible light) irradiation	16	18.94 ± 1.20
100 μm dye + 2-hr irradiation	16	11.00 ± 1.02

Source: J.S. D'Arrigo, "Aromatic Proteinaceous Surfactants Stabilize Long-Lived Gas Microbubbles from Natural Sources," *Journal of Chemical Physics,* 75, 1981, pp. 962–968.

Table 11. Effect of 100 μm methylene blue dye on bubble production in agarose gels (0.2% w/w, 0.27 ml) containing distilled water and pH 4.0 (1 mm $NaHCO_3$) buffer.

Gel Treatment	*No. of Trials*	*Number of Bubbles (Mean ± SEM)*
Nonirradiated	16	17.94 ± 1.27
100 μm dye; nonirradiated	16	14.25 ± 1.18
2-hr (intense visible light) irradiation	16	20.00 ± 2.04
100 μm dye + 2-hr irradiation	16	11.56 ± 1.00

Source: J.S. D'Arrigo, "Aromatic Proteinaceous Surfactants Stabilize Long-Lived Gas Microbubbles from Natural Sources," *Journal of Chemical Physics,* 75, 1981, pp. 962–968.

Accordingly, the reduction in bubble production almost certainly stemmed from destruction of gas microbubbles, i.e., from methylene blue-sensitized photo-oxidation of amino acid residues within the proteinaceous surfactants surrounding microbubbles. It is useful to point out that Knowles and Gurnani have reported that in acidic aqueous solution methylene blue causes extensive photo-oxidation of the aromatic amino acid tryptophan.[191] While other aromatic amino acids also displayed an appreciable sensitivity to the photochemical action of methylene blue, it was not nearly as strong as the photosensitivity of tryptophan. This same finding has been reported by different investigators working at other pH's.[192-194] In view of all the above findings, it appeared quite probable that the naturally occurring surfactants that surround and stabilize long-lived microbubbles are proteinaceous compounds that contain and for which surface activity depends upon aromatic amino acid residues that include tryptophan.

To evaluate this more specific hypothesis, additional experiments were carried out using the tryptophan-specific reagent 2-hydroxy-5-nitrobenzyl bromide. This reagent is known to be highly selective for tryptophan in proteins and peptides at acidic pH values.[195-199] While it has been reported that 2-hydroxy-5-nitrobenzyl bromide reacts with sulfhydryl groups of proteins at higher pH's, this reaction is so slight that the reagent can safely be used to modify tryptophan residues in proteins and peptides. It was therefore further assuring to find in the present experimental study (Table 12) that 1.0 M 2-hydroxy-5-nitrobenzyl bromide did, in fact, have a striking effect at pH 5.0 on bubble production in agarose gels containing distilled water. Specifically, the reagent at pH 5.0 reduced bubble production, as compared with the control buffered at the same pH, by over 95%, whereas at pH 8.0 the bubble reduction was far less (i.e., 60%). The much smaller but still quite noticeable effect of the reagent at pH 8.0 is very likely not due to a tryptophan-specific chemical reaction but rather to physical absorption of 2-hydroxy-5-nitrobenzyl bromide on the surface of the proteinaceous surfactants surrounding microbubbles.[199] The observed pH dependence of the 95% bubble reduction achieved with 2-hydroxy-5-nitrobenzyl bromide renders it virtually certain that the proteinaceous surfactants stabilizing long-lived gas microbubbles in agarose powder contain tryptophan residues. These residues appear to be essential for the normal surface-active behavior of these proteinaceous compounds.

In addition, 2-hydroxy-5-nitrobenzyl bromide was observed to have a similarly strong, pH-dependent effect on bubble production in agarose gels containing aqueous soil extract (Table 13). In this case, 0.7 M 2-hydroxy-5-nitrobenzyl bromide at pH 5.0 reduced bubble production, as compared with the control buffered at the same pH, by over 95%. At pH 8.0 the bubble reduction was 65%. Hence, the proteinaceous surfactants stabilizing long-lived microbubbles in aqueous soil extract must also contain tryptophan residues in their structure. At first this finding seemed difficult to reconcile with the work of Kojima[200] and, later, with that of Bremner, who summarizes his findings on the tryptophan content of soils as follows: "Tryptophan, which is destroyed by acid hydrolysis but not alkali hydrolysis, could not be detected in alkali hydrolysates of soils."[201] However, Barman and Koshland have pointed out that "basic hydrolysis occasionally gives accurate value [regarding tryptophan content] with model peptides and with some proteins, but poor results are usually obtained for inexplicable reasons.... The fact that alkaline or acid hydrolysis destroys tryptophan and produces unreliable assays indicates that a method [such as the use of 2-hydroxy-5-nitrobenzyl bromide] involving direct analysis of the protein is desirable."[202] Moreover, a pyridine derivative identified as α-picoline-γ-carboxylic acid, which Schreiner and Shorey have suggested may be formed in soils by decomposition of tryptophan, has been isolated both from a Hawaiian soil by Shorey and from other soils by Schreiner and Shorey.[203,204]

In view of the foregoing arguments, it is concluded that the naturally occurring, largely hydrophobic surfactants that surround and stabilize long-lived gas microbubbles are proteinaceous compounds that contain and whose surface activity depends upon aromatic amino acid residues, particularly tryptophan. This conclusion is in complete accord with the well-documented surface-active properties of a wide variety of proteins and certain glycoproteins at air-water interfaces.[205-244] Lastly, the major finding in these particular

Table 12. Effect of 1.0 M 2-hydroxy-5-nitrobenzyl bromide (HNBB) on bubble production in buffered agarose gels (0.3% w/w, 0.27 ml) containing distilled water.

Gel Treatment	*No. of Trials*	*Number of Bubbles (Mean ± SEM)*
pH 5.0 (0.1 M Na_3 citrate) buffer	16	9.50 ± 1.14
pH 5.0 buffer + 1.0 M HNBB	16	0.31 ± 0.15
pH 8.0 (0.1 M Na_3 citrate) buffer	16	11.69 ± 1.47
pH 8.0 buffer + 1.0 M HNBB	16	4.00 ± 0.77

Source: J.S. D'Arrigo, "Aromatic Proteinaceous Surfactants Stabilize Long-Lived Gas Microbubbles from Natural Sources," *Journal of Chemical Physics,* 75, 1981, pp. 962–968.

Table 13. Effect of 0.7 M 2-hydroxy-5-nitrobenzyl bromide (HNBB) on bubble production in buffered agarose gels (0.3% w/w, 0.27 ml) containing aqueous soil extract.

Gel Treatment	*No. of Trials*	*Number of Bubbles (Mean ± SEM)*
pH 5.0 (0.1 M Na_3 citrate) buffer	16	28.88 ± 1.93
pH 5.0 buffer + 0.7 M HNBB	16	1.25 ± 0.21
pH 8.0 (0.1 M Na_3 citrate) buffer	16	38.06 ± 2.81
pH 8.0 buffer + 0.7 M HNBB	16	15.31 ± 2.23

Source: J.S. D'Arrigo, "Aromatic Proteinaceous Surfactants Stabilize Long-Lived Gas Microbubbles from Natural Sources," *Journal of Chemical Physics,* 75, 1981, pp. 962–968.

experiments of protein-stabilized gas microbubbles in filtered aqueous extracts of forest soil may provide an explanation for the widespread occurrence of these long-lived microbubbles in nature: specifically, humic substances that are known to bind proteinaceous material reversibly (that presumably may include surfactant-stabilized microbubbles) and thereby form complexes resistant to decomposition "are among the most widely distributed natural products on the Earth's surface, occurring in soils, lakes, rivers, and in the sea."[245]

ISOLATION AND INITIAL CHARACTERIZATION OF THE GLYCOPROTEIN SURFACTANT SURROUNDING NATURAL MICROBUBBLES

Current experimental work in this laboratory is aimed at the systematic development of an efficient method for isolating the proteinaceous surfactant, which serves to stabilize natural microbubbles, from both commercial agarose powder and from forest soil samples collected locally. An acidified-ethanol extraction-Millipore filtration procedure has very recently been used to isolate this rather hydrophobic proteinaceous surfactant from commercial agarose powder (D'Arrigo, unpublished results). Successful isolation of the surfactant was evidenced in part by the fact that the treated agarose powder was found to have lost 99.6% of its bubble-producing capacity in subsequent decompression tests (Table 14). Sodium dodecyl sulfate polyacrylamide gel electrophoresis in the presence of sulfhydryl-reducing agents has shown the dissolved proteinaceous surfactant to be homogeneous and have a molecular weight <20,000 daltons (D'Arrigo, unpublished results). Furthermore, the single proteinaceous band observed on the SDS-polyacrylamide gels was stained successfully with either Coomassie blue dye for peptides or periodic acid-Schiff reagent for glycopeptides.[246] The glycoprotein nature of the microbubble surfactant was confirmed during subsequent automated amino acid analyses during which small amounts of glucosamine and galactosamine were readily detected in acid hydrolysates of the isolated microbubble surfactant (D'Arrigo, unpublished data).

Table 14. Bubble production following rapid decompression to atmospheric pressure of agarose gels (0.27 ml) saturated with N_2 at 85 psig.

Gel Type	*No. of Trials*	*Number of Bubbles (Mean ± SEM)*
Agarose (Lot No. 60808) before protein extraction	12	24.67 ± 2.95
Agarose (Lot No. 60808) after protein extraction	36	0.08 ± 0.05

The actual amino acid composition data obtained to date on the purified glycoprotein surfactant from commercial agarose powder are presented in Table 15. When the surfactant surrounding natural microbubbles in aqueous soil extract is also isolated successfully, its amino acid composition ought to resemble the rather unusual amino acid ratios listed in Table 15 for the glycoprotein already isolated from agarose powder. Accordingly, it was quite encouraging to find in the only preliminary experiment completed to date that even a simple 80% ethanol-Millipore filtered extraction of the aqueous soil extract itself removed rather pure glycoprotein material, containing both glucosamine and galactosamine, which has a total amino acid composition (Table 15, right column) that closely resembles that of the glycoprotein surfactant previously isolated from commercial agarose powder. After successful isolation of the larger quantities of this specific glycoprotein surfactant available in forest soils is confirmed, future work in this laboratory will be directed toward complete amino acid sequence analysis glycoprotein surfactant surrounding natural microbubbles.[247]

Table 15. Amino acid composition* of microbubble glycoprotein surfactant isolated from agarose powder and forest soil.

Amino Acid	*Surfactant from Agarose*	*Surfactant from Soil*
Lysine	5 (6,5,3,4)	7
Histidine	1 (1,1,1,2)	3
Arginine	2 (2,1,2,1)	3
Aspartic acid	9 (7,9,12,8)	10
Threonine	4 (4,3,4,5)	4
Serine	11 (13,8,8,16)	9
Glutamic acid	9 (11,8,9,6)	13
Proline	1 (~0, ~0,3,2)	3
Glycine	15 (15,16,13,14)	22
Alanine	7 (6,8,7,7)	7
Cysteine**	1 (~0, ~0,2,1)	1
Valine	5 (4,7,4,4)	5
Methionine	~0 (0,0,<1,<1)	1
Isoleucine	3 (2,3,4,2)	4
Leucine	5 (5,5,6,3)	7
Tyrosine	1 (2,1,1,1)	2
Phenylalanine	2 (3,1,2,2)	2

*Amino acid values are derived from normalized data to yield nearest integer values with each determination for the six most reliable amino acids, i.e., aspartic acid, glutamic acid, glycine, alanine, leucine, and lysine. For the microbubble protein surfactant isolated from agarose powder, nearest integer to the means was calculated from four determinations, with the unaveraged data also given in parentheses. Hydrolysis is in 5.7 N HCl for 24 hr at 110°C. No corrections have been made for destruction or incomplete hydrolysis.

**Cysteine as ½ cystine.

SUMMARY

Recent experimental work using carbohydrate (agarose) gels derived from marine algae has provided much evidence that largely hydrophobic surfactants surround and stabilize the long-lived gas microbubbles present in such aqueous gels. Later work involving three different types of protein-specific chemical tests has further revealed that these biological surfactants, which were also found to be present (in much higher concentrations) in forest soils, are proteinaceous compounds whose surface activity depends upon aromatic amino acid residues. Subsequent isolation of the microbubble surfactant in soluble form, followed by sodium dodecyl sulfate polyacrylamide gel electrophoresis, demonstrated that the surfactant was homogeneous, had a molecular weight <20,000 daltons, and was stained by both protein- and carbohydrate-specific reagents. Automated amino acid analyses of the glycoprotein surfactant revealed the presence of at least 17 different amino acids and small amounts of glucosamine and galactosamine.

ACKNOWLEDGMENTS

The author is grateful to Mr. Neil S. Reimer for his excellent assistance with amino acid analysis and gel electrophoresis and to Dr. Kerry T. Yasunobu for the use of his amino acid analyzer. In addition, I wish to thank Ms. Kai-Fei Chang for technical assistance with the decompression experiments. Portions of this work were supported by NASA Contract No. NAS 9-16048 and a University of Hawaii Research Corporation Leahi Trust Grant.

REFERENCES

1 DEAN. R.B. 1944. The formation of bubbles. J. Appl. Physics. *15*:446–451.

2 HARVEY, E.N., D.K. BARNES, W.D. MCELROY, A.H. WHITELEY, D.C. PEASE, and K.W. COOPER, 1944. Bubble formation in animals. I. Physical factors. J. Cell. Comp. Physiol. *24*:1–22.

3 PEASE, D.C., and L.R. BLINKS. 1947. Cavitation from solid surfaces in the absence of gas nuclei. J. Phys. Colloid Chem. *51*:556–567.

4 FOX, F.E., and K.F. HERZFELD. 1954. Gas bubbles with organic skin as cavitation nuclei. J. Acoust. Soc. Am. *26*:984–989.

5 IYENGAR, K.S., and E.G. RICHARDSON 1958. Measurements on the air nuclei in natural water which give rise to cavitation. Brit. J. Appl. Physics *9*:154–158.

6 KNAPP, R.T. 1958. Cavitation and nuclei. Trans. Am. Soc. Mech. Engin. *80*:1315–1324.

7 STRASBERG, M. 1959. Onset of ultrasonic cavitation in tap water. J. Acoust. Soc. Am. *31*:163–176

8 MANLEY, D.M.J.P. 1960. Change size of air bubbles in water containing a small dissolved air content. Brit. J. Appl. Physics *11*:38–42.

9 TURNER, W.R. 1961. Microbubble persistence in fresh water. J. Acoust. Soc. Am. *33*:1223–1233.

10 LANGTON, N.H., and P. VAUGHAN. 1962. Cavitation and the ultrasonic degradation of high polymers. Brit. J. Appl. Phys. *13*:478–482.

11 HASAN, M.M., and K.S. IYENGAR. 1963. Size and growth of cavitation bubble nuclei. Nature (London) *199*:995–996.

12 MESSINO', D., D. SETTE and F. WANDERLINGH. 1963. Statistical approach to ultrasonic cavitation. J. Acoust. Soc. Am. *35*:1575–1583.

13 WEBSTER, E. 1963. Cavitation. Ultrasonics *1*:39–48.

14 BERND, L.H. 1964. Study of the surface films of gas nuclei (as related to cavitation and tensile strength in water). Part I. 52/p. ONR Contract No. NONR-4296(00), AD 608094, G.E.-ATL.

15 FLYNN, H.G. 1964. Physics of acoustic cavitation in liquids, p. 57–172. *In* W.P. Mason (ed.), Physical acoustics: Principles and methods. Academic Press, New York.

16 BERND, L.H. 1966. Cavitation, tensile strength, and the surface films of gas nuclei, p. 77–108. *In* R.D. Cooper and S. W. Doroff (ed.), Sixth symposium on naval hydrodynamics. Physics of fluids, maneuverability and ocean platforms, ocean waves, and ship-generated waves and wave resistance. Office of Naval Research, Washington, D.C.

17 BERND, L.H. 1967. Study of the surface films of gas nuclei (as related to cavitation and tensile strength in water). Part IV. Dissolving of gas nuclei in turbulent flow, 50 p. ONR Contract NONR-4296(00), AD 829881, G.E.-RDC.

18 MESSINO', C.D., D. SETTE, and F. WANDERLINGH. 1967. Effects of solid impurities on cavitation nuclei in water. J. Acoust. Soc. Am. *41*:573–583.

19 SETTE, D., and F. WANDERLINGH. 1967. Thermodynamic theory of bubble nucleation induced in liquids by high-energy particles. J. Acoust. Soc. Am. *41*:1074–1082.

20 DARNER, C.L. 1970 Sonic cavitation in water. NRL Report 7131, ADA031182, NRL-USRD.

21 HAYWARD, A.T.J. 1970. The role of stabilized gas nuclei in hydrodynamic cavitation inception. J. Physics D *3*:574–579.

22 SIROTYUK, M.G. 1970. Stabilization of gas bubbles in water. Sov. Physics Acoust. *16*:237–240.

23 ACOSTA, A.J., and B.R. PARKIN. 1975. Cavitation inception–a selective review. J. Ship. Res. *19*:193–205.

24 KIMOTO, H. 1977. Cavitation in purified water, p. 1–2. *In* R.L. Waid (ed.), Cavitation and polyphase flow forum–1977. American Society of Mechanical Engineers, New York.

25 D'ARRIGO, J.S. 1978. Improved method for studying the surface chemistry of bubble formation. Aviat., Space & Environ. Med. *49*:358–361.

26 D'ARRIGO, J.S. 1979. Physical properties of the nonionic surfactants surrounding gas cavitation nuclei. J. Chemical Physics *71:*1809–1813.

27 D'ARRIGO, J.S. 1980. Structural features of the nonionic surfactants stabilizing long-lived bubble nuclei. J. Chemical Physics *72:*5133–5138.

28 DUCKWORTH, M., and W. YAPHE. 1971. Preparation of agarose by fractionation from the spectrum of polysaccharides in agar. Anal. Biochem. *44:*636–641.

29 DUCKWORTH, M., and W. YAPHE. 1971. The structure of agar. Part I. Fractionation of a complex mixture of polysaccharides. Carbohyd. Res. *16:*189–197.

30 GUISELEY, K.B., and D.W. RENN. 1975. Agarose: Purification, properties, and biomedical applications. FMC Corp., Rockland.

31 BEREZINA, N.P., and N.V. NIKOLAEVA-FEDOROVICH. 1967. Problem of the purity of water in electrochemical studies. Sov. Electrochem. *3:*1–4.

32 BURNHAM, A.K., G.V. CALDER, J.S. FRITZ, G.A. JUNK, H.J. SVEE, and R. WILLIS. 1972. Identification and estimation of neutral organic contaminants in potable water. Anal. Chem. *44:*139–144.

33 JUNK, G.A., and S.E. STANLEY. 1975. Organics in drinking water. Part I. Listing of identified chemicals. IS No. 3671. Ames Laboratory, ERDA.

34 TRUSSELL, A.R., and M.D. UMPHRES. 1978. An overview of the analysis of trace organics in water. J. Am. Water Works Assoc. *70:*595–603.

35 MURTAUGH, J.J., and R.L. BUNCH. 1965. Acidic components of sewage effluents and river water. J. Water Pollut. Control Fed. *37:*410–415.

36 JUNK, G.A. 1977. Organic contaminants in water, p. 137–141. *In* F. Coulston and E. Mrak (ed.), Water quality; Proceedings of an international forum. Academic Press, New York.

37 BAKER, C.B., J.W. HOLL, and R.E.A. ARNDT. 1976. Influence of gas content and polyethylene oxide additive upon confined jet cavitation in water, p. 6–8. *In* R.L. Waid (ed.), Cavitation and polyphase flow forum–1976. American Society of Mechanical Engineers, New York.

38 CRUM, L.A. 1977. Measurements of the growth of air bubbles by rectified diffusion. 25 p. ONR Contract No. NR384-923. Technical Report No. 377 U.S. Naval Academy.

39 MANLEY, op cit.

40 BERND, 1964, op cit.

41 BERND, 1966, op cit.

42 BERND, 1967, op cit.

43 SIROTYUK, op cit.

44 WILLIAMS, P.M. 1961. Organic acids in Pacific Ocean waters. Nature (London) *189:*219–220.

45 WILLIAMS, P.M., and A. ZIRINO. 1964. Scavenging of 'dissolved' organic matter from sea-water with hydrated metal oxides. Nature (London) *204:*462–464.

46 DUURSMA, E.K. 1965. The dissolved organic constituents of sea water, p. 433–475. *In* J.P. Riley and G. Skirrow (ed.), Chemical oceanography. Academic Press, New York.

47 WANGERSKY, P.J. 1965. The organic chemistry of sea water. Am. Sci. *53*:358–374.

48 KALLE, K. 1966. The problem of the Gelbstoff in the sea. Oceanogr. Mar. Biol. Ann. Rev. *4*:91–104.

49 BARBER, R.T. 1968. Dissolved organic carbon from deep waters resists microbial oxidation. Nature (London) *220*:274–275.

50 KHAYLOV, K.M. 1968. Dissolved organic macromolecules in sea water. Geokhimiya *5*:595–603.

51 SKOPINTSEV, B. 1972. On the age of stable organic matter–aquatic humus in oceanic waters, p. 205–207. *In* D. Dyrssen and D. Jagner (ed.), The changing chemistry of the oceans. Wiley, New York.

52 PARSONS, T.R., and J.D.H. STRICKLAND. 1962. Oceanic detritus. Science *136*:313–314.

53 BAYLOR, E.R., and W.H. SUTCLIFFE, JR. 1963. Dissolved organic matter in seawater as a source of particulate food. Limnol. Oceanogr. *8*:369–371.

54 RILEY, G.A. 1963. Organic aggregates in seawater and the dynamics of their formation and utilization. Limnol. Oceanogr. *8*:372-381.

55 MENZEL, D.W., and J.H. RYTHER. 1964. The composition of particulate organic matter in the western North Atlantic. Limnol. Oceanogr. *9*:179–186.

56 MENZEL, D.W. 1966. Bubbling of sea water and the production of organic particles: a re-evaluation. Deep-Sea Res. *13*:963–966.

57 MENZEL, D.W. 1967. Particulate organic carbon in the deep sea. Deep Sea Res. *14*:229–238.

58 JOHNSON, B.D., and R.C. COOKE. 1980. Organic particle and aggregate formation resulting from the dissolution of bubbles in seawater. Limnol. Oceanogr. *25*:653–661.

59 WILSON, A.T. 1959. Surface of the ocean as a source of air-borne nitrogenous material and other plant nutrients. Nature (London) *184*:99–101.

60 BLANCHARD, D.C. 1964. Sea-to-air transport of surface active material. Science *146*:396–397.

61 GARRETT, W.D. 1967. The organic chemical composition of the ocean surface. Deep-Sea Res. *14*:221–227.

62 GARRETT, W.D. 1967. Stabilization of air bubbles at the air-sea interface by surface-active material. Deep-Sea Res. *14*:661–672.

63 BARGER, W.R., and W.D. GARRETT. 1970. Surface active organic material in the marine atmosphere. J. Geophys. Res. *75*:4561–4566.

64 BLANCHARD, D.C. 1975. Bubble scavenging and the water-to-air transfer of organic material in the sea. Adv. Chem. Ser. *145*:360–387.

65 RAMSEY, W.L. 1962. Dissolved oxygen in shallow nearshore water and its relation to possible bubble formation. Limnol. Oceanogr. *7*:453–461.

66 MEDWIN, H. 1970. In situ acoustic measurements of bubble populations in coastal ocean waters. J. Geosphys. Res. *75*:599–611.

67 MEDWIN, H. 1977. In situ acoustic measurements of microbubbles at sea. J. Geophys. Res. *82*:971–976.

68 JOHNSON, B.D., and R.C. COOKE 1979. Bubble populations and spectra in coastal waters: a photographic approach. J. Geophys. Res. *84*:3761–3766.

69 D'ARRIGO, 1980, op cit.

70 DUCKWORTH and YAPHE, 1971, Preparation of agarose by fractionation from the spectrum of polysaccharides in agar, op cit.

71 DUCKWORTH and YAPHE, 1971, The structure of agar. Part I. Fractionation of a complex mixture of polysaccharides, op cit.

72 GUISELEY and RENN, op cit.

73 NISSENBAUM, A., and I.R. KAPLAN. 1972. Chemical and isotopic evidence for the in situ origin of marine humic substances. Limnol. Oceanogr. *17*:570–582.

74 STUERMER, D.H. and G.R. HARVEY. 1974. Humic substances from seawater. Nature (London) *250*:480–481.

75 KERR, R.A., and J.G. QUINN. 1975. Chemical studies on the dissolved organic matter in seawater. Isolation and fractionation. Deep-Sea Res. *22*:107–116.

76 STUERMER, D.H., and G.R. HARVEY. 1977. The isolation of humic substances and alcohol-soluble organic matter from seawater. Deep-Sea Res. *24*:303–309.

77 BLANCHARD, D.C., and E.J. HOFFMAN. 1978. Control of jet drop dynamics by organic material in seawater. J. Geophys. Res. *83*:6187–6191.

78 LEMESSURIER, D.H. 1972. Supersaturation and 'preformed nuclei' in the etiology of decompression sickness. Paper presented at the Second International Meeting on Aerospace Medicine, Melbourne (unpublished).

79 YOUNT, D.E., and R.H. STRAUSS. 1976. Bubble formation in gelatin: a model for decompression sickness. J. Appl. Physics *47*:5081–5089.

80 YOUNT, D.E., T.D. KUNKLE, J.S. D'ARRIGO, F.W. INGLE, C. YEUNG, and E.L. BECKMAN. 1977. Stabilization of gas cavitation nuclei by surface active compounds. Aviat., Space, & Environ. Med. *48*:358–361.

81 D'ARRIGO, 1978, op cit.

82 D'ARRIGO, J.S. 1977. An improved method for studying the physical chemistry of bubble formation. Biophys. J. *17*:302a.

83 VEIS, A. 1964. The macromolecular chemistry of gelatin. Academic Press, New York.

84 HJERTEN, S. 1961. Agarose as an anticonvection agent in zone electrophoresis. Biochem. Biophys. Acta *53*:514–517.

85 YOUNT and STRAUSS, op cit.

86 D'ARRIGO, 1978, op cit.

87 D'ARRIGO, J.S., and Y. MANO. 1979. Bubble production in agarose gels subjected to different decompression schedules. Undersea Biomed. Res. *6*:93–98.

88 D'ARRIGO, J.S. 1979. Evidence that nonionic surfactants surround gas cavitation nuclei. *In* Proceedings of the 1979 Aerospace Medical Association Meeting (unpublished; limited distribution).

89 D'ARRIGO, 1980, op cit.

90 DOSCHER, T.M., G.E. MYERS, and D.C. ATKINS, JR. 1951. The behavior of nonionic surface active agents in salt solutions. J. Colloid Sci. *6*:223–235.

91 LONG, F.A., and W.F. McDEVIT. 1952. Activity coefficients of nonelectrolyte solutes in aqueous salt solutions. Chem. Rev. *51*:119–169.

92 JELINEK, C.F., and R.L. MAYHEW. 1954. Nonionic surfactants: their chemistry and textile uses. Textile Res. J. *24*:765–778.

93 MACLAY, W.N. 1956. Factors affecting the solubility of nonionic emulsifiers. J. Colloid Sci. *11*:272–285.

94 BAILEY, F.E., JR., and R.W. CALLARD. 1959. Some properties of poly(ethylene oxide) in aqueous solution. J. Appl. Polym. Sci. *1*:56–62.

95 DURHAM, K. 1961. Properties of detergent solutions–amphipathy and absorption, p. 1–28. *In* K. Durham (ed.), Surface activity and detergency. Macmillan, London.

96 BECHER, P. 1962. Nonionic surface-active compounds. V. Effect of electrolyte. J. Colloid Sci. *17*:325–333.

97 KURIYAMA, K. 1962. Temperature dependence of micellar weight of nonionic surfactant in the presence of various additives. Part 2. Addition of sodium chloride and calcium chloride. Kolloid Z.Z. Polym. *181*:144–149.

98 SCHICK, M.J. 1962. Surface films of nonionic detergents. I. Surface tension study. J. Colloid Sci. *17*:801–813.

99 SCHICK, M.J. 1963. Physical chemistry of nonionic detergents. J. Am. Oil Chem. Soc. *40*:680–687.

100 BAILEY, F.E., and J.V. KOLESKE. 1967. Configuration and hydrodynamic properties of the polyoxyethylene chain in solution, p. 794–822. *In* M.J. Schick (9 ed.), Nonionic surfactants. Dekker, New York.

101 SAITO, H., and K. SHINODA. 1967. The solubilization of hydrocarbons in aqueous solutions of nonionic surfactants. J. Colloid Interface Sci. *24*:10–15.

102 MALIK, W.U., and S.M. SALEEM. 1968. Effect of additives on that critical micelle concentration of some polyethoxylated nonionic detergents. J. Am. Oil Chem. Soc. *45*:670–672.

103 DOREN, A., and J. GOLDFARB. 1970. Electrolyte effects on micellar solutions of nonionic detergents. J. Colloid Interface Sci. *32*:67–72.

104 BECHER, P. 1963. Nonionic surface-active compounds. V. Effect of electrolyte–Addendum. J. Colloid Sci. *18*:196–197.

105 SHINODA, K., and H. TAKEDA. 1970. The effect of added salts in water on the hydrophile–lipophile balance of nonionic surfactants: the effect of added salts on the phase inversion temperature of emulsions. J. Colloid Interface Sci. *32*:642–646.

106 RAY, A., and G. NEMETHY. 1971. Effects of ionic protein denaturants on micelle formation by nonionic detergents. J. Am. Chem. Soc. *93*:6787–6793.

107 POPESCU, F., and I. ANA. 1973. Influenta electrolitilor asupra temperaturii de iesire din solutie a agentilor activi de suprafata (AAS) neionici. Pet. Gaze *24*:88–90.

108 SCHOTT, H. 1973. Salting in of nonionic surfactants by complexation with inorganic salts. J. Colloid Interface Sci. *43*:150–155.

109 DEGUCHI, K., and K. MEGURO. 1975. The effects of inorganic salts and urea on the micellar structure of nonionic surfactant. J. Colloid Interface Sci. *50*:223–227.

110 FLORENCE, A.T., F. MADSEN, and F. PUISIEUX. 1975. Emulsion stabilization by non-ionic surfactants: The relevance of surfactant cloud point. J. Pharm. Pharmacol. *27*:385–394.

111 SCHOTT, H., and S.K. HAN. 1975. Effect of inorganic additives on solutions of nonionic surfactants. II. J. Pharm. Sci. *64*:658–664.

112 SCHOTT, H., and S.K. HAN. 1976. Effect of inorganic additives on solutions of nonionic surfactants. III. CMCs and surface properties. J. Pharm. Sci. *65*:975–978.

113 SCHOTT, H., and S.K. HAN. 1976. Effect of inorganic additives on solutions of nonionic surfactants. IV. Kraft points. J. Pharm. Sci. *65*:979–981.

114 NISHIKIDO, N., and R. MATUURA. 1977. The effect of added inorganic salts on the micelle formation of nonionic surfactants in aqueous solutions. Bull. Chem. Soc. Jpn. *50*:1690–1694.

115 SCHOTT, H., and S.K. HAN. 1977. Effect of symmetrical tetraalkylammonium salts on cloud point of nonionic surfactants. J. Pharm. Sci. *66*:165–168.

116 SCHICK, op cit.

117 MARON, S.H., and J.B. LANDO. 1974. Fundamentals of physical chemistry. Macmillan, New York.

118 D'ARRIGO, J.S. 1978. Screening of membrane surface charges of divalent cations: an atomic representation. Am. J. Physiol. *235*:109–117.

119 SCHRIER, E.E., and E.B. SCHRIER. 1967. The salting-out behavior of amides and its relation to the denaturation of proteins by salts. J. Phys. Chem. *71*:1851–1860.

120 KABARA, J.J., A.J. CONLEY, and J.P. TRUANT. 1972. Relationship of chemical structure and antimicrobial activity of alkyl amides and amines. Antimicrob. Agents Chemother. *2*:492–498.

121 KABARA, J.J., and G.V. HAITSMA. 1975. Aminimides. II. Antimicrobial effect of short chain fatty acid derivatives. J. Am. Oil Chem. Soc. *52*:444–447.

122 CHIPALKATTI, H.R., C.H. GILES, and D.G.M. VALLANCE. 1954. Adsorption at organic surfaces. Part I. Adsorption of organic compounds by polyamide and protein fibers from aqueous and nonaqueous solutions. J. Chem. Soc. *1954*:4375–4390.

123 ARSHID, F.M., C.H. GILES, S.K. JAIN, and A.S.A. HASSAN. 1956. Studies in hydrogen-bond formation. Part III. The reactivity of amines, amides, and azo-compounds in aqueous and non-aqueous solutions. J. Chem. Soc. *1956*:72–75.

124 ROBINSON, D.R., and W.P. JENCKS. 1965. The effect of concentrated salt solutions on the activity coefficient of acetyltetraglycine ethyl ester. J. Am. Chem. Soc. *87*:2470–2479.

125 EGAN, R.W., M.A. JONES, and A.L. LEHNINGER. 1976. Hydrophile-lipophile balance and critical micelle concentration as key factors influencing surfactant disruption of mitochondrial membranes. J. Biol. Chem. *251*:4442–4447.

126 GRIFFIN, W.C. 1949. Classification of surface-active agents by "HLB." J. Soc. Cosmet. Chem. *1*:311–326.

127 GRIFFIN, W.C. 1954. Calculation of HLB values of nonionic surfactants. J. Soc. Cosmet. Chem. *5*:249–256.

128 GREENWALD, H.L., G.L. BROWN, and M.N. FINEMAN. 1956. Determination of the hydrophile-lipophile character of surface active agents and oils by a water titration. Anal. Chem. *28*:1693–1697.

129 MOORE, C.D., and M. BELL. 1956. Non-ionic surface-active agents and their utilization in cosmetic preparations. Soap Perfum. Cosmet. *29*:893–896.

130 DAVIES, J.T. 1957. A quantitative kinetic theory of emulsion type. I. Physical chemistry of the emulsifying agent, p. 426–438. *In* J.H. Schulman (ed.), Proceedings of the 2nd International Congress of Surface Activity, vol. 1. Butterworths, London.

131 SCHOTT, H. 1971. Hydrophilic-lipophilic balance and distribution coefficients of nonionic surfactants. J. Pharm. Sci. *60*:648–649.

132 D'ARRIGO, 1979, Physical properties of the nonionic surfactants. . . ," op cit.

133 YOUNT, KUNKLE, D'ARRIGO, INGLE, YEUNG, and BECKMAN, op cit.

134 LONG and McDEVIT, op cit.

135 SCHOTT and HAN, op cit.

136 NISHIKIDO and MATUURA, op cit.

137 SCHICK, op cit.

138 BERND, 1964, op cit.

139 TRUSSELL and UMPHRES, op cit.

140 JUNK, op cit.

141 MOORE and BELL, op cit.

142 PACIFICO, C., and M.E. IONESCU. 1957. Surfactant survey. J. Am. Oil Chem. Soc. *34*:203–210.

143 STEVENS, C.E. 1957. Nonionic surfactants. J. Am. Oil Chem. Soc. *34*:181–185.

144 ROSEN, M.J., and H.A. GOLDSMITH. 1972. Systematic analysis of surface-active agents, 2nd ed. Wiley-Interscience, London.

145 LONGMAN, G.F. 1975. The analysis of detergents and detergent products. Wiley-Interscience, London.

146 D'ARRIGO, J.S. 1981. Aromatic proteinaceous surfactants stabilize long-lived gas microbubbles from natural sources. J. Chemical Physics *75*:962–968.

147 JOHNSON and COOKE, op cit.

148 MANLEY, op cit.

149 D'ARRIGO, 1981, op cit.

150 BLOCK, R.J., E.L. DURRUM, and G. ZWELG. 1958. A manual of paper chromatography and paper electrophoresis. Academic, New York.

151 GREENSTEIN, J.P., and M. WINITZ. 1961. Chemistry of the amino acids. Wiley, New York.

152 DUCKWORTH and YAPHE, 1971, Preparation of agarose by fractionation from the spectrum of polysaccharides in agar.

153 DUCKWORTH and YAPHE, 1971, The structure of agar. Part I. Fractionation of a complex mixture of polysaccharides.

154 GUISELEY and RENN, op cit.

155 SHOREY, E.C. 1906. Organic nitrogen in Hawaiian soils. Annu. Rep. Hawaii Agr. Exp. Sta. *1906*:37–59.

156 SCHREINER, O., and E.C. SHOREY. 1909. The isolation of harmful organic substances from soils. U.S. Dep. Agr. Bull. *53*:1–45.

157 JODIDI, S.L. 1911. The chemical nature of the organic nitrogen in the soil. J. Am. Chem. Soc. *33*:1226–1241.

158 JODIDI, L.S. 1912. The chemical nature of the organic nitrogen in the soil. Second paper. J. Am. Chem. Soc. *34*:94–99.

159 KELLEY, W.P. 1914. The organic nitrogen of Hawaiian soils. I. The products of acid hydrolysis. J. Am. Chem. Soc. *36*:429–434.

160 KELLEY. W.P. 1914. The organic nitrogen of Hawaiian soils. II. The effects of heat on soil nitrogen. J. Am. Chem. Soc. *36*:434–438.

161 KELLEY, W.P., and A.R. THOMPSON. 1914. The organic nitrogen of Hawaiian soils. III. The nitrogen of humus. *36*:438–444.

162 POTTER, R.S., and R.S. SNYDER. 1915. Amino acid nitrogen of soils and the chemical groups of amino acids in the hydrolyzed soil and their humic acids. J. Am. Chem. Soc. *37*:2219–2224.

163 WALTERS, E.H. 1915. The presence of proteoses and peptones in soils. J. Ind. Eng. Chem. *7*:860–863.

164 LATHROP, E.C. 1916. Protein decomposition in soils. Soil. Sci. *1*:509–532.

165 DAVIES, W.L. 1928. The proteins of different types of peat soils. J. Agr. Sci. *18*:682–690.

166 HOBSON, R.P., and H.J. PAGE. 1932. Studies on the carbon and nitrogen cycles in the soil. VII. The nature of the organic nitrogen compounds of the soil: "humic" nitrogen. J. Agr. Sci. *22*:497–515.

167 HOBSON, R.P., and H.J. PAGE. 1932. Studies on the carbon and nitrogen cycles in the soil. VIII. The nature of the organic nitrogen compounds of the soil: "non-humic" nitrogen. J. Agr. Sci. *22*:516–526.

168 KOJIMA, R.T. 1947. Soil organic nitrogen. I. Nature of the organic nitrogen in a muck soil from Geneva, New York. Soil Sci. *64*:157–165.

169 KOJIMA, R.T. 1947. Soil organic nitrogen. II. Some studies on the amino acids of protein material in a muck soil from Geneva, New York. Soil Sci. *64*:245–252.

170 BREMNER, J.M. 1950. The amino-acid composition of the protein material in soil. Biochem. J. *47*:538–542.

171 BREMNER, J.M. 1951. A review of recent work on soil organic matter. Part I. J. Soil Sci. *2*:67–82.

172 STEVENSON, F.J. 1954. Ion exchange chromatography of the amino acids in soil hydrolysates. Soil Sci. Soc. Am. Proc. *18*:373–377.

173 STEVENSON, F.J. 1956. Isolation and identification of some amino compounds in soils. Soil. Sci. Soc. Am. Proc. *20*:201–204.

174 STEVENSON, F.J. 1956. Effect of some long-time rotations on the amino acid composition of the soil. Soil Sci. Soc. Am. Proc. *20*:204–208.

175 SOWDEN, F.J. 1958. The forms of nitrogen in the organic matter of different horizons of soil profiles. Can. J. Soil. Sci. *38*:147–154.

176 YOUNG, J.L., and J.L. MORTENSEN. 1958. Soil nitrogen complexes. I. Chromatography of amino compounds in soil hydrolysates. Ohio Agr. Exp. Sta. Res. Circ. *61*:1–18.

177 BREMNER, J.M. 1965. Organic nitrogen in soils, p. 93–120. *In* W.V. Bartholomew and F.E. Clark (ed.), Soil nitrogen. American Society of Agronomy, Madison, Wisconsin.

178 BRIONES, A.M. 1969. Nature and distribution of organic nitrogen in tropical soils. Ph.D. thesis, University of Hawaii, Honolulu.

179 SIMONART, P., L. BATISTIC, and J. MAYAUDON. 1967. Isolation of protein from humic acid extracted from soil. Plant Soil *27*:153–161.

180 BIEDERBECK, V.O., and E.A. PAUL. 1973. Fractionation of soil humate with phenolic solvents and purification of the nitrogen-rich portion with polyvinylpyrrolidone. Soil Sci. *115*:357–366.

181 MARTIN, J.P., A.A. PARSA, and K. HAIDER. 1978. Influence of intimate association with humic polymers on biodegradation of $[^{14}C]$ labeled organic substrates in soil. Soil Biol. Biochem. *10*:483–486.

182 CRANWELL, P.A., and R.D. HAWORTH. 1975. The chemical nature of humic acids, p. 13–18. *In* D. Povoledo and H.L. Golterman (ed.), Humic substances: their structure and function in the biosphere. Centre for Agricultural Publishing and Documentation, Wageningen, Netherlands.

183 VERMA, L., J.P. MARTIN, and K. HAIDER. 1975. Decomposition of carbon-14-labeled proteins, peptides, and amino acids; free and complexed with humic polymers. Soil Sci. Soc. Am. Proc. *39*:279–284.

184 HOBSON and PAGE, op cit.

185 STEVENSON, 1954, op cit.

186 WHITE, A., P. HANDLER, and E.L. SMITH. 1964. Principles of biochemistry, 3rd ed. McGraw-Hill, New York.

187 KNOWLES, A., and S. GURNANI. 1972. A study of the methylene blue-sensitized oxidation of amino acids. Photochem. Photobiol. *16*:95–108.

188 WEIL, L., W.G. GORDON, and A.R. BUCHERT 1951. Photooxidation of amino acids in the presence of methylene blue. Arch. Biochm. Biophys. *33*:90–109.

189 WEIL, L. 1965. On the mechanism of the photo-oxidation of amino acids sensitized by methylene blue. Arch. Biochem. Biophys. *110*:57–68.

190 GUY, R.D., D.R. NARINE, and S. DESILVA. 1980. Organo-cation speciation. I. A. comparison of the interactions of methylene blue and paraquat with bentonite and humic acid. Can. J. Chem *58*:547–554.

191 KNOWLES and GURNANI, op cit.

192 WEIL, L., A.R. BUCHERT, and J. MAHER. 1952. Photooxidation of crystalline lysozyme in the presence of methylene blue and its relation to enzymatic activity. Arch. Biochem. Biophys. *40*:245–252.

193 WEIL, L., S. JAMES, and A.R. BUCHERT. 1953. Photooxidation of crystalline chymotrypsin in the presence of methylene blue. Arch. Biochem. Biophys. *46*:266–278.

194 SPIKES, J.D., and R. LIVINGSTON. 1969. The molecular biology of photodynamic action: sensitized photoautoxidations in biological systems, p. 29–110. *In* L.G. Augenstein, R. Mason, and M. Zelle (ed.), Advances in radiation biology, vol. 3. Academic, New York.

195 HORTON, H.R., H. KELLY, and D.E. KOSHLAND, JR. 1965. Environmentally sensitive protein reagents: 2-methoxy-5-nitrobenzyl bromide. J. Biol. Chem. *240*:722–724.

196 BARMAN, T.E., and D.E. KOSHLAND, JR. 1967. A colorimetric procedure for the quantitative determination of tryptophan residues in proteins. J. Biol. Chem. *242*:5771–5776.

197 VALLEE, B.L., and J.F. RIORDAN. 1969. Chemical approaches to the properties of active sites of enzymes. Annu. Rev. Biochem. *38*:733–784.

198 STARK, G.R. 1970. Recent developments in chemical modification and sequential degradation of proteins. Adv. Protein Chem. *24*:261–308.

199 KARKHANIS, Y.D., D.J. CARLO, and J. ZELTNER 1975. A simplified procedure to determine tryptophan residues in proteins. Anal. Biochem. *69*:55–60.

200 KOJIMA, 1947, Soil organic nitrogen. I. Nature of the organic nitrogen in a muck soil from Geneva, New York, op cit.

201 BREMNER, op cit.

202 BARMAN and KOSHLAND, op cit.

203 SHOREY, op cit.

204 SCHREINER and SHOREY, op cit.

205 GORTER, E. 1934. Spreading in a monomolecular film. A method for studying biologic problems. Am. J. Dis. Child. *47*:945–957.

206 NEURATH, H. 1936. The influence of denaturation of the spreading of proteins on a water surface. J. Phys. Chem. *40*:361–368.

207 NEURATH, H., and H.B. BULL. 1938. The surface activity of proteins. Chem. Rev. *23*:391–435.

208 PANKHURST, K.G.A. 1941. Foam formation and foam stability: The effect of the adsorbed layer. Trans. Faraday Soc. *37*:496–505.

209 PERRI, J.M., JR., and F. HAZEL. 1946. Lime hydrolysis of soybean protein. Ind. Eng. Chem. *38*:549–554.

210 PERRI, J.M., JR., and F. HAZEL. 1947. Effect of electrolytes on the foaming capacity of alpha soybean protein dispersions. J. Phys. Colloid Chem. *51*:661–666.

211 DAVIS, S.G., C.R. FELLERS, and W.B. ESSELEN, JR. 1949. Foam fractionation procedures in the isolation of fruit proteins. Food Tech. *3*:198–201.

212 THUMAN, W.C., A.G. BROWN, and J.W. McBAIN. 1949. Studies of protein foams obtained by bubbling. J. Am. Chem. Soc. *71*:3129–3135.

213 CUMPER, C.W.N. , and A.E. ALEXANDER. 1950. The surface chemistry of proteins. Trans. Faraday Soc. *46*:235–253.

214 CUMPER, C.W.N. 1953. The stabilization of foams by proteins. Trans. Faraday Soc. *49*:1360–1369.

215 DAVIES, J.T. 1954. Some factors influencing the orientation of ϵ-amino groups in monolayers of proteins and amino acid polymers. Biochem. J. *56*:509–513.

216 GOLDACRE, R.J. 1958. Surface films, their collapse on compression, the shapes and sizes of cells and the origin of life, p. 278–298. *In* J.F. Danielli, K.G.A. Pankhurst, and A.C. Riddiford (ed.), Surface phenomena in chemistry and biology. Pergamon, New York.

217 MALCOLM, B.R. 1962. Conformation of synthetic polypeptide and protein monolayers at interfaces. Nature (London) *195*:901–902.

218 NAKAMURA, R. 1964. Studies on the foaming property of the chicken egg white. Part VII. On the foaminess of the denatured ovalbumin. Agr. Biol. Chem. *28*:403–407.

219 OLDFIELD, J.F.T., and J.V. DUTTON. 1968. Surface active constituents in beet sugar crystallization. Int. Sugar J. *70*:7–9.

220 MALCOLM, B.R. 1968. Molecular structure and deuterium exchange in monolayers of snythetic polypeptides. Proc. R. Soc. London Ser. A *305*:363–385.

221 JONES, D.D., and M. JOST. 1970. Isolation and chemical characterization of gas-vacuole membranes from *Microcystis aeruginosa* Kuetz. emend. Elenkin. Arch. Mikrogiol. *70*:43–64.

222 BIKERMAN, J.J. 1973. Foams. Springer, New York.

223 KING, R.J., D.J. KLASS, E.G. GIKAS, and J.A. CLEMENTS. 1973. Isolation of apoproteins from canine surface active material. Am. J. Physiol. *224*:788–795.

224 HEARD, B., R. WHITE, and E. KING. 1974. Effects of serum proteins on pulmonary surfactants *in vitro*: evidence of a protective mechanism in pulmonary edema. Chest *66*:13S–16S.

225 ADAMSON, A.W. 1976. Physical chemistry of surfaces, 3rd ed. Wiley, New York.

226 PIERCE, J.S. 1978. The role of positive and negative factors in head retention (Guinness Overseas Ltd.) Inst. Brewing Aust. N.Z. Sect. Proce. *15*:51–65.

227 WHITEAR, A.L. 1978. Basic factors that determine foam stability. Inst. Brewing Aust. N.Z. Sect. Proc. *15*:67–75.

228 KALISCHEWSKI, K., W. BUMBULLIS, and K. SCHÜGERL. 1979. Foam behavior of biological media. I. Protein foams. Eur. J. Appl. Microbiol. Biotechnol. *7*:21–31.

229 KALISCHEWSKI, K., and K. SCHÜGERL. 1979. Investigation of protein foams obtained by bubbling. Colloid Polym. Sci. *257*:1099–1110.

230 WANISKA, R.D., and J.E. NINSELLA. 1979. Foaming properties of proteins: Evaluation of a column aeration apparatus using ovalbumin. J. Food Sci. *44*:1398–1402.

231 PHILLIPS, M.C. 1981. Protein conformation at liquid interfaces and its role in stabilizing emulsions and foams. Food Tech. *35*:50–57.

232 PASSERO, M.A., R.W. TYE, K.H. KILBURN, and W.S. LYNN. 1973. Isolation and characterization of two glycoproteins from patients with alveolar proteinosis. Proc. Natl. Acad. Sci. *70*:973–976.

233 BHATTACHARYYA, S.N., M.A. PASSERO, R.P. DiAUGUSTINE, and W.S. LYNN 1975. Isolation and characterization of two hydroxyproline-containing glycoproteins from normal animal lung lavage and lamellar bodies. J. Clin. Invest. *55*:914–920.

234 KING, R.J., J. RUCH, E.G. GIKAS, A.C.G. PLATZKER, and R.K. CREASY. 1975. Appearance of apoproteins of pulmonary surfactant in human amniotic fluid. J. Appl. Physiol. *39*:735–742.

235 BHATTACHARYYA, S.N., M.C. ROSE, M.G. LYNN, C. MACLEOD, M. ALBERTS, and W.S. LYNN. 1976. Isolation and characterization of a unique glycoprotein from lavage of chicken lungs and lamellar organelles. Am. Rev. Resp. Dis. *114*:843–850.

236 BHATTACHARYYA, S.N., S. SAHU, and W.S. LYNN. 1976. Structural studies on a glycoprotein isolated from alveoli of patients with alveolar proteinosis. Biochem. Biophys. Acta *427*:91–106.

237 BHATTACHARYYA, S.N., and W.S. LYNN. 1977. Structural studies on the oligosaccharides of a glycoprotein isolated from alveoli of patients with alveolar proteinosis. J. Biol. Chem. *252*:1172–1180.

238 BHATTACHARYYA, S.N., and W.S. LYNN. 1977. Studies on structural relationship between two glycoproteins isolated from alveoli of patients with alveolar proteinosis. Biochem. Biophys. Acta *494*:150–161.

239 KING, F.J., H. MARTIN, D. MITTS, and F.M. HOLMSTROM. 1977. Metabolism of the apoproteins in pulmonary surfactant. J. Appl. Physiol. *42*:483–491.

240 BHATTACHARYYA, S.N., and W.S. LYNN. 1978. Isolation and characterization of a pulmonary glycoprotein from human amniotic fluid. Biochem. Biophys. Acta *537*:329–335.

241 KATYAL, S.L., and G. SINGH. 1979. An immunologic study of the apoproteins of rat lung surfactant. Lab. Invest. *40*:562–567.

242 BHATTACHARYYA, S.N., and W.S. LYNN. 1979. Structural characterization of a glycoprotein isolated from alveoli of patients with alveolar proteinosis. J. Biol. Chem. *254*:5191–5198.

243 BHATTACHARYYA, S.N., and W.S. LYNN. 1980. Structural studies on a collagen-like glycoprotein isolated from lung lavage of normal animal. Biochem. Biophys. Acta *626*:451–458.

244 LEHTO, V.-P., I. KANTOLA, T. TERVO, and L.A. LAITINEN. 1981. Ruthenium red staining of blood-bubble interface in acute decompression sickness in rats. Undersea Biomed. Res. *8*:101–111.

245 OGNER, G., and M. SCHNITZER. 1971. Chemistry of fulvic acid, a soil humic fraction, and its relation to lignin. Can. J. Chem. *49*:1053–1063.

246 GLOSSMANN, H., and D.M. NEVILLE, JR. 1971. Glycoproteins of cell surfaces. J. Biol. Chem. *246*:6339–6346.

247 D'ARRIGO, 1981, op cit.

19 THE ACETONE-BUTANOL FERMENTATION OF *CLOSTRIDIUM ACETOBUTYLICUM*

Duane C. Ulmer

Department of Biotechnology, Zurich

Antonio R. Moreira

International Flavors and Fragrances Inc.

James C. Linden

Colorado State University

The escalating cost of petroleum products during the 1970s has caused renewed interest in obtaining fuels and chemicals by alternative technologies. The conversion of cellulosic residues to fermentable sugars for the production of ethanol has undoubtedly received most of the attention to date. Ethanol produced from so-called "renewable resources" can be used as an extension of existing fuel supplies as well as be a chemical feedstock for the production of various other chemical products.

Butanol, like ethanol, can be produced by fermentation processes. The commercialized production of butanol is carried out by the fermentation of cane molasses with the microorganism *Clostridium acetobutylicum*.[1] Acetone and ethanol, as well as the organic acids acetate and butyrate, are also produced in the fermentation broth, and large amounts of carbon dioxide and hydrogen are released during fermentation. Butanol, a longer-chain hydrocarbon than ethanol, has chemical properties closer to those of petroleum. Some of the chemical properties of butanol, in relationship to methanol, ethanol, and other hydrocarbons, are listed in Table 1. Two notable characteristics are butanol's low vapor pressure and low miscibility with water.

Anaerobic gas- and acid-producing bacteria, such as *C. acetobutylicum*, are also of interest for enhanced oil recovery. The injection of bacteria and nutrients into a reservoir with subsequent fermentation would cause partial repressurization of the reservoir due to the release of gases such as carbon dioxide and hydrogen.[2] The organic acids produced would also react with reservoir carbonates to form additional carbon dioxide. Acetone and the aliphatic alcohols produced by the *C. acetobutylicum* may also aid in oil release.

Table 1. Characteristics of Chemically Pure Fuels

Fuel	*Molecular Weight (lb/mole)*	*Specific Gravity*	*Boiling Point (°C)*	*Vapor Pressure at 100°F (psi)*	*Combustion Energy (Btu/lb)*	*Latent Heat (Btu/lb)*	*Solubility (parts in 100 parts H_2O)*	*Stoichiometric Air-Fuel Ratio*
Methanol	32	0.79	65	4.6	10,260	503	∞	6.5
Ethanol	46	0.79	78	2.2	13,160	396	∞	9
Butanol	74	0.81	117	0.3	15,770	186	9	11.2
Octane	114	0.70	210	1.72	20,750	155	insoluble	15.2
Hexadecane	240	0.79	287	3.46	20,320	—	insoluble	15

A major drawback of the acetone-butanol fermentation is the relatively dilute concentration of solvents obtained in the range of 2 weight percent.[3,4] The solvents, most notably butanol, are reportedly toxic to the bacterial cells at levels produced during the fermentation. That is the reason for the relatively low concentrations produced.[4] Since the products are toxic at relatively low levels, the concentration of fermentable sugars in the initial medium is also relatively low, 6.0–6.5% (w/v).

Attempts at solving the ethanol and butanol toxicity problem have centered on mutation, adaptation, and selection for tolerant strains[4,5] and on random nutrient and chemical addition to the nutrient broth.[6] These manipulations have given only slightly higher yields. This study in acetone-butanol fermentation focuses on obtaining a better understanding of the basic cause of the alcohol toxicity. The concentrations of ethanol and butanol that are toxic to the producing *Clostridium* are the same as those that affect energy-driven sugar transport in a variety of other cells.[7] Since sugar transport through the cell membrane is tied closely to cell viability, it is proposed that herein lies the basis for alcohol toxicity in such fermentations. Also included are characteristics of the acetone-butanol process, the microorganism, and some effects of ethanol and butanol on cell growth and sugar transport during glucose fermentation by *C. acetobutylicum*.

BIOCHEMISTRY OF ACETONE-BUTANOL FERMENTATIONS

C. acetobutylicum belongs to the group of *Clostridium* comprising the butyric acid bacteria. They are characterized by the formation of acetic and butyric acids, carbon dioxide, and hydrogen. The products of the fermentation of glucose by some species of butyric acid bacteria are given in Table 2. It is interesting to note that the gases, carbon dioxide and hydrogen, produced during the *C. acetobutylicum* fermentation weigh over 1.5 times as much as the acids and neutral solvents produced.[8] Lenz and Moreira have calculated that over 97% of the energy content of the glucose fermented is conserved in the chemicals produced (Table 3).[9]

Table 2. Fermentation products from three butyric acid *Clostridia*.

Fermentation Product	*C. butyricum*	*C. acetobutylicum*	*C. butylicum*
Butyrate	76	4	17
Acetate	42	14	17
Butanol	—	56	59
Acetone	—	22	—
Isopropanol	—	—	12
Ethanol	—	7	—
Acetylmethylcarbinol	—	6	—
CO_2	188	221	204
H_2	235	135	78

moles/100 moles of glucose fermented.
Source: H.J. Rehm, *Industrielle Mikrobiologie* (Berlin: Springer-Verlag, 1980).

Table 3. Energy Analysis of the Acetone-Butanol Fermentation

Product of Fermentation	*Product Yield*[a]	*Sp Heat of Combustion* $\Delta \hat{H}_c$[b]	ΔH_c, *J (Btu)*	*Energy Yield (%)*
Butyric acid	4	2.18×10^9 (9.36×10^5)	8.72×10^9 (8.26×10^6)	
Acetic acid	14	8.75×10^8 (3.76×10^5)	1.23×10^{10} (1.16×10^7)	
Hydrogen	135	2.87×10^8 (1.23×10^5)	3.87×10^{10} (3.66×10^7)	
Ethanol	7	1.37×10^9 (5.88×10^5)	9.59×10^9 (9.59×10^6)	
Butanol	56	2.68×10^9 (1.15×10^6)	1.50×10^{11} (1.42×10^8)	
Acetone	22	1.79×10^9 (7.69×10^5)	3.94×10^{10} (3.73×10^7)	
Acetoin	6	2.25×10^9 (9.66×10^5)	1.35×10^{10} (1.28×10^7)	
$\Sigma\Delta H_c$			2.72×10^{11} (2.58×10^8)	97[d]

[a]Units: kg-mol/100 kg-mol of glucose fermented.
[b]In J/kg-mol (Btu/lb-mol).
[c]Estimated from heat of formation data.
[d]Energy yield (%) = $[\Sigma\Delta H_c/(100 \times \Delta H_c \text{(glucose)})] \times 100$.

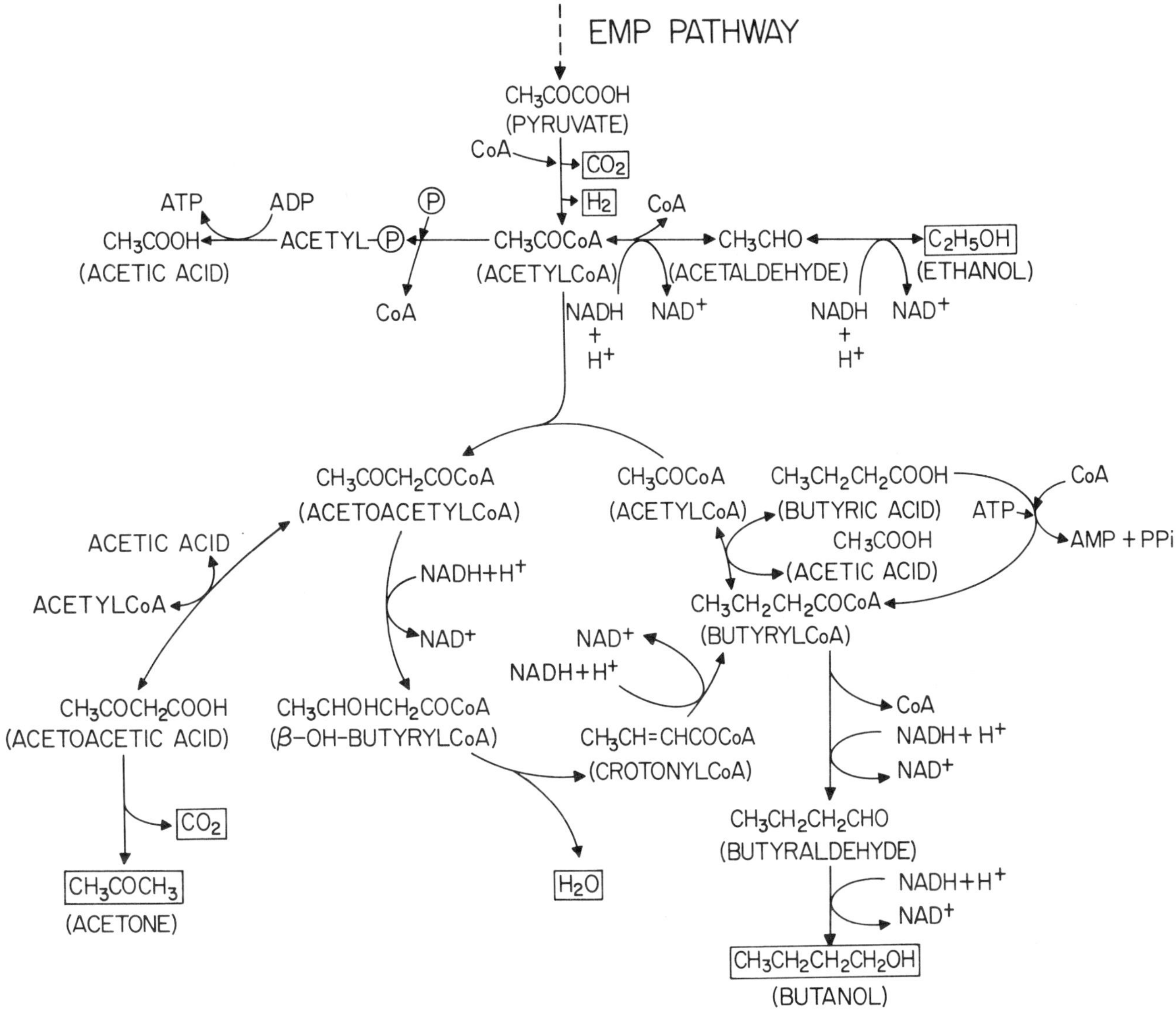

Fig. 1. Pathway of glucose fermentation by *C. acetobutylicum*

The overall pathway of glucose fermentation by *C. acetobutylicum*, based on available literature, is outlined in Figure 1. Sugars are converted to pyruvate via the Embden-Meyerhof-Parnas (EMP) pathway. Pyruvate is then broken down to acetyl-CoA with the release of carbon dioxide and hydrogen. This breakdown of pyruvate is referred to as a Clostridial type, as it differs from a similar process in enteric bacteria where formate is formed rather than CO_2 and H_2.

Acetyl-CoA plays a central role in the acetone-butanol fermentation and in other *Clostridium* fermentations because it serves as a precursor to ethanol, acetate, and all of the fermentation solvents. The formation of acetone and butanol occurs in what may be divided into two separate processes. Initially, two acetyl-CoA molecules combine to form acetoacetyl-CoA, initiating a cycle that produces butyrate. During this cyclic mechanism, the two moles of $NADH_2$ formed prior to pyruvate are reoxidized to NAD^+.

At some point during the accumulation of butyrate in the medium, a new enzyme system is activated, leading to the formation of acetone and butanol and disrupting the normal cyclic mechanism. The exact reason for the shift to production of neutral solvents is not understood clearly but is often attributed to the drop in the pH of the medium to 4.0.[10] The shift to formation of neutral products would thus prevent a further decrease in the pH. However, numerous reports have indicated that successful solvent production can be achieved without the pH of the medium dropping below 5.0.[11,12,13] The regulation for the changeover from acid production to the production of neutral solvents therefore appears to be much more complex than can be explained by changes in pH alone.

C. acetobutylicum possesses a transferase system that diverts acetoacetyl-CoA from the normal cycle to produce acetoacetate that can then be decarboxylated to acetone. The diversion of acetyl-CoA to acetone alone would be detrimental to the cell since the two steps for reoxidizing $NADH_2$ would have been eliminated. The regeneration of NAD^+ is accomplished by reducing the butyric acid produced to butanol through three consecutive reactions.

The fermentation pattern of glucose metabolism is thus the production of organic acids via the EMP pathway until the accumulation of acids disrupts the cyclic mechanism and the production of neutral solvents starts. The production of neutral solvents corresponds to the ''breakpoint'' in industrial fermentations as the pH begins to climb and the titratable acidity drops. The gases are produced throughout the fermentation. Table 4 shows the course of product formation in a fermentation of glucose in a complex medium where the initial pH of 6.2 was allowed to fall to pH 5.0 and was then controlled thereafter so that it would not fall below 5.0.

CHARACTERISTICS OF *Clostridium Acetobutylicum* AND ACETONE-BUTANOL FERMENTATION

Bergey's manual describes *C. acetobutylicum* as a straight rod, 0.6–0.9 by 2.4–4.7 μm; often granulose positive.[14] It is motile with peritrichous flagella and is gram positive with the cell wall containing DL-diaminopimelic acid. Spores are oval. It will ferment the following carbohydrate substrates; arabinose, galactinol, fructose, galactose, glucose, glycogen, lactose, maltose, mannose, salicin, starch, sucrose, trehalose, and xylose. Optimum temperature for growth is 37°C.

Spivey and Peterson & Fred describe variations in the morphology of the culture during different stages of the fermentation process, depending upon the strain of organism used and the formulation of the medium.[15,16] Cho and Doy have described some of the ultrastructures of *C. acetobutylicum.*[17]

Isolation. Beesch has described several methods for isolating acetone-butanol producing bacteria from materials such as soil, manure, roots of leguminous plants, cereals, decayed wood, cornstalks, sewage, and river bottom mud.[18,19] A species of *Clostridium* is ubiquitous in nature; sources for isolating such bacteria are limited only by the investigator's imagination.

The simplest medium for isolation is a sterile 4 percent (w/v) corn mash placed in culture tubes. After inoculation, the corn mash tubes are placed in boiling water for two minutes to destroy any vegetative forms present. The tubes are then cooled and incubated at 37°C for several days. If acetone-butanol organisms are present, the tubes will show considerable gas production and the culture will have a butylic-type odor. Cultures that have appeared to ferment well are plated on malt agar and are incubated anaerobically for 48–72 hours. Colonies can be picked from three plates and tested for solvent production (types of solvents and ratios produced), pH, time of fermentation, and other characteristics.

Calam has described methods for isolating *C. acetobutylicum* from broad beans, potatoes, or maize, using a potato medium and a molasses-rice bran medium.[20] As a general rule, the investigator should use an isolation medium that will resemble the medium to be utilized for solvent production.

Table 4. Batch fermentation data at pH 5.0 for *Clostridium acetobutylicum* in 5.8% (w/v) glucose medium.

Time (hr)	*Glucose (g/l)*	*Cell Mass (g/l)*	*Acetone (g/l)*	*Butanol (g/l)*	*Ethanol (g/l)*	*Acetic Acid (g/l)*	*Butyric Acid (g/l)*
8.5	49.3	0.4	0.08	0.35	—	0.95	0.54
11.0	46.8	1.1	0.14	0.41	—	1.97	1.26
14.0	37.8	2.5	0.07	0.45	—	2.60	3.31
17.0	28.7	3.1	1.63	2.12	0.14	4.50	4.17
21.0	8.1	3.5	1.63	6.04	0.34	4.58	2.48
24.0	3.4	3.8	2.71	7.76	0.55	3.08	1.87
33.0	0.6	3.5	3.26	7.51	—	1.58	1.55
39.0	0.4	3.1	4.07	11.10	0.62	1.71	1.33

Source: Moreira, Ulmer, and Linden, ''Butanol Toxicity in the Butylic Fermentation,'' Biotechnology Bioengineering Symposium 11, 1981, pp. 567–579.

Storage. Since the acetone-butanol organisms produce true spores, the cultures are maintained as spores. Industrially this is usually accomplished using soil as a carrier.[21] Briefly, this involves adding a culture that has completed its fermentation (about 72 hours) to sterile sand or soil until it is moderately wet. The flasks are then dried in a desiccator under reduced pressure.

No problems have been encountered in storing cultures in a 5 percent (w/v) cornmeal medium. This medium was comprised of cornmeal obtained from a local supermarket with 0.05 percent (w/v) cysteine added to ensure anaerobiosis. The mash was prepared by boiling the cornmeal for one hour in the appropriate volume of distilled water, adjusting the pH to 6.0, dispensing in screw cap tubes, and autoclaving at 121°C for 15 minutes. The culture has been recovered from corn tubes stored longer than eight months and undoubtedly would be viable for several years.

Nutrient Requirements. Lampeen and Peterson found biotin and p-aminobenzoic acid to be the main growth factors necessary for growth and fermentation by *C. acetobutylicum.*[22] In industrial practice the fermentable material usually contains all of the necessary trace nutrients. Fermentation of molasses generally requires the addition of a nitrogen and phosphate source. For more details concerning nutrient addition during industrial processes, see Beesch and Rehm.[23,24] On a laboratory scale, yeast extract can be added to ensure all of the vitamin requirements are met or, alternatively, a synthetic medium can be devised to contain the necessary growth factors.

Cultivation Medium. The nutrition requirements of *C. acetobutylicum* are few; thus, it can be grown on a variety of media. For solvent production and other experimental tests, a complex soluble medium was used.[25] The soluble medium contained (grams/liter): glucose, 60.0; KH_2PO_4, 0.75; K_2HPO_4, 0.75; $MgSO_4$, 0.20; $MnSO_4$-H_2O, 0.01; $FeSO_4$-$7H_2O$, 0.01; NaCl, 1.00; cysteine-HCl, 0.50; yeast extract, 5.00; asparagine-H_2O, 2.00; $(NH_4)_2SO_4$, 2.00. Various synthetic media have also been utilized to culture *C. acetobutylicum.*[26,27]

The pH for the butanol organisms may range from 4.0 to 7.0. A typical starting pH of the fermentation broth is 6.0 to 6.5. The final pH depends on the medium employed but usually is in the range of 5.2 to 6.2 During fermentation of a molasses medium, the pH will decrease to about 5.2 before the breakpoint is achieved.[28,29] The pH of a cornmeal medium, however, may decrease to as low as 4.0 before reaching a final value close to 4.3.[30]

Heat Shocking. In creating an inoculum for acetone-butanol fermentation, the cultures generally are subjected to a heat shock.[31,32] This treatment consists of placing tubes in a water bath at 70–100°C for as long as 90 seconds and then rapidly cooling them. The heat shock will vary, depending on the size and nature of the tube and the character of the medium employed. The heat shock is given to eliminate any vegetative cells or weak spores present in the tube and to stimulate germination of the remaining spores. In general, heat shock treatment has been beneficial for selecting active solvent producers.

Carbohydrate Sources. The original industrial process for acetone-butanol production utilized corn as a substrate; at present, materials such as invert or blackstrap molasses are used.[32,33] The fermentable sugar level of these substrates is initially set at 6.0–6.5 percent (w/v). Other carbohydrate sources that can be used include wheat, rice, Jerusalem artichokes, potatoes, and beet molasses.[34,35] Carbohydrates derived from waste products also have been fermented successfully to produce solvents. These include the production of acetone-butanol using the sugars released by acid hydrolysis of corn cobs, acid hydrolyzates of wood, and waste sulfite liquor.[36,37,38] These processes suffer from the relatively low level of sugar concentrations (1–3 percent w/v) that are present in the medium.

Soni et al. have used *C. saccaroper butylacetonicum* to produce butanol from concentrated enzymatic hydrolyzates of bagasse and rice straw.[39] To obtain 14.8 g/l butanol, it was necessary to subject the hydrolyzate to ammonium sulfate precipitation, followed by activated carbon treatment and addition of 1.25 percent (w/v) ferrous sulfate.

Maddox was able to ferment sulfuric acid whey filtrate supplemented with 0.5 percent (w/v) yeast extract to produce 1.2 percent (w/v) butanol.[40] The ratio of butanol-acetone-ethanol products was approximately 10:1:1. Lindberg and Moreira have also obtained butanol levels as high as 1.4 percent (w/v) from cheese whey with a butanol-acetone-ethanol ratio of approximately 25:1:1.[41]

These results indicate that a variety of substrates may be utilized for acetone-butanol production. However, it would be necessary to test each substrate for optimum conditions to produce the highest yields and the best solvent ratios.

Effect of Oxygen. Although *C. acetobutylicum* is an anaerobic microorganism requiring a redox potential of −250 mV for excellent growth and solvent production, short-term exposure to oxygen is not lethal to the cells. O'Brien and Morris found that submerged cultures aerated to maintain a redox potential of −50 to 0 mV (dissolved O_2 < 1 μmolar) did not affect the growth of the microorganism.[42] Exposure to oxygen (dissolved O_2 of 40–50 molar) for four to six hours was not lethal but decreased the rate of glucose consumption, halted growth, and prevented net synthesis of DNA, RNA, and protein. The effects of oxygen could be reversed quickly when anaerobic conditions were reestablished. Conditions of the fermentations were such that no solvents were produced, even under anaerobic conditions.

Experience has shown that reasonable precautions to exclude oxygen from the medium are sufficient to obtain good growth and solvent production.

Jones et al. have described the production of elongated fruiting body-like structures by *C. acetobutylicum* grown anaerobically on potato agar for two days and then exposed to aerobic conditions.[43] The strains of *C. acetobutylicum* ATCC 824 and 10132 failed to produce these structures. A suggested function of such structures was to act as a protective barrier against oxygen.

Solvent Ratios and Yields. Solvent yields are about 26.5 percent (w/w) based on dry corn with a solvent ratio of butanol-acetone-ethanol of 6:3:1.[44] Thirty percent yield of solvents from the sugars fermented in molasses with an approximate 6:3:1 ratio of butanol-acetone-ethanol is considered normal. However, variations in solvent ratios can be accomplished by isolating different cultures.[45] Differences among isolates include fermentation of different carbohydrates, different solvent ratios, and different utilizable sugar concentrations.

The yield and ratio of solvents can also be modified to some extent by varying the temperature of the culture or the composition of the nutrients and by adding of various chemicals.

Recycling some of the remaining stillage from the fermentation of blackstrap molasses enhances the yield of solvents 15 percent.[44] This increase in yield was attributed to the presence of soluble proteins, proteases, peptones, amino acids, and vitamins present in the stillage that would exert a stimulating effect. Addition of calcium carbonate to the medium caused a decrease in the yield of acetone and butanol in proportion to the amount of carbonate added, while the amounts of butyric and acetic acid were increased.

Copper ions have been found to be quite toxic to *C. acetobutylicum*.[46] Fermentations are usually inhibited by 40 ppm copper and are almost completely inhibited by the addition of 50 ppm copper.

Sucrose, levulose, and xylose yield normal ratios of solvents, but arabinose yields a 5:4:1 ratio of butanol-acetone-ethanol. Growth on lactose present in cheese whey has been found to yield a solvent ratio of 25:1:1 of butanol-acetone-ethanol.[47]

Autolysin Production. The production of solvents during the acetone-butanol fermentation is affected by product inhibition and may also be inhibited by the production of an autolysin.[48,49] During the fermentation of a molasses medium, it was found that an autolysin was released toward the end of the exponential growth phase and was accompanied by lysis of the culture and inhibition of the production of solvents. The autolysin has been characterized as a glycoprotein with a molecular weight of 28,000. It had no effect on deoxyribonucleic acid, ribonucleic acid, or protein synthesis and did not result in a loss of intracellular adenosine triphosphate. The exact role of the autolysin in the fermentation process is open to speculation; it may be a factor accounting for the low level of solvents achieved.

Recently, Allcock et al. have isolated an autolysin-deficient mutant to determine if such a microorganism can enhance the concentrations of solvents obtained during a fermentation.[50] The mutant produced 70 percent less autolysin and had a cell wall that was more resistant to the wild-type autolysin unless treated with penicillin. The mutant cells could tolerate up to 31 g/l butanol without cell death or any effect on growth rate. The mutant grew in up to 16 g/l of solvents; however, the highest solvent production obtained was 20 g/l of solvents. Hence, the level of solvent produced was not improved over that from the wild-type culture.

Continuous Culture. The use of continuous culture has recently been employed to study the nutritional and environmental factors that affect solvent production by *C. acetobutylicum*. The results showed that no solvents were produced under conditions employed or appeared only after addition of butyrate to the broth.[51,52,53] These results lead to the conclusion that solvent production by *C. acetobutylicum* is not feasible in a chemostat. However, Lueng and Wang reported that butanol productivity could be increased in continuous culture fermentation over that obtained in batch culture.[54] Above a maximum dilution rate, the fermentation favored butyrate production.

Immobilized Cells. One report in the literature has used immobilization techniques for vegetative cells and spores of *C. acetobutylicum* for solvent production.[55] Spores and vegetative cells were immobilized in calcium alginate gel and were studied under what were termed nongrowth conditions. Immobilization of cells in the solvent-producing stage caused a rapid decrease in the relative butanol production rate. In less than two days the production rate was less than 25 percent of the original rate. When suspended cells in a batchwise process were employed, it was necessary to use a medium of 10 g/l glucose and 5 g/l butyric acid to obtain a butanol concentration of 1.08 g/l.

Immobilization of spores followed by outgrowth of viable cells within the gel and continued incubation for 24 hours under nongrowth conditions produced 1.81 g/l butanol from a 10-g/l glucose solution. Although the authors concluded that a continuous process based on immobilization of spores is feasible, extensive investigation is needed to establish economical working conditions for such a process.

Byproducts. The carbon dioxide and hydrogen produced during the fermentation may be considered byproducts rather than a waste stream. The carbon dioxide produced can be recovered by absorption and then desorbed, purified, and compressed to liquid or solid CO_2.[56] With an appropriate catalyst it should be possible to convert the CO_2 and H_2 to methanol economically, creating another marketable product.

The stillage that is produced contains riboflavin, B complex vitamins, and protein, making the residue a valuable source of animal feed.[56]

Research Implications. The acetone-butanol fermentation was one of the first pure microbiological processes used on an industrial scale. The process presently used has changed little from that employed by Weizmann in 1915. With the production of solvents from petrochemical sources, the fermentation process has ceased to operate except in South Africa, where a combination of economic, political, and geographical conditions have kept this process viable on an industrial scale. There has therefore been little incentive to investigate conditions that would enhance the solvent yield other than empirically changing fermentation conditions. The use of more modern technology, such as continuous culture and cell immobilization, has failed to enhance the productivity of the fermentation. Presently there is still much to be learned regarding the basic physiology of the *C. acetobutylicum*, the regulation of solvent production in batch processes, and the mechanism by which fermentation products inhibit growth.

Recent studies have shown that the levels of *n*-butanol obtained in a typical fermentation of *C. acetobutylicum* are sufficient to inhibit cell growth rate, transport of nutrients, and membrane-bound ATPase activity.[58] The inhibitory effect of butanol on these cellular functions could be one explanation for the low level of solvents produced during the fermentation. Ethanol also caused a reduction in the growth rate and uptake of nutrients; however, it required approximately a tenfold greater concentration than butanol to effect the same degree of inhibition. The difference in concentrations required for these effects has been attributed to differences in the hydrophobicity of the alcohols, as shown by the data in Table 5.

Solvents greatly affect the lipid bilayer of cell membranes; they dissolve in the bilayers and may increase bilayer fluidity. Modifying cell membranes to tolerate higher concentrations of alcohols, i.e., adjusting the membrane fluidity by altering the molecular species of the phospholipid component in the membrane, may stabilize nutrient transport. The increased tolerance of *C. acetobutylicum* cells to butanol may allow greater levels of solvents to be attained in the fermentation broth and hence reduce the costs associated with solvent recovery.

Table 5. Concentrations of aliphatic alcohols that limit various *C. acetobutylicum* cellular functions to half maximal rates.

	Molar Concentration		
Half Maximal Rates	*EtOH*	*BuOH*	*HxOH*
Cell growth	1.10	0.15	0.015
Alanine uptake	0.91	0.13	0.006
3-0-Methyl glucose uptake	1.00	0.10	0.007
Membrane ATPase activity	<1.1	0.09	<0.01

Source: Linden, Ulmer, and Moreira, "A Mechanism for Aliphatic Alcohol-induced Toxicity in *Clostridium acetobutylicum*," American Chem. Soc., New York, August 1981.

Another approach to enhance the level of solvents produced is being undertaken by a group in South Africa.[59] This group is focusing on obtaining various mutants, the overall objective being to establish a means of looking at the genetic information of *C. acetobutylicum*. With a basic understanding of the genetic information available, it may be possible to manipulate the genes to create a strain capable of greater solvent production.

As the cost of petroleum products escalates, fermentation technology for solvent production will regain worldwide emphasis. The development of more solvent-tolerant *C. acetobutylicum* strains, of efficient in situ solvent extraction techniques, or of means to utilize the gaseous byproducts may allow the fermentation process for solvent production to become economically viable.

REFERENCES

1 SPIVEY, J.J. 1978. The acetone/butanol/ethanol fermentation. Proc. Biochem. *13*:2–4, 25.

2 JOHNSON, A.C. 1979. Microbial oil release technique for enhanced recovery. *In* Conference on Microbiological Processes Useful in Enhanced Oil Recovery, Final Report. Aug. 29–Sept. 1, San Diego, California. Submitted by Science Applications Inc., East Brunswick, New Jersey.

3 WALTON, M.T., and J.L. MARTIN. 1979. Production of butanol acetone by fermentation, p. 187–210. *In* H.J. Peppler and D. Perlman (ed.), Microbial technology. I. Microbial processes, 2nd ed. Academic Press Inc., New York.

4 STEEL, R. 1958. Biochemical engineering: unit processes in fermentation, p. 125–148. Macmillan Co., New York.

5 HERRERO, A.H., and R.F. GOMEZ. 1980. Development of ethanol tolerance in *Clostridium thermocellum*: effect of growth temperature. Appl. Environ. Microbiol. *39*:571–577.

6 PETERSON, W.H., and E.B. FRED. 1932. Butyl-acetone fermentation of corn meal. Interrelations of substrate and products. Ind. Eng. Chem. *24*:237–242.

7 SULLIVAN, K.H., M.K. JAIN, and A.L. KOCK. 1974. Activation of the -galactoside transport system in *E. coli* ML-308 by n-alkanols. Biochem. Biophys. Acta. *352*:287–297.

8 REHM, H.J. 1980. Butanol-Aceton und weitere primare Produkte aus *Clostridien* p. 291–305. *In* Industrielle Mikrobiologie. Springer-Verlag, Berlin, Germany.

9 LENZ, T.G., and A.R. MOREIRA. 1980. Economic evaluation of the acetone-butanol fermentation. Ind. Eng. Chem. Prod. Res. Dev. *19*: 478–483.

10 DOELLE, H.W. 1975. Bacterial metabolism, 2nd ed, p. 587. Academic Press, New York.

11 REHM, op cit.

12 BARBER, J.M., F.T. ROBB, J.R. WEBSTER, and D.R. WOODS. 1979. Bacteriocin production by *Clostridium acetobutylicum* in an industrial fermentation process. Appl. Environ. Microbiol. *37*:433–437.

13 MOREIRA, A.R., D.C. ULMER, and J.C. LINDEN. 1981. Butanol toxicity in the butylic fermentation. Biotechnol. Bioeng. Symp. No. *11*: 567–579.

14 BUCHANAN, R.E., and N.E. GIBBONS (ed.), 1974. Bergey's manual of determinative bacteriology, 8th ed, p. 556–560. The Williams and Wilkins Co., Baltimore.

15 SPIVEY, op cit.

16 PETERSON and FRED, op cit.

17 CHO, K.Y., and C.H. DOY. 1973. Ultrastructure of the obligately anaerobic bacteria *Clostridium kluyveri* and *Cl. acetobutylicum*. Aust. J. Biol. Sci. *26*: 547–558.

18 BEESCH, S.C. 1952. Acetone-butanol fermentation of sugars. Ind. Eng. Chem. *44*:1677–1682.

19 BEESCH, S.C. 1953. A microbiological process report. Acetone-butanol fermentation of starches. Appl. Microbiol. *1*:85–95.

20 CALAM, C.T. 1980. Isolation of *Clostridium acetobutylicum* strains producing butanol and acetone. Biotechnol. Letr. *3*:111–116.

21 WALTON and MARTIN, op cit.

22 LAMPEN, J.O., and W.H. PETERSON. 1943. Growth factor requirements of *Clostridia*. Arch. Biochem. *2*:443–449.

23 BEESCH, 1952, op cit.

24 REHM, op cit.

25 MOREIRA, ULMER, and LINDEN, op cit.

26 LAMPEN and PETERSON, op cit.

27 GOTTSCHAL, J.C., and J.G. MORRIS. 1981. Nonproduction of acetone and butanol by *Clostridium acetobutylicum* during glucose — and ammonium — limitation in continuous culture. Biotechnol. Letr. *3*:525–530.

28 SPIVEY, op cit.

29 BARBER et al., op cit.

30 PETERSON and FRED, op cit.

31 BEESCH, 1953, op cit.

32 SPIVEY, op cit.

33 BEESCH, 1952, op cit.

34 FOUAD, M., A.A. ASOU-ZEID, and M. YASSEIN. 1976. The fermentative production of acetone-butanol by *Clostridium acetobutylicum*. Acta. Biol. Acad. Sci. Hung. *27*:107–117.

35 WENDLAND, R.T., E.I. FULMER, and L.A. UNDERKOFLER. 1941. Butylacetonic fermentation of Jerusalem artichokes. Ind. Eng. Chem. *33*:1078–1081.

36 LANGLYKKE, A.F., J.M. VAN LANEN, and D.R. FRASER. 1948. Butyl alcohol from xylose saccharification liquors from corn cobs. Ind. Eng. Chem. *40*:1716–1719.

37 LEONARD, R.H., W.H. PETERSON, and G.J. RITTER. 1947. Butanol-acetone fermentation of wood sugar. Ind. Eng. Chem. *11*:1443–1445.

38 WILEY, A.J., M.J. JOHNSON, E. McCOY, and W.H. PETERSON, 1941. Ind. Eng. Chem. *33*:606–610.

39 SONI, B.K., K. DAS, and T.K. GHOSE. 1982. Bioconversion of agrowastes into acetone butanol. Biotechnol. Letr. *4*:19–22.

40 MADDOX, I.S. 1980. Production of n-butanol from whey filtrate using *Clostridium acetobutylicum* N.C.T.B. 2951. Biotechnol. Letr *2*:493–498.

41 LINDBERG, S.C., and A.R. MOREIRA. 1982. Acetone-butanol fermentation of cheese whey. Paper presented at the AIChE Winter Meeting, Orlando, Florida, February 28–March 3.

42 O'BRIEN, R.W., and J.G. MORRIS. 1971. Oxygen and the growth and metabolism of *Clostridium acetobutylicum*. J. Gen. Microbiol. *68*:307–318.

43 JONES, D.T., J.R. WEBSTER, and D.R. WOODS. 1980. The formation of simple fruiting body-like structures associated with sporulation under aerobic conditions in *Clostridium acetobutylicum*. J. Gen. Microbiol. *116*:195–200.

44 BEESCH, 1953, op cit.

45 BEESCH, 1952, op cit.

46 LEONARD, PETERSON, and RITTER, op cit.

47 LINDBERG and MOREIRA, op cit.

48 BARBER et al., op cit.

49 WEBSTER, J.R., S.J. REID, N.T. JONES, and D.R. WOODS. 1980. The formation of simple fruiting body-like structures associated with sporulation under aerobic conditions in *Clostridium acetobutylicum*. J. Gen. Microbiol. *116*:195–200.

50 ALLCOCK, E.R., S.J. REID, D.T. JONES, and D.R. WOODS. 1981. Autolytic activity and an autolysis-deficient mutant of *Clostridium acetobutylicum*. Appl. Environ. Microbiol. *42*:929–935.

51 GOTTSCHAL and MORRIS, op cit.

52 ANDERSCH, W., H. BAHL, and G. GOTTSCHALK. 1982. Acetone-butanol production by *Clostridium acetobutylicum* in an ammonium-limited chemostat at low pH values. Biotechnol. Letr. *4*:29–32.

53 BAHL, H., W. ANDERSCH, K. BRAIN, and G. GOTTSCHALK. 1982. Effect of pH and butyrate concentrate on the production of acetone and butanol by *C. acetobutylicum* grown in continuous culture. Eur. J. Appl. Microbiol. Biotechnol. *14*:17–20.

54 LEUNG, J.C.Y., and D.I.C. WANG. 1981. Production of acetone and butanol by *Clostridium acetobutylicum* in continuous culture using free cells and immobilized cells, p. 348–352. *In* Proceedings of 2nd World Congress of Chemical Engineering, vol 1. Montreal, October 4–9.

55 HAGGSTROM, L., and N. MOLIN. 1980. Calcium alginate immobilized cells of *Clostridium acetobutylicum* for solvent production. Biotechnol. Letr. *2*:241–246.

56 SPIVEY, op cit.

57 BEESCH, 1953, op cit.

58 LINDEN, J.C., D.C. ULMER, and A.R. MOREIRA. 1981. A mechanism for aliphatic alcohol-induced toxicity in *Clostridium acetobutylicum*. Symposium on Fuels and Chemicals from Biomass. American Chemical Society National Meeting, New York, August 23–28.

59 ALLCOCK et al., op cit.

20 TOWARD MICROBIALLY ENHANCED OIL RECOVERY FROM NORTH SEA RESERVOIRS

V. Moses, J.P. Robinson, and D.G. Springham
Queen Mary College, London

Melanie J. Brown, M. Foster, J. Hume, C.W. May, T.S. McRoberts, and A. Western
QMC Industrial Research Ltd., London

The depth of the oil-bearing strata in North Sea reservoirs, coupled with their offshore location in frequently inhospitable weather conditions, makes drilling expensive and results in wide well spacing, with linear distances of 1 km not uncommon between well bottoms. Any system of microbially enhanced oil recovery intended for those reservoirs must take into account not only factors such as permeability and bed or flood water mineralization, which are important in all reservoirs, but must also accommodate the high temperatures and pressure associated with all North Sea fields. The wide well spacings, in particular, and the very long liquid displacement times (5–7 years) they imply make it quite impractical to consider microbial enhancement procedures that depend on feeding the organisms by organic nutrient injection, as in the molasses approach. There is only one source of organic carbon that might be made available to the microbial inoculum: the crude oil itself. The ideal substrate would be a minor component of the crude oil, adequate to support enough growth to allow the generation of the necessary quantities of those microbial products that would effect the mobilization of the bulk of the crude. However, the impossibility of providing molecular oxygen in such extensive areas of the reservoir and for such long periods makes it imperative that microbial growth on the oil fraction take place anaerobically.

The question of whether constituents of crude oil can be metabolized in the absence of molecular oxygen has been investigated by the use of anaerobic enrichment techniques. Many samples from diverse environments likely to contain crude oil-degrading bacteria have been screened for their in vitro ability to degrade Forties crude and other substrates anaerobically. Cultures capable of metabolizing crude oil in the presence of nitrate have been maintained for more than eight months by periodic feeding with more oil. The release of carbon dioxide was dependent on the presence both of crude oil and of nitrate. The addition of fresh medium to a substrate-depleted anoxic culture has shown that carbon dioxide and nitrite formation were both independent of molecular oxygen. A demulsifying agent (the nature of which was not disclosed to us) that had been added to the oil did not support gas production in the absence of oil.

Sediment samples from an oil separator incubated with Forties crude for 25 weeks produced methane and carbon dioxide in amounts significantly higher than in control enrichments. Methanogenic decomposition of crude oil, cyclodecane, naphthenic acids, and hexadecane has been obtained in sediment samples from a waste water lagoon after adaptation periods ranging from 30 days for crude oil to 127 days for hexadecane.

Preliminary attempts to isolate crude oil-utilizing bacteria on an oil-agar medium incubated under anaerobic conditions have yielded mixed populations that do not form axenic colonies when restreaked. Microscopic examination of the anaerobic crude oil enrichments revealed complex mixed cultures in all cases. It is possible that consortia of the type described for methanogenic benzoate degradation are also involved in methanogenic hydrocarbon decomposition, and this may be the case in nitrate-reducing cultures as well. Nevertheless, it is clear that these populations are able to oxidize and utilize at least some of the components of crude oil in the absence of molecular oxygen. Whether, in practical terms, it would be possible to design a microbially enhanced oil recovery system — dependent on its ability to grow anaerobically on crude oil if the population were a consortium of different bacterial types — remains to be seen. One particular difficulty with doing so might arise from the need to use temperature- and pressure-tolerant strains in the deeper reservoirs. Each member of the consortium would itself need to withstand the high ambient temperature and pressure. Alternatively, it may ultimately prove possible to characterize both the biochemistry of the anaerobic processes and their genetic determinants so that all of the required reactions might be combined into a single strain, able simultaneously to use oil anaerobically in anoxic cultures and to tolerate high temperature and pressure.

The environmental conditions in the deep reservoirs of the North Sea are unfavorable for microbial growth. Their depth results in temperatures of about 90°C and above and pressures approaching 300 atm. Water flooding and pressure maintenance would, of course, use sea water, and any proposed microbial inoculum would need to function satisfactorily at such a salinity. The immediate objective is to inquire whether this combination of factors might not exclude bacterial systems altogether. While some of the bacterial cultures able to grow anaerobically on crude oil will do so in the presence of sea water, the main attention has been directed to the effects of high temperature on bacterial growth.

The usually accepted limit for growth under laboratory conditions is 86°C, although growth can occur up to about 100°C in the boiling water of natural hot springs, and there are reports that this has been achieved in one laboratory.[1] The possibility of raising the upper temperature limit observed for a freshly isolated bacterial strain is interesting. Presumably growth becomes impossible at some temperature because of the thermal instability on one or more cellular components. If these were proteins, the implication would be that the rate of synthesis could not be raised high enough to compensate for the rate of thermal denaturation. Such a hurdle might be overcome in two ways: (*i*) by allowing cells to adjust by a series of adaptive changes to an increased level of derepression of the synthesis of the limiting component, or (*ii*) by mutational change to increase its thermal stability. Both are worth exploring. If the temperature limitation imposed by the most temperature-sensitive factor could be increased by a few degrees, the maximum growth temperature of the organism would presumably increase until the next thermostability obstacle was encountered. It might prove possible to circumvent this by repeating the adaptive or mutational change and so progress upward in a series of steps until too many individual limitations, all acting together, prevented any further rise in growth temperature.

This might be regarded as being comparable to the progressive adaptations exhibited by bacteria grown in the presence of trifluoroleucine or increasing concentrations of deuterium oxide.[2,3] Gradually increasing the concentrations of either of these materials enables the cells to grow at levels that were totally inhibitory on first contact. The adaptations were, however, unstable; bacteria adapted to grow at almost 100% D_2O totally lost their ability to do so after no more than two generations of growth again in H_2O, even though one-quarter of their nonexchangeable hydrogen atoms would have been deuterated.

One of the many problems of developing microbiological populations functional in situ in the deep reservoirs is the likely interactive effect of high pressure, temperature, and salinity. Little published data extend into the region of interest for enhanced oil recovery. The genetic control of tolerance to these external factors, while of great practical significance, is virtually unexplored; but there is some evidence to suggest that tolerance may be controlled genetically in a manner simpler than might at first seem to be the case.[4,5] This raises again the exciting prospect of using gene manipulation to combine into a single bacterial strain the properties necessary to allow anaerobic growth on crude oil and those that permit growth to occur in the downhole conditions of the deep reservoirs.

ACKNOWLEDGMENTS

Part of this work has been carried out under contract to the U.K. Department of Energy.

REFERENCES

1 HEINEN, W., and A.M. LAUWERS. 1981. Growth of bacteria at 100°C and beyond. Archives of Biochemistry *129*:127–128.

2 MANN, L.R.M., and V. MOSES. 1971. Properties of *Escherichia coli* grown in deuterated media. Folia Microbiologica *16*:267–284.

3 ORGEL, L.E. 1964. Adaptation to wide-spread disturbance of enzyme function. Journal of Molecular Biology *9*:208–212.

4 FRIEDMAN, S.M., and T. MOJICA-A, 1977. Transformants of *Bacillus subtilis* capable of growth at elevated temperatures, p. 117–126. *In* S.M. Friedman (ed.), Biochemistry of thermophily. Conference papers, published 1978. Academic Press: New York and London.

5 LINDSAY, J.A., and E.H. CREASER 1975. Enzyme thermostability is a transformable property between *Bacillus* spp. Nature *255*:650–652.

21 CELL-SURFACE HYDROPHOBICITY AND ADHERENCE TO SURFACES WITH SPECIAL REFERENCE TO ORAL BACTERIA

E. Weiss, M. Rosenberg, H. Judes, and E. Rosenberg

Tel Aviv University

Cell surface hydrophobicity is an important factor in the adherence and subsequent proliferation of microorganisms on solid surfaces and at interfaces.[1] Hydrophobic interactions play a role in the adherence of bacteria to nonwettable plastics,[2] phagocytes and buccal epithelial cells,[3,4] and in the partitioning of cells at liquid-hydrocarbon and liquid-air interfaces.[5-12] Adherence of petroleum-degrading bacteria to their hydrophobic substrate is a prerequisite for growth under certain conditions.[13]

While the hydrophobic properties of many laboratory strains have been tested using a variety of qualitative and quantitative approaches,[14-18] few studies have undertaken the screening of fresh microbial isolates for cell surface hydrophobicity.[19,20,21] The adherent properties of bacterial isolates may change with prolonged laboratory cultivation.[22] To our knowledge, there have been no previous reports on the cell-surface hydrophobicity of bacteria adhering to the tooth surface. A quantitative technique for measuring bacterial hydrophobicity based on adherence to liquid hydrocarbons was recently described.[23,24] The simplicity of this method lends itself to the rapid screening of numerous microbial samples. In the present study the hydrophobic surface characteristics of 155 fresh microbial isolates obtained from the surfaces of extracted teeth and incubated matrices were examined. Over 70% of these isolates exhibited pronounced cell surface hydrophobicity, as determined by adherence to hexadecane.

MATERIALS AND METHODS

Isolation of microorganisms. Adherent microorganisms were isolated from four freshly extracted teeth, obtained from four patients, and from two Tofflemire stainless steel matrix bands cut to a size of 0.5 cm × 0.5 cm × 0.038 mm (Teledyne, Elk Grove Village, Illinois). The latter were sterilized and subsequently incubated within an acrylic bite plate in the mouth of a healthy, low-caries (low tooth decay) volunteer for 2 hr.[25]

Following their removal, teeth and matrices were washed extensively with 20 ml saline in sterile petri dishes, were transferred to test tubes containing 5 ml saline, and were vortexed for 2 min to remove loosely bound cells. Teeth and matrices were then placed in wells of sterile Costar dishes (Costar, Cambridge, Mass.) containing 5 ml saline and were exposed to sonic oscillation at 0°C for 15 sec at 20 watts power in a Braun Labtronic 1510 sonifier (standard probe). Appropriate dilutions were spread onto brain-heart infusion agar (Difco) and were incubated under aerobic conditions at 37°C. Colonies were chosen randomly from these plates for further study.

Incubation conditions. Following two passages on brain-heart infusion agar, isolated colonies were inoculated into test tubes containing 5 ml brain-heart infusion broth (Difco) and were grown for 48 hr at 37°C with slow reciprocal shaking. Cells were harvested, washed twice, and suspended in PUM buffer (22.2 g of $K_2HPO_4 \cdot 3H_20$, 7.26 g of KH_2PO_4, 1.8 g of urea, 0.2 g of $MgSO_4 \cdot 7H_20$, and distilled water to 1,000 ml; pH 7.1) to optical densities of 1.0–1.5 at 400 nm (Gilford model 240 spectrophotometer).

Adherence to hexadecane. The general method for measuring bacterial adherence to liquid hydrocarbons has been described previously.[26,27] To 1.2 ml of bacterial suspension, various volumes of hexadecane were added. After preincubation at 30°C, the two phases were vortexed for 120 sec on a Thermolyne (Sybron) mixer. After the phases were allowed to separate, the aqueous phase was carefully removed and its light absorbance at 400 nm was measured. The proportion of adherent cells was determined by comparing the drop in absorbance following the mixing procedure with the absorbance of the suspension prior to the assay.

RESULTS

The four teeth used in the experiments are described in Table 1.

The number of adherent microorganisms released from the teeth and stainless steel matrices by sonic oscillation, as determined by colony-forming units on brain-heart infusion agar, is shown in Table 2. The two carious teeth, A and B, yielded approximately 70 times more cells than the noncarious teeth, C and D. The number of microorganisms isolated from each of the matrices was of the same order of magnitude as the number of cells released from noncarious teeth C and D. Different colonies appearing on the plates were selected at random, cloned, and subsequently examined for their ability to adhere to hexadecane.

To illustrate the technique, Figure 1 shows adherence of two of the strains as a function of increasing hexadecane volume. While adherent strain M2-14 (isolated from matrix 2) was increasingly removed from the aqueous phase by greater volumes of hexadecane, the turbidity of the suspension of cells of D-26 (from tooth D) was not significantly affected. Following the mixing procedure, adherent cells rose with the hydrocarbon droplets to form a stable upper cream, consisting of cell-coated droplets. In the case of nonadherent strains, the hydrocarbon droplets coalesced following the mixing procedure and little or no decrease in turbidity of the aqueous phase was observed. For the results presented in Table 2, cell suspensions were mixed in the presence of 0.2 ml hexadecane. Strains were scored adherent positive if more than 75% of the cells were removed to the aqueous-hydrocarbon interface following the assay procedure. With few exceptions the isolates examined were either totally adherent (over 90%) or not at all adherent (less than 10%).

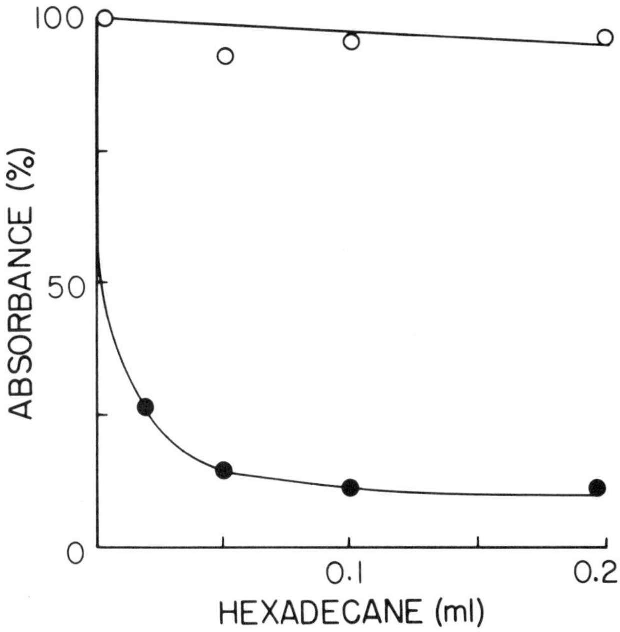

Fig. 1. Strains M2-14 (●) and D-26 (○) were compared for adherence to hexadecane. Initial optical absorbance: M2-14, 1.35; D-26, 1.23

Table 1. Description of extracted teeth used for isolating adherent bacteria.

Tooth[a]	*Condition*	*Reason for Extraction*
A	Untreated large caries	Caries
B	Untreated large caries	Caries
C	Intact	Pericorinitis
D	Root canal treatment and composite filling	Periodontitis

[a]Each of the four teeth was a third molar extracted from an adult male.

Table 2. Screening of oral bacteria for cell surface hydrophobicity.

Source of Microorganisms	*Total Colony-Forming Units*	*Isolates Examined*	*Isolates Adherent to Hexadecane*[a]
Tooth A	1.8×10^6	32	32
Tooth B	2.3×10^6	23	18
Tooth C	2.6×10^4	31	31
Tooth D	3.3×10^4	17	1
Matrix 1	1.2×10^4	29	13
Matrix 2	2.2×10^4	23	17

[a]Isolates were designated as adherent if more than 75% of the cells were removed following mixing in the presence of 0.2 ml hexadecane.

In teeth A and C, all of the isolates examined were hydrophobic. In tooth B, 78% of the isolates adhered to hexadecane, whereas only one of the 17 isolates from tooth D was adherent. A high frequency of hydrophobic bacteria was also isolated from the stainless-steel matrices that were incubated for only 2 hr in a bite plate in the mouth of a healthy volunteer (Table 2).

Although the hydrophobic bacteria isolated in the present report have not yet been classified, certain generalizations can be made: (*i*) all of the isolates obtained from tooth A were streptococci with similar cell and colony morphology;

(*ii*) the single isolate from tooth D that adhered to hexadecane was a gram-positive bacillus; (*iii*) hydrophobic isolates from teeth B and C and the two stainless-steel matrices comprised a wide variety of morphological types, including filamentous forms, coryneforms, bacilli, streptococcal and other coccal forms; and (*iv*) all the streptococci examined adhered to hexadecane.

DISCUSSION

Various studies have demonstrated that the tooth surface is normally covered by a hydrophobic film of organic matter termed "acquired pellicle."[28] While it has been postulated that the low surface energy imparted by this organic layer may act to impede bacterial attachment, it has also been suggested that hydrophobic interactions may actually participate in adherence of certain oral species to the tooth surface.[29,30,31]

A surprisingly high overall proportion of bacteria tested (72%) exhibited pronounced ability to adhere to the test hydrocarbon.

Bacterial isolates studied in this report probably represent a small fraction of the bacteria that adhere to and colonize the tooth surface, both because of the removal procedure (mild sonic oscillation) and cultivation conditions (aerobic, rich medium). Nevertheless, the diversity of adherent types found demonstrates for the first time that various species of hydrophobic bacteria can be isolated directly from the tooth surface. In addition, data obtained using the stainless-steel matrices demonstrate that hydrophobic microorganisms may become attached to surfaces following a brief incubation period in the oral cavity.

The demonstration of hydrophobic bacteria at the saliva-tooth interface raises a number of interesting questions that are now amenable to experimental examination: (*i*) Do hydrophobic microorganisms, similar to those isolated here, play a role in aiding or abetting formation of plaque, dental caries, and periodontal disease? (*ii*) Are similar mechanisms involved in adherence of these oral species to the tooth surface and to hexadecane? (*iii*) What intraoral factors influence the affinity of hydrophobic bacteria for the tooth-saliva interface? Further investigations into the possible role of hydrophobic interactions in mediating adherence of bacteria to the tooth surface are underway.

SUMMARY

Adherent bacteria were released from the surfaces of four freshly extracted teeth by mild sonic oscillation and were screened for cell surface hydrophobicity on the basis of their ability to adhere to hexadecane. Of the 103 tooth isolates examined, 82 adhered to the test hydrocarbon. Hydrophobic bacteria could similarly be isolated from stainless-steel dental matrix bands following brief incubation in the mouth of a volunteer; 30 to 52 isolates examined adhered to hexadecane. Among those strains that adhered to hexadecane, streptococci were the most frequent type isolated. Various other morphological types were also observed, including bacilli, coryneforms, and filamentous bacteria. The high overall proportion of hydrophobic bacteria found in this study (72%) suggests that cell surface hydrophobicity may play a role in adherence of certain oral species to the tooth surface.

ACKNOWLEDGMENTS

We would like to thank Yardena Mazor for excellent technical assistance and Yair Aharonowitz for critical reading of the manuscript.

REFERENCES

1 MARSHALL, K.C. 1976. Interfaces in microbial ecology. Harvard University Press, Cambridge, Mass.

2 ROSENBERG, M. 1981. Bacterial adherence to polystyrene: a replica method of screening for bacterial hydrophobicity. Applied and Environmental Microbiology *42*:375–377.

3 VAN OSS, C.J. 1978. Phygocytosis as a surface phenomenon. Annual Review of Microbiology *32*:19–39.

4 ROSENBERG, M., A. PERRY, E.A. BAYER, D.L. GUTNICK, E. ROSENBERG, and I. OFEK. 1981. Adherence of *Acinetobacter calcoaceticus* RAG-1 to human epithelial cells and to hexadecane. Infection and Immunity *33*:29–33.

5 MARSHALL, op cit.

6 MUDD, S., and E.B.H. MUDD. 1924. The penetration of bacteria through capillary spaces. IV. A kinetic mechanism in interfaces. Journal of Experimental Medicine *40*:633–645.

7 ROSENBERG, M., D. GUTNICK, and E. ROSENBERG. 1980. Adherence of bacteria to hydrocarbons: a simple method for measuring cell-surface hydrophobicity. FEMS Microbiology Letters *9*:29–33.

8 ROSENBERG, M., E. ROSENBERG, and D. GUTNICK. 1980. Bacterial adherence to hydrocarbons, p. 541–542. *In* R.C.W. Berkeley, J.M. Lynch, J. Melling, P.R. Rutter, and B. Vincent (ed.), Microbial adhesion to surfaces. Ellis Horwood, Chichester, England.

9 ROSENBERG, op cit.

10 ROSENBERG, M., and E. ROSENBERG. 1981. Role of adherence in growth of *Acinetobacter calcoaceticus* RAG-1 on hexadecane. Journal of Bacteriology (in press).

11 ROSENBERG, PERRY, BAYER, GUTNICK, ROSENBERG, and OFEK, op cit.

12 DAHLBÄCK, B., M. HERMANSSON, S. KJELLEBERG, and B. NORKRANS. 1981. The hydrophobicity of bacteria–an important factor in their initial adhesion at the air-water interface. Archives of Microbiology *128*:267–270.

13 ROSENBERG and ROSENBERG, op cit.
14 KJELLEBERG, S., C. LAGERCRANTZ, and T. LARSSON. 1980. Quantitative analysis of bacterial hydrophobicity studied by the binding of dodecanoic acid. FEMS Microbiology Letters *7*:41–44.
15 MAGNUSSON, K.E., and G. JOHANSSON. 1977. Probing the surface of *Salmonella typhimurium* SR and R bacteria by aqueous biphasic partitioning in systems containing hydrophobic and charged polymers. FEMS Microbiology Letters *2*:225–228.
16 MUDD and MUDD, op cit.
17 SMYTH, C.J., P. JONSSON, E. OLSSON, O. SÖDERLIND, J. ROSENGREN, S. HJERTEN, and T. WADSTRÖM. 1978. Differences in hydrophobic surface characteristics of porcine enteropathogenic *Escherichia coli* with or without K88 antigen as revealed by hydrophobic interaction chromatography. Infection and Immunity *22*:462–472.
18 VAN OSS, op cit.
19 DAHLBÄCK, HERMANSSON, KJELLEBERG, and NORKRANS, op cit.
20 FARIS, A., T. WADSTRÖM, and J.H. FREER. 1981. Hydrophobic adsorptive and hemagglutinating properties of *Escherichia coli* possessing colonization factor antigens (CFA/I or CFA/II), type 1 pili, or other pili. Current Microbiology *5*:67–72.
21 SMYTH, JONSSON, OLSSON, SÖDERLIND, ROSENGREN, HJERTEN, and WADSTRÖM, op cit.
22 GIBBONS, R.J. 1980. Adhesion of bacteria to the surfaces of the mouth, p. 351–388. *In* R.C.W. Berkeley, J.M. Lynch, J. Melling, P. Rutter, and B. Vincent (ed.), Microbial adhesion to surfaces. Ellis Horwood, Chichester, England.
23 ROSENBERG, GUTNICK, and ROSENBERG, op cit.
24 ROSENBERG, ROSENBERG, and GUTNICK, op cit.
25 GLANTZ, P.O. 1977. Adhesion to teeth. International Dental Journal *27*:324–332.
26 ROSENBERG, GUTNICK, and ROSENBERG, op cit.
27 ROSENBERG, ROSENBERG, and GUTNICK, op cit.
28 GLANTZ, P.O. 1969. On wettability and adhesiveness. Odontologisk Revy, *20*, Supplement *17*:1–132.
29 GLANTZ, op cit.
30 GIBBONS, op cit.
31 RUTTER, P. 1979. Accumulation of organisms on the teeth, p. 139–164. *In* D.C. Ellwood, J. Melling, and P. Rutter (ed.), Adhesion of microorganisms to surfaces. Academic Press, London.

22 MICROBIAL DEGRADATION OF ATHABASCA BITUMEN

P.I. Forrester, D. Pearson, and J. Roth

Gemini Biochemical Research Ltd., Calgary

The widespread decreased production of conventional oil has forced oil companies to seek other sources of petroleum. One of these, tar sand, is a mixture of approximately 85% sand and 15% bitumen with low percentages of water and clay minerals. Bitumen is an extremely viscous oil with a specific gravity of approximately 1. The origin of the tar sands is in dispute, but current evidence supports the view that these deposits represent the end product obtained from microbial degradation of a conventional crude oil.[1]

Figure 1 shows the tar sand deposits of Northern Alberta. There are estimated to be from 800 billion to 1.2 trillion barrels of oil locked up in these deposits.[2,3] Roughly 10% of this total is recoverable by surface mining technology. The remaining 90% is buried so deeply that some sort of in situ operation will be required to extract the oil. At present, there are two commercial-scale plants in operation, producing 60,000–120,000 bbl per day. Both of these plants use surface mining technology followed by hot-water treatment to separate the bitumen from the sand. Two other plants are in the late planning stage, one to use surface mining technology and the other to use an in situ method.

One of the main environmental concerns with both the hot-water process and in situ steam stimulation is that large quantities of bitumen-contaminated water are produced (roughly 2.5 bbl of water per barrel of oil). This water also contains clay minerals that will not settle, thus causing the water to be impounded for extremely long times before it can be recycled or returned to the environment. These large volumes of bitumen-contaminated water pose a potential threat to the environment. Microbial activity may have a role to play in purifying this water. Gemini Biochemical Research Ltd. has two projects in progress to study this activity. One project studies the use of microorganisms to clean up water produced by in situ plants.[4] The other, to be presented here, studies the microbial activity in the tailings ponds from surface mining plants.

The objectives of this project are:

1. To monitor the levels of dispersed bitumen in the ponds.
2. To fractionate the bitumen and monitor changes, if any, in these fractions with time.
3. To study the microbial populations of the ponds.
4. To select for organisms which can degrade the bitumen.
5. To study conditions required to maximize the degradation.

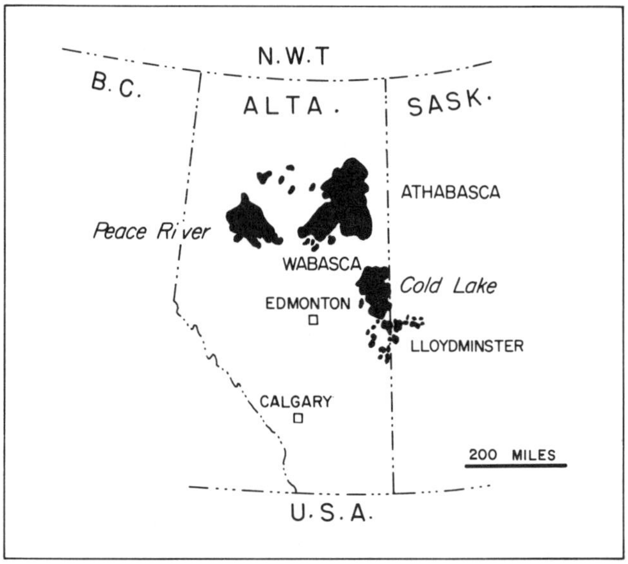

Fig. 1. Map of Alberta showing the location and extent of the oil sands deposits

MATERIALS AND METHODS

Materials. All ingredients for microbiological media were obtained from Difco Laboratories, St. Louis. Solvents for chromatography and fractionation were HPLC grade and were obtained from Fisher Scientific and CanLab. All other reagents were the best grade available and were obtained from local suppliers.

Sampling techniques. Samples of approximately 1 l were obtained by submerging polyethylene bottles into the top 20–30 cm of the pond. In all samples attempts were made to avoid including floating bitumen in the bottles so that analysis would only include the bitumen suspended in the pond and not the floating bitumen. It was noticed that at the time of the second sampling the amounts of floating bitumen were greatly reduced due to the installation of the floating booms at the inlet points.

Analytical techniques. Tailings pond samples (200 ml) were extracted twice with equal volumes of benzene either by vortexing or shaking in a separatory funnel. The extracts were combined and the benzene was removed on a rotary evaporator. To quantify the total amount of bitumen present, the residue was transferred to tared glass fiber filter papers with a minimal volume of benzene, usually 10 ml. After evaporating the benzene in a fume hood, the amount of bitumen was determined gravimetrically. When the bitumen was to be fractionated, the following procedure was utilized. Twenty milliliters of n-pentane were added to the residue obtained by evaporation of the benzene. After sitting for 20 minutes, the suspension was filtered through a medium-porosity fritted glass Buchner funnel. The insoluble material was rinsed with an additional 10 ml of n-pentane. The filtrate was termed the n-C_5 soluble extract and the material collected on the filter was termed the asphaltene fraction. The pentane was removed by rotary evaporation and n-C_5 soluble extract was transferred to a tared glass vial with 5 ml of n-pentane. After evaporation of the pentane, the n-C_5 soluble extract was weighed.

The asphaltene fraction was dissolved from the filter with 20 ml of benzene, the benzene was removed by rotary evaporation, and the asphaltene fraction was transferred to a tared vial with 5 ml of benzene. After evaporation of the benzene, the asphaltenes were weighed.

The n-C_5 soluble fraction was fractionated on a silica-alumina column (0.5 cm × 10 cm; top half alumina, bottom half silica). The nonpolar fraction was eluted with 5 ml of n-pentane, the aromatic fraction with 5 ml of methylene chloride:n-pentane (3:1 v/v), and the polar fraction with 5 ml of methylene chloride:methanol (3:2 v/v). The fractions were collected in tared vials and were quantitated gravimetrically after evaporation of solvent.

Microbiological techniques. (1) Pure strains of microorganisms were isolated from tailings pond samples by serial dilutions and plate spreading techniques.

(2) Epifluorescent microscopy was performed by a modification of the technique described by Geesey & Costerton (1979).[5] One milliliter of sample was filtered through a millipore filter (0.45 μ, 2.5-cm diameter) under a slight vacuum. Two milliliters of acridine orange solution (50 mg/1,000 ml in 0.1 M potassium phosphate buffer, pH 7.5, prefiltered before use) was added to the filter for 2 minutes. The excess dye was filtered off and the filter was rinsed with 5 ml filtered sterile distilled water. The filters were air dried in covered petri dishes then transferred and fixed to glass microscope slides with nonfluorescing immersion oil. The slides were examined using a Zeiss microscope with fluorescent attachments using an HBO 50-W mercury arc burner and P 520 filter set.

(3) Microbial degradation of bitumen was monitored by growing pure cultures in Y.M. broth in which the glucose was replaced with 0.5 ml n-dodecane. After stationary phase was reached, usually in 2-3 days, 50 mg of bitumen was added. This was considered to be day zero. The bitumen was extracted at varying times and was characterized as described above.

RESULTS AND DISCUSSION

Bitumen content. The bitumen content of the pond at the various sampling times is shown in Table 1. A steady decline in bitumen content was noted during the sampling period.

Table 1. Bitumen content of the tailings pond.

Surface Samples					
Sampling Time	*Bitumen (ppm)*	*SATS (%)*	*AROM (%)*	*Polar (%)*	*ASPH (%)*
July 1979	399	17.2	28.2	22.1	32.4
May 1980	267	22.8	39.1	22.2	16.0
August 1980	183	32.2	36.7	14.1	17.0
October 1980	191	21.1	42.5	19.1	17.4
April 1981	157	21.6	52.0	26.4	15.2

Subsurface Samples		
Depth	*October 1980*	*April 1981*
1 m	205	107
3 m	196	125
5 m	194	139
7 m	298	129
10 m	391	120
13 m	1,065	2,356
15 m	205	NS
18 m	15,626	NS

NS–no sample

This is due to two reasons: (*i*) when the plant was first operational (1977) the efficiency of extraction of the bitumen probably was much lower than is now experienced and thus more bitumen was discharged into the pond; (*ii*) booms were installed across the discharge areas to contain large mats of floating bitumen for recovery.

The bitumen content of subsurface samples was relatively constant down to a depth of approximately 13 m. Large, submerged, floating mats of bitumen were encountered at this depth. In the October 1980 samples the bottom of the pond was found at approximately 18 m. In the April 1981 samples the bottom was found at 13 m.

Sediment analysis. The results of sediment analysis on both surface and subsurface samples are shown in Table 2. The concentration of suspended solids is fairly constant to a depth of 7–10 m, at which point there is a dramatic change in the amount of material $>3\ \mu$ and $>11\ \mu$ in size. There is no evidence of large-scale mixing of the deeper material with the upper levels as occurs in the spring with most natural bodies of water.

Table 2. Sediment analysis.

	Sampling Time					
	October 80		*April 81*		*July 81*	
Depth	$>3\mu$	$>11\mu$	$>3\mu$	$>11\mu$	$>3\mu$	$>11\mu$
Surface	0.37	0.060	0.45	0.18	0.18	0.12
1 m	0.28	0.067	0.46	0.19	0.16	0.15
3 m	0.34	0.100	0.46	0.20	0.18	0.13
5 m	0.40	0.077	0.48	0.18	0.17	0.12
7 m	0.91	0.138	0.52	0.25	0.18	0.18
10 m	ND	8.81	0.56	0.16	8.40	8.44
13 m	ND	14.64	6.44	6.23	19.17	19.44
15 m	ND	16.77	NS	NS	NS	NS
18 m	ND	23.96	NS	NS	NS	NS

All values expressed as percent
ND–not determined
NS–no sample

Microbial populations. Two methods were used to measure microbial populations: viable aerobic microorganisms were estimated by growth on plate count agar and total microbial populations by epifluorescent microscopy. The preliminary data obtained by these methods are illustrated in Table 3. It appears as if the number of viable organisms increases over the summer months and then decreases as organisms die off during the winter months. The data from the epifluorescent microscopy show a reverse trend. No explanation can be given at present for these seemingly contradictory results.

The sudden drop-off in viable organisms observed between 7 and 10 m (October 1980) is believed to be due to a sharp decrease in oxygen tension observed at this depth

Table 3. Microbial populations of the tailings pond.

Surface Samples

Sampling Date	*Plate Counts*	*Epifluorescence*
May 1980	ND	2.28×10^8
August 1980	8.80×10^6	3.52×10^7
October 1980	2.36×10^7	2.55×10^7
April 1981	3.28×10^6	3.62×10^8
July 1981	4.31×10^6	7.80×10^7

Subsurface Samples

	October 1980		*April 1981*		*July 1981*	
Depth	*Plate Counts*	*Epifluor*	*Plate Counts*	*Epifluor*	*Plate Counts*	*Epifluor*
1 m	6.40×10^7	2.13×10^7	1.98×10^6	4.00×10^8	3.73×10^6	10.70×10^7
3 m	0.46	2.08	2.29	3.61	4.1	13.70
5 m	6.65	1.41	0.41	3.27	5.83	8.15
7 m	3.17	1.01	0.56	0.90	2.47	8.38
10 m	0.15	4.16	2.11	1.78	26.3	5.47
13 m	0.02	1.74	1.19	0.80	0.29	7.42
15 m	0.07	2.08	NS	NS	NS	NS
18 m	0.01	2.26	NS	NS	NS	NS

ND–not done
NS–no sample

(M. McKinnon, Syncrude Canada, personal communication). The oxygen discontinuity occurred at a greater depth in the other sets of samples. Thus, the test levels of organisms were more constant with depth and changes in numbers were more difficult to associate with oxygen discontinuity.

Microbial degradation of bitumen. As a preliminary experiment to determine if the bitumen in the pond was susceptible to microbial degradation by organisms present in the pond, samples of tailings pond water were incubated aerobically in the presence of added mineral salts. After two weeks of incubation, the bitumen was extracted. Results of this experiment are shown in Table 4.

Table 4. Incubation of tailings pond samples supplemented with mineral salts.

Bitumen Concentration (ppm)		*Percent Degradation*
Before Incubation[a]	*After Incubation*[a]	
320	186	42
308	80	74
332	150	55

[a]Samples incubated for 2 weeks

These preliminary results suggest that microbes degrading bitumen are present in the tailings pond samples. Thus, two methods were tried to obtain cultures of bitumen degraders.

First, enrichment cultures were obtained on a number of hydrocarbons and bitumen components. The ability of these cultures to degrade bitumen is shown in Table 5. As can be seen from this table, only enrichment cultures obtained on dodecane and naphthalene could degrade bitumen. Figure 2 illustrates the organisms present in some of the enrichment cultures.

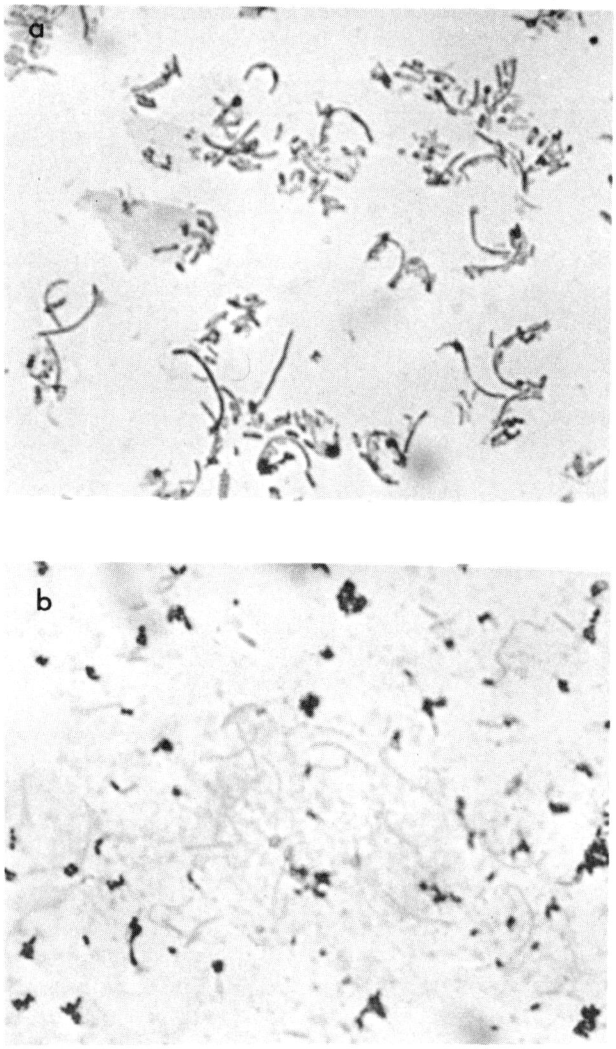

Fig. 2. Photomicrographs of (*a*) naphthalene enrichment culture and (*b*) bezothiophene enrichment culture

Table 5. Enrichment cultures.

Enriched on	[a]*Concentration (%)*	[b]*Bitumen Degradation (%)*
Dodecane	0.1	2.8
	0.2	8.9
	0.5	37.6
	1.0	47.4
Naphthalene	0.1	1.5
	0.2	3.8
	0.5	3.7
	1.0	17.1
Phenol	0.1	0
Benzoic acid	0.01	0
m-cresol	0.01	0
Benzothiophene	0.1	0
Dibenzothiophene	0.1	0

[a]percent by weight added to minimal salts medium
[b]percent degraded estimated by extraction with methylene chloride

A second method to obtain bitumen degraders was to obtain single-colony isolates and test for their ability to degrade bitumen. Approximately 40 single-colony isolates were obtained. The bitumen-degrading abilities of the best ones are shown in Table 6 in which all cultures degraded to greater than 42% bitumen. If a comparison is made between the degraded and undegraded bitumen, it can be seen that all fractions of the bitumen can be degraded and that some strains preferentially degrade certain fractions. For example, strains S103-79, S119-79, and S120-79 preferentially degrade the saturate fraction. Strains S101-79 and S120-79 degrade asphaltenes. These experiments were conducted in

Table 6. Degradation of bitumen by pure strains.

Strain #	*Degradation[a] (%)*	*Fractionation of residual bitumen*			
		P/N (%)	*AROM (%)*	*Polar (%)*	*ASP (%)*
S101-79	51	18.0	54.7	13.9	8.2
S102-79	57	10.7	50.3	10.7	13.2
S103-79	67	6.6	28.3	13.3	52.4
S104-79	73	16.4	31.3	14.9	38.8
S109-79	48	19.7	35.8	16.1	33.1
S113-79	42	9.1	39.4	20.3	22.2
S116-79	67	22.6	48.2	18.3	21.3
S119-79	61	5.9	31.7	21.5	12.7
S120-79	64	6.0	60.1	20.8	4.9
Undegraded bitumen		16.3	43.0	14.3	26.4

[a]After two weeks incubation in rich medium

rich medium that contained dodecane, malt extract, yeast extract, and peptone in mineral salts medium. Table 7 shows the constituents required for bitumen degradation. The best degradation is obtained when all ingredients are present, especially the complex organic components. It should be noted that all strains lost their ability to degrade bitumen on continued transfer.

SUMMARY

The increased exploitation of Alberta's tar sands has made the study of the microbiology of bitumen a worthwhile pursuit. Large amounts of bitumen-contaminated water are produced by surface mining operations and by in situ production plants. Over the past two years over 40 strains of microorganisms have been isolated from the tailings ponds at the surface mining operations. These strains were partially characterized and tested for their ability to degrade bitumen. Bitumen degraders were further studied for growth requirements and the nature of the changes in bitumen components during degradation. In addition, enrichment cultures on a number of hydrocarbons and nonhydrocarbons (dodecane, naphthalene, phenol, benzoic acid, m-cresol, and bitumen) were produced. Finally, preliminary data on the fluctuations of the microbial populations and chemical components of one of the tailings pond were presented. In summary:

(*i*) High levels of viable microorganisms are present in the tailings pond from tar sands operations.

(*ii*) Under the appropriate conditions bitumen can be biodegraded.

(*iii*) Certain strains of organisms preferentially attack particular fractions of the bitumen.

(*iv*) Ability to biodegrade bitumen is readily lost on continued transfer of microbes on rich media.

Table 7. Effect of media composition on bitumen degradation.

Medium	*Bitumen Degradation (%)[a]*	
	Gravimetric	*Colorometric*
dH_2O	3.1	7.3
MM	6.7	6.9
MM + do	5.0	8.9
MM + ME	0.7	2.8
MM + Pep	13.0	15.5
MM + YE	12.0	17.5
MM + do + ME	14.0	21.9
MM + do + Pep	24.3	31.6
MM + do + YE	22.5	25.4
MM + ME + Pep	19.4	21.5
MM + ME + YE	20.5	29.4
MM + Pep + YE	42.7	48.0
MM + do + ME + Pep	36.8	44.0
MM + do + ME + YE	12.3	23.9
MM + do + Pep + YE	45.8	54.0
MM + ME + Pep + YE	43.7	45.2
MM + do + ME + Pep + YE	56.5	49.7

dH_2O–distilled water; MM–minerals salts medium; do–dodecane (1mg/ml); ME–malt extract (0.3%); Pep–peptone (0.5%); YE–yeast extract; [a]Bitumen extracted after 1 week incubation

REFERENCES

1 BAILEY, N.J.L., A.M. JOBSON, and M.A. ROGERS. 1973. Bacterial degradation of crude oil: Comparison of field and experimental data. Chem. Geol. *11*:203–22.

2 KENDALL, G.H. 1977. Importance of reservoir description in evaluating in situ recovery methods for Cold Lake heavy oil. Part I. Reservoir description. Bull. Can. Pet. Geol. *25*:314–327.

3 LIM, Y.C., and J.R. THOMAS. 1980. Oil sands technology: conventional oil from unconventional Canadian oil deposits. Proceedings, 30th Can. Chem. Eng. Conference *11*:1060–1091.

4 FORRESTER, P.I., K. MAHADEV, and J. ROTH. 1981. Upgrading brackish water: A study on the use of microorganisms to clean up produced water. Preliminary report submitted to AOSTRA.

5 GEESEY, G.G., and J.W. COSTERTON. 1979. Microbiology of a northern river: Bacterial distribution and relationship to suspended sediment and organic carbon. Can. J. Microbiol. *25*:1058–1062.

23 COMPARATIVE STUDIES ON MICROBIAL DEGRADATION OF AROMATICS AND SATURATES IN CRUDE OIL

P.M. Fedorak, J.M. Foght, and D.W.S. Westlake

University of Alberta

The use of microbes in microbial enhanced oil recovery involves knowledge of the many biochemical changes they impart to oil. This study examines the bioconversions and aromatics in oil. A microbial culture's ability to metabolize a chosen substrate can be determined by monitoring (*i*) growth on the substrate, (*ii*) the loss of the substrate from the culture medium, or (*iii*) the production of a specific metabolic end product.[1] In studying the degradation of hydrocarbons in crude oil and petroleum products, all three approaches have been used. For example, Davies and Westlake screened 60 fungal isolates for their ability to grow on seven different crude oils and on pure hydrocarbons (n-tetradecane, toluene, and naphthalene).[2] Besides monitoring growth, those authors measured the loss of substrates from the oil by means of gas chromatography (GC). Conventional packed GC columns have been used extensively to detect the loss of n-alkanes and isoprenoids from microbial cultures challenged with petroleum.[3-6] However, as a result of the complex nature of petroleum, analyses using these packed columns have been unable to provide information on the degradation of the other fractions of oil (aromatics, polar compounds, and asphaltenes). Studies with pure aromatic substrates and synthetic mixtures of aromatics have shown that they are susceptible to microbial attack.[7,8,9]

^{14}C-labeled substrates also have been used to demonstrate the biodegradation of hydrocarbons (P.M. Fedorak, J.M. Foght, and D.W.S. Westlake, Water Res., in press).[10,11] This extremely sensitive and specific approach allows measurement of the production of $^{14}CO_2$ as a result of substrate mineralization. Although a wide variety of labeled substrates has been studied (e.g., n-hexadecane, pristane, cyclohexane, toluene, naphthalene, phenanthrene, anthracene, and benzanthracene), the range of compounds that can be used is limited mainly by commercial availability and low specific activity. Thus, a method that allows a more comprehensive analysis of the oil is required to understand better the order and extent of microbial degradation of petroleum hydrocarbons.

Capillary GC methods provide much improved resolution of the compounds in both the saturate and aromatic fractions of oil and allow comparative study of the degradation of these compounds by detecting the loss of various peaks from the "fingerprint" of the oil. Only recently has this technique been used to study the degradation of the water-soluble fraction of gas oil by mixed and pure bacterial cultures and the degradation of the aromatics and saturates in crude oil.[12-15]

The data presented here demonstrate that mixed populations from environmental sources are able to degrade saturate and aromatic compounds, show the relative susceptibilities of selected ^{14}C-labeled saturate and aromatic hydrocarbons to microbial mineralization, elaborate on the order and extent of microbial attack on the saturate and aromatic fractions of crude oil as determined by a capillary GC method, and compare the abilities of pure cultures of fungi and bacteria to degrade these two fractions of oil.

MATERIALS AND METHODS

Sources of inocula. Marine water and sediment samples were obtained from Point Partridge State Park, East Fidalgo Bay near Anacortes, and Port Angeles harbor, all in the state of Washington. These sites are described further by Westlake and Cook.[16] Garden soil was obtained from the University of Alberta greenhouses. For pure culture studies, fungi were isolated by direct plating of serially diluted marine samples onto saline Sabouraud's agar containing streptomycin and penicillin. Bacterial isolates were purified from marine enrichment cultures that had been maintained by monthly transfers with Prudhoe Bay crude oil as the sole carbon source.

Growth conditions. Two-hundred-ml aliquots of sea water were placed in 500-ml Erlenmeyer flasks and were incubated at 8°C with shaking at 250 rpm (3.8-cm eccentricity). Some cultures were supplemented with 1.0% (v/v) filter-sterilized nitrogen/phosphorus solution (N,P: 59 g PO_4^{3-}; 100 g NH_4NO_3; 1,000 ml water; pH 7.3). Ten g of soil were added to flasks containing 200 ml salts solution (0.5 g K_2HPO_4; 1 g NH_4Cl; 2 g Na_2SO_4; 2 g KNO_3; 0.2 g $MgSO_4 \cdot 7H_2O$; trace $FeSO_4 \cdot 7H_2O$; 1,000 ml water) and these were incubated with shaking at 20°C. Pure cultures of marine bacteria were grown in artificial sea water (ASW: 23.4 g NaCl; 0.75 g KCl; 7.0 g $MgSO_4 \cdot 7H_2O$; 1,000 ml water; pH 7.3) at 8°C with shaking. All flasks received 0.1% (v/v) Prudhoe Bay crude oil as the sole carbon source.

The fungal isolates were grown at 20°C on slants of basal marine agar (l g proteose peptone #3; l g yeast extract; 20 g purified agar; 1,000 ml ASW) with 0.1 ml Prudhoe Bay crude oil added to the surface of the medium.

In all cases, sterile controls containing Prudhoe Bay oil were incubated with the test cultures in order to account for nonmicrobial loss of hydrocarbons (weathering).

Mineralization studies. Radiorespirometric determinations of mineralization of Prudhoe Bay oil supplemented with one of four ^{14}C-labeled hydrocarbons and with N,P were carried out as described previously in Westlake and Cook. Replicate 50-ml samples and controls in 125-ml flasks were sacrificed at intervals, and the $^{14}CO_2$ plus volatilized ^{14}C-substrate were collected on a filter paper soaked with a CO_2-trapping agent. Lag times were estimated by comparing test flasks with mercuric chloride-poisoned sample flasks serving as volatility controls. When test flask $^{14}CO_2$ dpm exceeded by 5,000 the ^{14}C-volatile substrate dpm (corresponding to approximately 2% of the total activity added), the time expired was recorded as the lag time for initiation of mineralization of the particular ^{14}C-hydrocarbon (Figure 1). The general scheme of the radiorespirometric experiments is given in Figure 2. All radiochemicals ([1-^{14}C]naphthalene, n-[1-^{14}C]hexadecane, [9-^{14}C]phenanthrene, and [9-^{14}C]anthracene) were obtained from Amersham Corp.

Analysis of residual oil. The basic design of experiments is given in Figure 3, and details are described by Fedorak and Westlake.[17,18] After incubation, cultures were acidified with 2 ml of concentrated HCl and were supplemented with chrysene, an extraction standard. Extraction of the residual oil with methylene chloride was followed by filtration through anhydrous Na_2SO_4. The solvent-oil phase was concentrated in hexanes and was applied to a prepared silica gel column. Elution of the column with increasing proportions of methylene chloride in pentane allowed

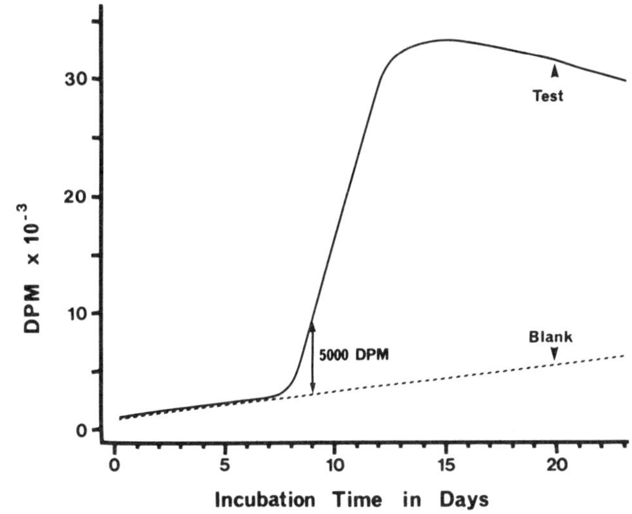

Fig. 1. Criterion used to define lag time for mineralization of ^{14}C-labeled substrates. The arbitrary difference of 5,000 DPM between the test flasks and the control flasks represents approximately 2% of the total substrate activity added to the culture. In this example, the lag time is 9 days

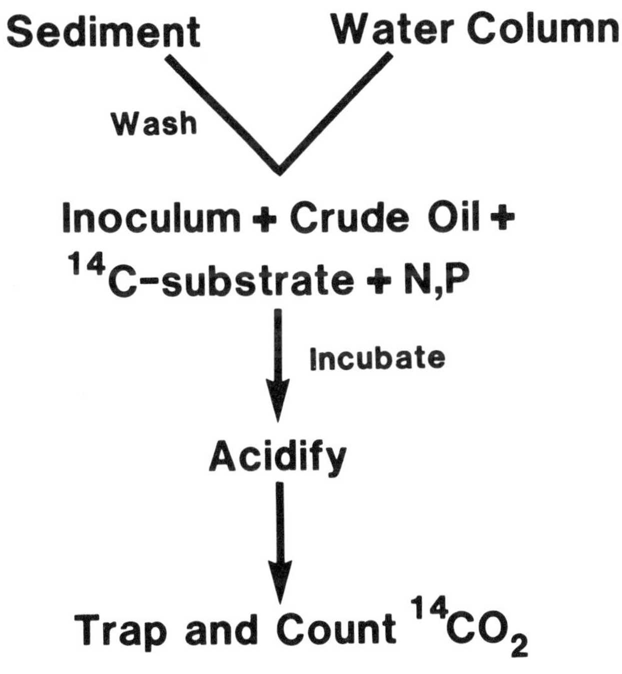

Fig. 2. General experimental scheme for mineralization studies using ^{14}C-labeled hydrocarbons added to Prudhoe Bay crude oil

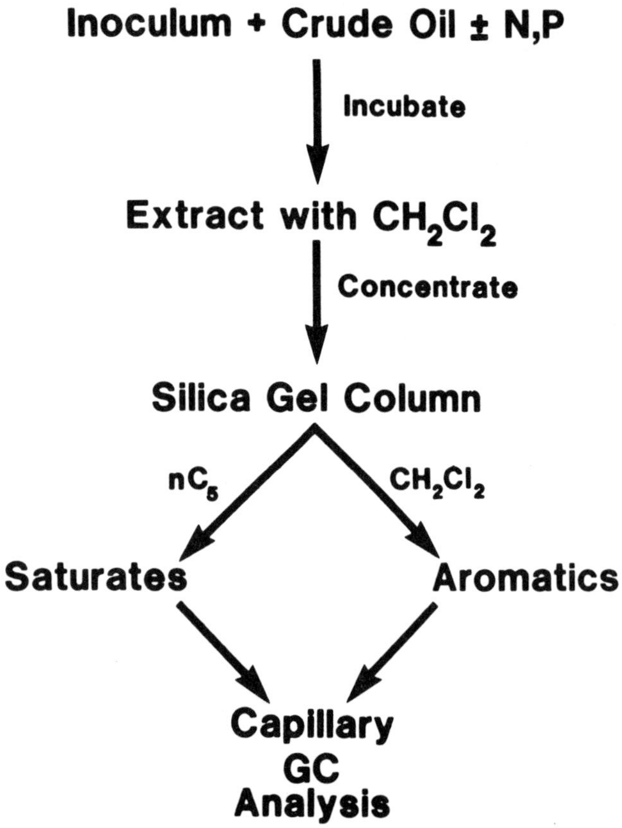

Fig. 3. General experimental scheme for studying the microbial degradation of compounds in the saturate and aromatic fractions of Prudhoe Bay crude oil using capillary GC methods

separation of the saturate and aromatic fractions, which were then concentrated in hexanes. Hexamethylbenzene (hmb) was added prior to capillary GC analysis.

RESULTS AND DISCUSSION

The saturate and aromatic GC profiles of Prudhoe Bay oil prior to incubation are shown in Figure 4 and the identities of the peaks are given in Table 1. Figure 5 shows the microbial degradation of the aromatic and saturate fractions of Prudhoe Bay oil after incubation for 27 days at 8°C with ocean water taken from an unpolluted beach in the state park at Point Partridge. The profiles from the sterile (weathering) controls are shown on the top, those from the unsupplemented culture (NO, N,P) are in the middle, and those from the supplemented culture (+N,P) are at the bottom. With no N,P added, the aromatic profile was greatly altered, but there was little change apparent in the saturate profile when compared with the control. With N,P added there was a marked change in the saturate chromatogram. The n-alkanes were completely removed, leaving only the isoprenoids (pr and ph) and other unidentified peaks. The aromatic profile from the supplemented culture differed little from that of the unsupplemented culture.

Table 1. Identity of selected hydrocarbons in the GC analysis of Prudhoe Bay oil.

Designation	*Compound(s)*
1	Naphthalene
2	2-methylnaphthalene
3	1-methylnaphthalene
hmb	Hexamethylbenzene (GC marker)
x	3-methylbiphenyl (tentative identification)
5	Dibenzothiophene
6	Phenanthrene
7	Anthracene
chr	Chrysene (GC marker)
9	C_3,C_4,C_5, benzenes
10	C_2-naphthalenes
11	C_3-naphthalenes
12	Methylphenanthrenes
13	C_2-phenanthrenes
C_{10}, C_{12}, etc.	Even-numbered n-alkanes
pr	Pristane
ph	Phytane

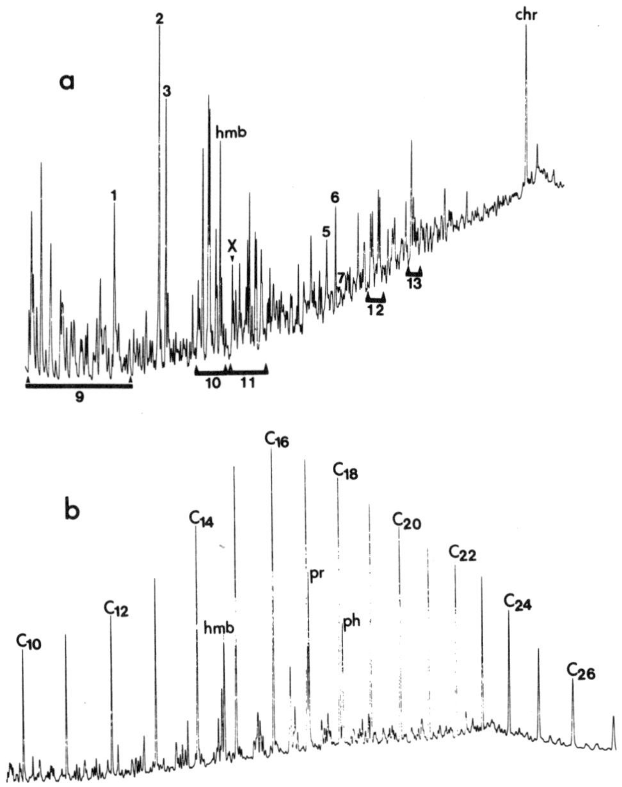

Fig. 4. Glass capillary GC analyses of (*a*) the aromatic fraction and (*b*) the saturate fraction of Prudhoe Bay oil prior to incubation. See Table 1 for identification of compounds

Figure 6 shows the results when water from a commercial harbor at Port Angeles was used as an inoculum. Organic contamination from pulping activities and hydrocarbons from marine traffic were apparent at this site. With no N,P added, extensive degradation of the aromatic fraction with little or no degradation of the saturate fraction was observed again. With N,P added, there was slightly more degradation observed in the aromatic fraction and complete degradation of the n-alkanes and isoprenoids in the saturate fraction.

A comparison of the profiles in Figures 5 and 6 indicates that the addition of N,P stimulated the degradation of the saturated fraction but had a lesser effect on the aromatic degradation. This trend had also been observed with other environmental samples, including marine sediments, greenhouse soil, and river water.

A different manner of attack is demonstrated by the population from East Fidalgo Bay (Figure 7), the site of two large oil refineries that receive much of their oil from tankers. In the presence of N,P there was extensive degradation of both fractions. However, unlike the degradation seen in Figures 5 and 6, this population was capable of degrading a portion of the n-alkanes and only some of the smaller aromatic compounds, such as naphthalene (peak 1), methylnaphthalenes (peaks 2 and 3), C_2-naphthalenes (region 10), and phenanthrene (peak 6), without the addition of N,P.

Figures 5–7 show that mixed populations from environmental sources can readily degrade the aromatic compounds under consideration, and that the nutrient status of the medium and the source of the inoculum influence which groups of compounds are most readily used.

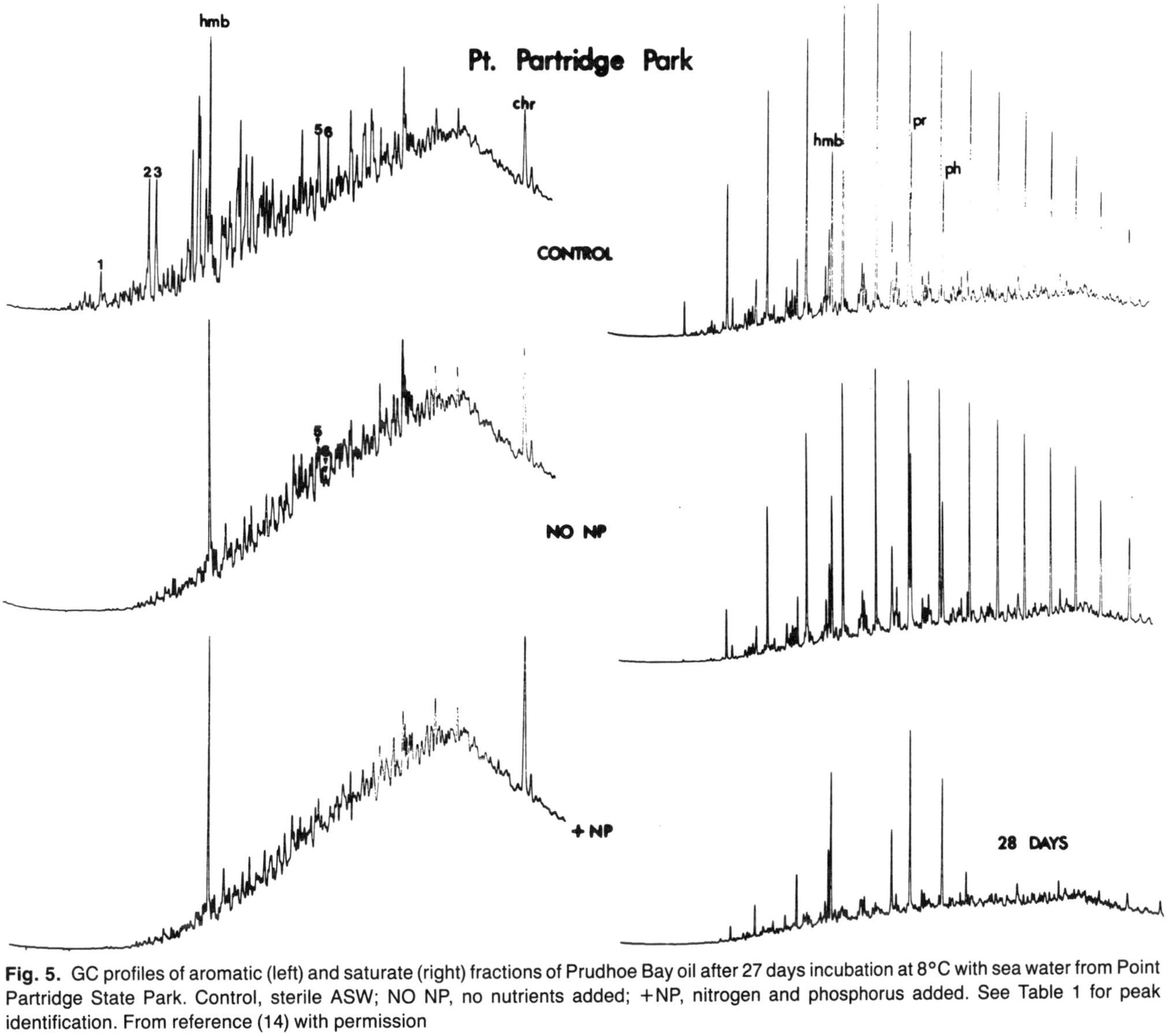

Fig. 5. GC profiles of aromatic (left) and saturate (right) fractions of Prudhoe Bay oil after 27 days incubation at 8°C with sea water from Point Partridge State Park. Control, sterile ASW; NO NP, no nutrients added; +NP, nitrogen and phosphorus added. See Table 1 for peak identification. From reference (14) with permission

The order in which these aromatic and saturate compounds are attacked was also studied. In experiments using N,P-supplemented marine samples grown at 8°C in the presence of crude oil supplemented with one of four different ^{14}C-labeled hydrocarbons, the lag time for initiation of mineralization was measured (Table 2). Regardless of sample type, seasonal variation in population, or site type, naphthalene was always the first compound attacked (i.e., shortest lag time) and anthracene was the most recalcitrant. The relative susceptibilities of hexadecane and phenanthrene were dependent upon the microbial population present in the sample. Completeness of degradation of the ^{14}C compounds was reflected in the extent of mineralization (data not presented). Results generally showed that phenanthrene was utilized most rapidly and to the greatest extent, with anthracene usually the most persistent of the four compounds tested.[19]

Because of the limited availability of ^{14}C-labeled hydrocarbons for mineralization studies, the technique of capillary GC was used to expand these observations and to elucidate further the order of degradation in both the saturate and aromatic fractions.

Figures 8 and 9 show the aromatic and saturate profile of oil incubated with greenhouse soil supplemented with N,P. Control and test flasks were analyzed at various times. After 2 days of incubation at 20°C, there was a marked reduction in the amount of naphthalene (peak 1) and 2-methylnaphthalene (peak 2) recovered from the growth flask

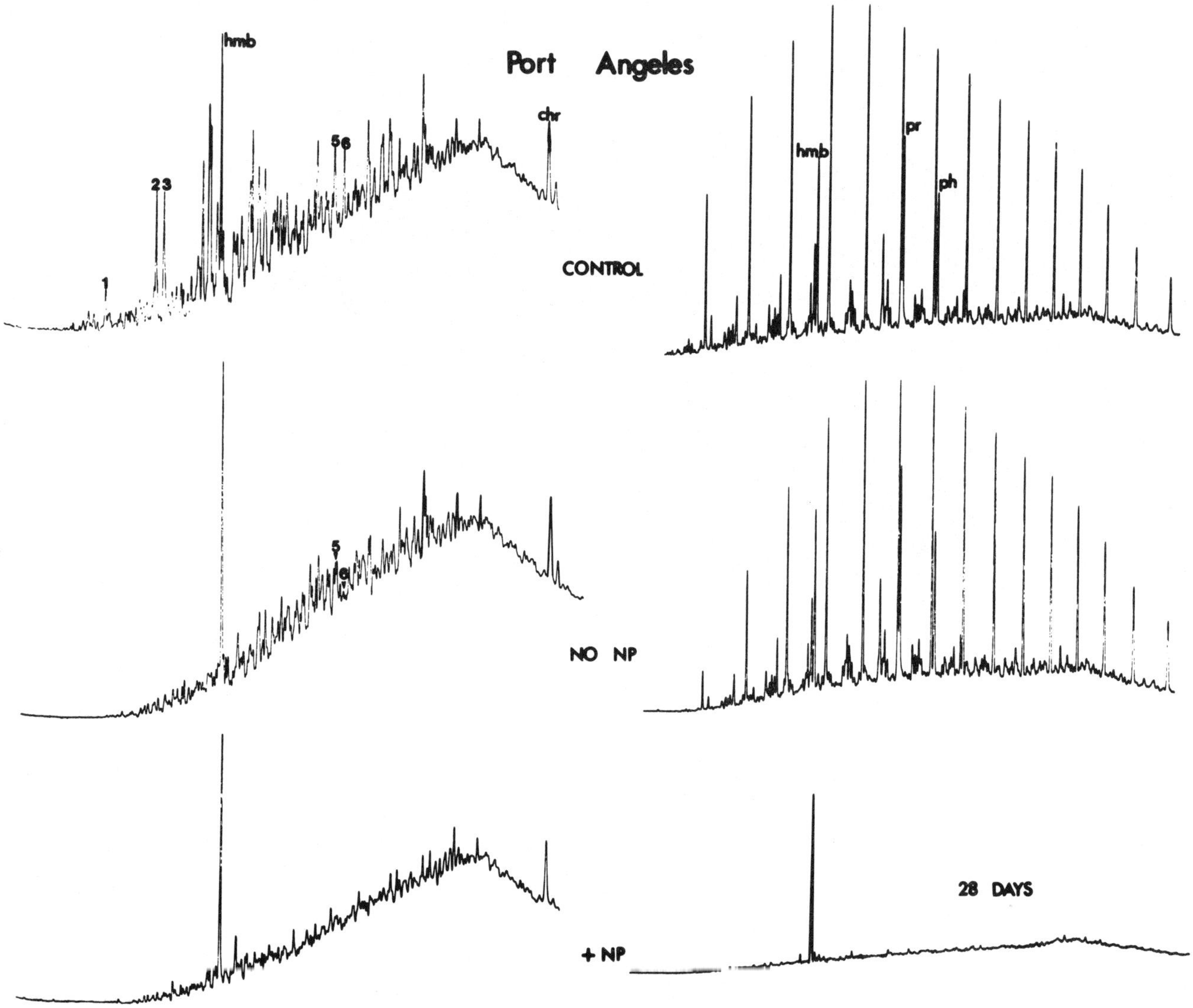

Fig. 6. GC profiles of aromatic (left) and saturate (right) fractions of Prudhoe Bay oil after 27 days' incubation at 8°C with sea water from Port Angeles Harbor. See Table 1 for peak identification. From reference 14 with permission

Table 2. Time required to initiate mineralization of selected ^{14}C substrates by marine microorganisms in mixed cultures incubated at 8°C.

Sample Type	Month of Sampling	Site Type	Lag Time (Days)			
			Naphthalene	Hexadecane	Phenanthrene	Anthracene
Water	April	Pristine	3	9	11	15
		Polluted	3	6	8	11
Water	June	Pristine	4	10	10	11
		Polluted	2	6	6	8
Subtidal sediment	June	Pristine	8	11	9	13
		Polluted	5	9	7	9
Beach	January	Pristine	9	18	ND[a]	ND
		Polluted	5	7	ND	ND

[a]Not determined

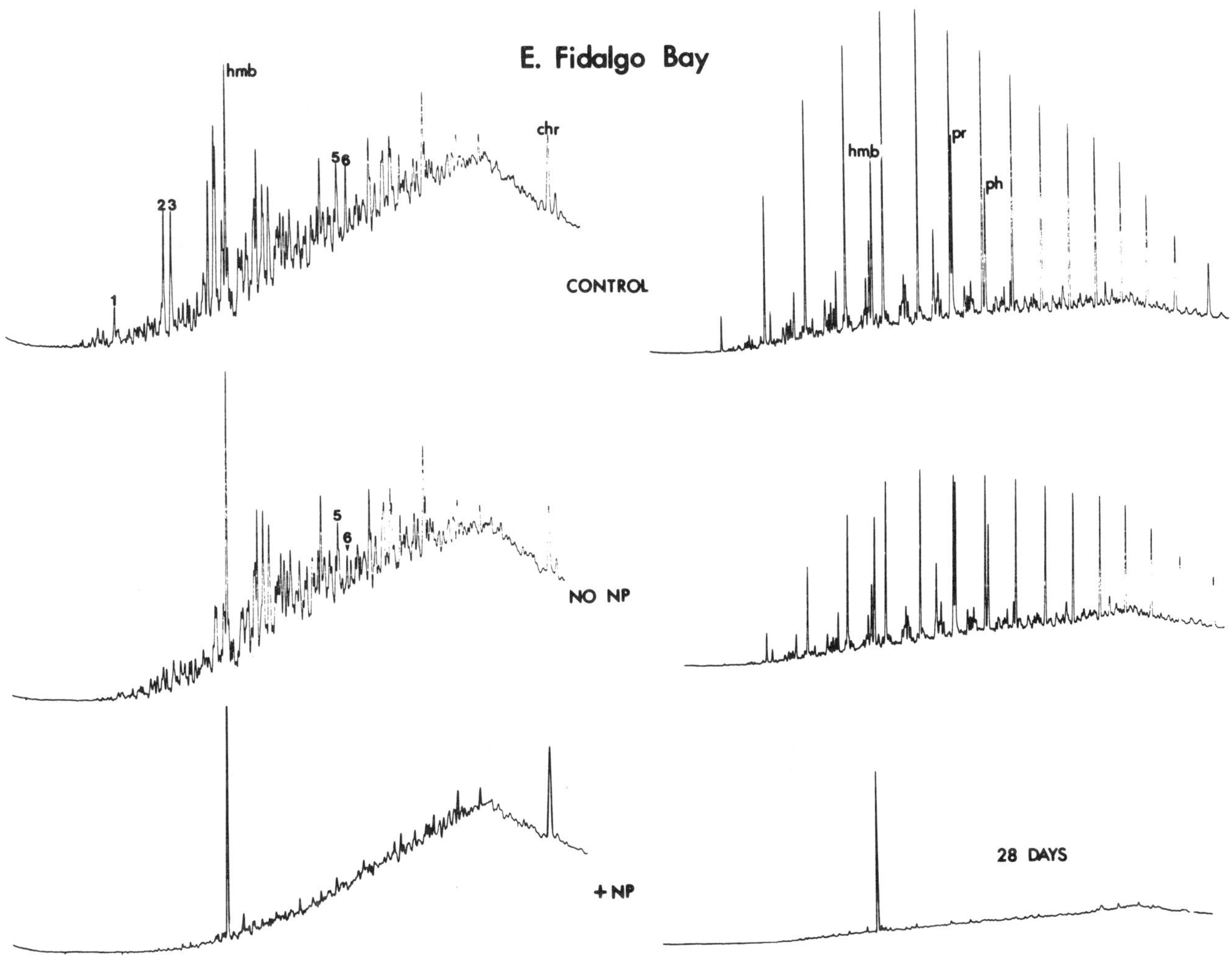

Fig. 7. GC profiles of aromatic (left) and saturate (right) fractions of Prudhoe Bay oil after 27 days' incubation at 8°C with sea water from East Fidalgo Bay. See Table 1 for peak identification. From reference 14 with permission

compared with the control flask. A comparison of the n-C_{17} to pristane ratios in the control and sample extracts indicates that there was some degradation of the n-alkanes (approximately 5%). After 3 days of incubation, there was an overall decrease in the n-alkane peaks, while in the aromatic fraction a decrease in the amount of 1-methylnaphthalene (peak 3) occurred. At day 3, the C_2-naphthalenes (region 10), dibenzothiophene (peak 5) and phenanthrene (peak 6) were present. However, by day 4 (Figure 9) these had been removed from the oil. In the saturate fraction by day 4, most of the n-alkanes had been removed, leaving only the isoprenoids and other unidentified compounds. Between days 4 and 5 there was a loss of the C_3-naphthalenes (region 11) and methylphenanthrenes (region 12). By day 9 there was some change in the C_2-phenanthrenes (region 13), and by day 11 these had been degraded. In the saturate fraction between days 7 and 11 there was a slow, continuous degradation of the isoprenoids (pr and ph) and other unidentified compounds.

Figure 10 summarizes the hydrocarbon degradation sequence observed. Naphthalene and 2-methylnaphthalene were the first hydrocarbons to be extensively degraded. Once the degradation of the n-alkanes commenced, they all disappeared relatively quickly. At about the same time, dibenzothiophene and phenanthrene were lost from the oil. The degradation of parent, unsubstituted aromatic compounds occurred first, followed by the degradation of the substituted compounds in the order of their increasing

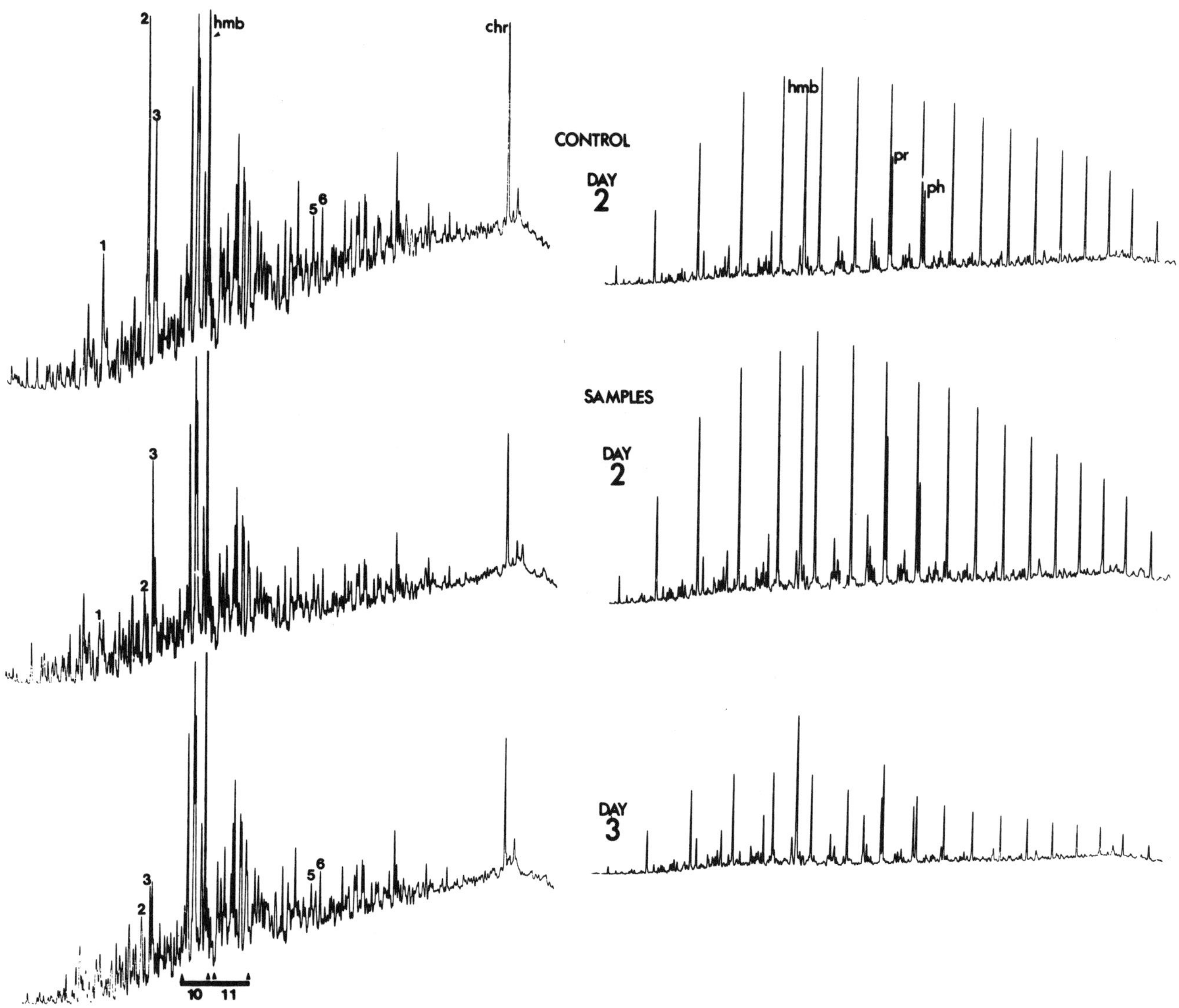

Fig. 8. GC profiles of aromatic (left) and saturate (right) fractions of Prudhoe Bay oil after 2 and 3 days' incubation. Control, autoclaved soil and salts solution; samples, 2 g soil as inoculum. From reference 15 with permission

complexity. For example, naphthalene was degraded before the C_2-naphthalenes, which were degraded before the C_3-naphthalenes; and phenanthrene was degraded before the methylphenanthrenes, which were degraded before the C_2-phenanthrenes. The degradation of the isoprenoids appeared to start shortly after the loss of the n-alkanes, but their rate of degradation was relatively slow. This sequence supports and expands the data gathered using ^{14}C-labeled substrates and indicates that many aromatic compounds are more susceptible to microbial degradation than previously reported.

Also, single isolates from oil-degrading enrichment cultures were obtained and these were grown individually on oil. The capillary GC method was used to determine which compounds in the aromatic and saturate fractions were being utilized. The first organisms studied in this manner were marine fungal isolates (Table 3), which had previously been shown to attack n-alkanes (unpublished data). Cultures of nine isolates were grown on slants with Prudhoe Bay crude oil as a carbon source, and after 70 days of incubation at 20°C six of the isolates completely degraded the n-alkanes while the other three partially removed the n-alkanes. None of the isolates degraded the isoprenoids pristane and phytane. In the aromatic fraction, the amount of 2-methylnaphthalene present was very slightly reduced by six of the isolates, but there was no evidence of degradation of any other aromatics.

In another study, 16 isolates from marine bacterial enrich-

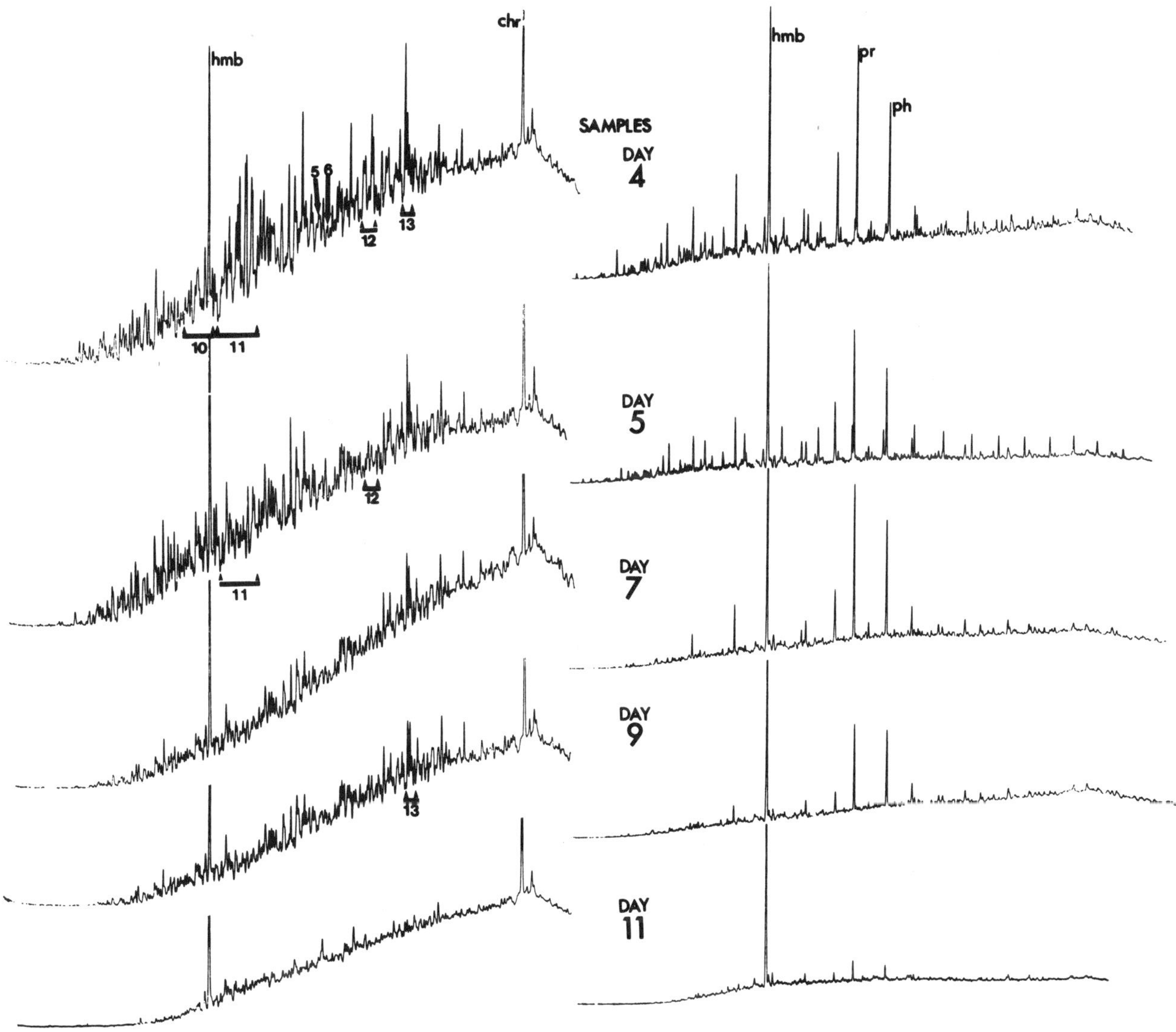

Fig. 9. GC profiles of aromatic (left) and saturate (right) fractions of Prudhoe Bay oil after incubation for 4–11 days with 2 g soil as inoculum. From reference 15 with permission

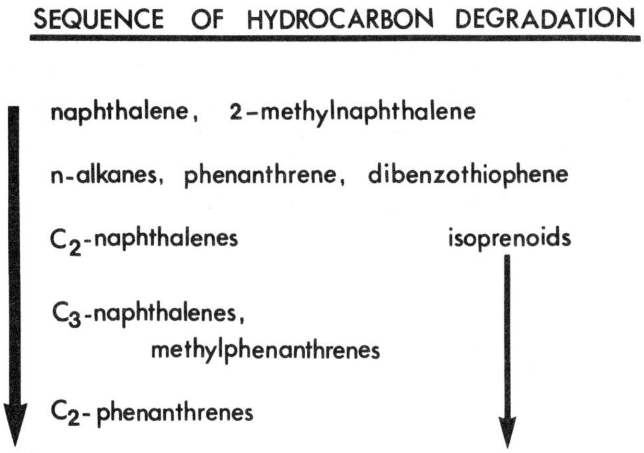

Fig. 10. Summary of observed hydrocarbon degradation sequence. Direction of arrow indicates increasing incubation time required before microbial attack

ments were grown at 8°C for 28 days in the presence of Prudhoe Bay crude oil (Table 4). Six of these isolates gave detectable changes in the GC profiles of the oil. One isolate completely degraded the n-alkanes and the isoprenoids, but neither made any change in the aromatic profile. One isolate degraded only the methylnaphtalenes. Another extensively degraded the aromatic fraction, including the methylnaphtalenes, the C_2-naphthalenes, dibenzothiophene, phenanthrene, and some of the methylphenanthrenes, and also degraded the aromatic designated X (Figure 4a.), which tentatively has been identified as 3-methylbiphenyl based on the retention time of an authentic standard. Two other isolates degraded only this compound and made no other detectable changes in the aromatic or saturate profiles. The isolates described in Table 4 survived nine monthly transfers in pure culture. The latter seven isolates, which caused no detectable changes in the aromatic or saturate fractions, were able to survive the extensive dilution of the transfers. These cultures must have grown at the expense of other

Table 3. Marine fungal isolates on Prudhoe Bay crude oil. Degradation after 70 days at 20°C.

Number of Isolates	*n-Alkanes*	*Isoprenoids*	*2-Methyl-naphthalene*	*Other Aromatics*
2	++[a]	−[b]	−	−
1	+[c]	−	−	−
4	++	−	sl[d]	−
2	+	−	sl	−

[a]++ = complete removal of peaks from chromatogram
[b]− = no change
[c]+ = partial removal
[d]sl = slight change in relative peak height

unresolvable compounds in the oil. Three of the 16 isolates did not survive repeated transfer in pure culture.

To date, approximately 50 microbial isolates from various enrichment cultures have been screened to compare their activities on the aromatic and saturate fractions, and the following trends are apparent: *(i)* many isolates survive repeated transfers but show no changes in the aromatic or saturate profiles; *(ii)* those isolates that degrade some of the aromatic compounds produce no changes in the saturate fraction; and *(iii)* those isolates which extensively degrade the n-alkanes and isoprenoids do not degrade the aromatic compounds studied. The only exceptions noted were those fungi which appear to degrade only 2-methylnaphthalene and saturated n-alkanes (Table 3).

To gain a greater understanding of the microbial degradation or transformation of crude oil, more elaborate analytical methods must be used to determine which compounds are being modified or removed from the oil and which compounds are not susceptible to attack. In most GC profiles of oil exposed to bacterial cultures, there is an envelope of unresolved compounds. Little or no information has been obtained regarding the identity of these compounds or their susceptibility to microbial metabolism. These envelopes are present even when using high-efficiency capillary GC.

Table 4. Marine bacterial isolates on Prudhoe Bay oil. Degradation after 28 days at 8°C.

Number of Isolates	*n-Alkanes*	*Isoprenoids*	*2- and 1-Methylnaphthalene*	*C_2-Naphthalene, Dibenzothiophene, Phenanthrene, Methylphenanthrenes*	*Single aromatic X*
1	+[a]	−[b]	−	−	−
1	+	+	−	−	−
1	−	−	+	−	−
1	−	−	+	+	+
2	−	−	−	−	+
7	−	−	−	−	−

[a]Peak(s) completely removed from chromatogram.
[b]No change in peak(s) compared to control oil.

Application of ^{13}C and ^{1}H nuclear magnetic resonance spectroscopy may provide information on changes occurring in these envelopes. The use of element-specific GC detectors, such as those for sulfur and nitrogen, would allow the detection of many compounds in oil that are usually present in low concentrations in conventional crudes but are more abundant in heavy oils. A greater understanding of the compounds that make up the polar fraction of oil and improved analytical methods are needed to enable study of their biodegradability.

Data presented in Table 4 show that many marine bacterial isolates survived repeated transfer in pure culture with crude oil as their sole carbon source, yet the analytical methods used failed to reveal which compounds were being metabolized. The fact that three of the sixteen isolates did not survive repeated transfer in pure culture may indicate their dependence upon other organisms in the mixed maintenance culture to provide them with specific nutrients or to detoxify components in the oil. A greater knowledge of the interactions between microorganisms and oil is relevant to understanding their roles in degrading petroleum pollutants in the environment, in microbe-assisted corrosion of pipelines, in the biogenesis of oils, and in the recovery of petroleum from reserves.

SUMMARY

The sequential degradation of saturates and aromatics in crude oil was studied using pure and mixed cultures of bacteria and fungi from environmental sources, including ocean water, marine sediments, and soil. Measurement of $^{14}CO_2$, evolved from mineralization of selected ^{14}C-labeled hydrocarbons in the presence of oil incubated with marine cultures, was used to determine the order of initiation of attack, which usually occurred as follows: naphthalene, phenanthrene, hexadecane, and anthracene. Glass capillary gas chromatographic analysis of these samples confirmed the radiorespirometric data and was also used to study the degradation of a wider range of saturate and aromatic compounds. The time required to initiate microbial attack on aromatic compounds increased with molecular complexity. The source of the environmental sample, the constituents of the microbial population, and the concentrations of available nitrogen and phosphorus determined the sequence of attack of aromatics and saturates.

ACKNOWLEDGMENTS

We acknowledge the technical assistance of Ms. K. Semple and Ms. C. Adamyk. Data in Table 4 were collected by Ms. R.S. Clare as part of her undergraduate thesis project in the Department of Microbiology, University of Alberta.

This work was supported in part by the United States Marine Ecosystems Analysis Puget Sound Project (Environmental Protection Agency Interagency Agreement No. D6-E693-EV, program element No. EHE 625-A) and in part by Natural Sciences and Engineering Research Council of Canada operating grant No. A-3687.

REFERENCES

1 WESTLAKE, D.W.S. 1979. Heterotrophic activity determined by substrate disappearance, p. 156–163. *In* J.W. Costerton, and R.R. Colwell (ed.), Native aquatic bacteria: enumeration, activity and ecology. ASTM STP 695. American Society for Testing and Materials.

2 DAVIES, J.S., and D.W.S. WESTLAKE. 1979. Crude oil utilization by fungi. Can. J. Microbiol. *25*:146–156.

3 ATLAS, R.M. 1975. Effects of temperature and crude oil composition on petroleum biodegradation. Appl. Microbiol. *30*:396–403.

4 WESTLAKE, D.W.S., F.D. COOK, and A.M. JOBSON. 1978. Microbial degradation of petroleum hydrocarbons. United States Environmental Protection Agency, Washington, D.C.

5 WESTLAKE, D.W.S., A. JOBSON, R. PHILLIPPE, and F.D. COOK. 1974. Biodegradability and crude oil composition. Can. J. Microbiol. *20*:915–928.

6 WILLIAMS, G.R., E. CUMINS, A.C. GARDENER, M. PALMIER, and T. RUBIDGE. 1981. The growth of *Pseudomonas putida* in AVTUR aviation turbine fuel. J. Appl. Bacteriol. *50*:551–557.

7 CRIPPS, R.E., and R.J. WATKINSON. 1978. Polycyclic aromatic hydrocarbons: metabolism and environmental aspects, p. 113–134. *In* R.J. Watkinson (ed.), Developments in biodegradation of hydrocarbons. 1.

8 HOPPER, D.J. 1978. Microbial degradation of aromatic hydrocarbons, p. 85–112. *In* R.J. Watkinson (ed.), Developments in biodegradation of hydrocarbons. 1.

9 SOLI, G., and E.M. BENS. 1972. Bacteria which attack petroleum hydrocarbons in a saline medium. Biotechnol. Bioeng. *14*:319–330.

10 ROUBAL, G., and R.M. ATLAS. 1978. Distribution of hydrocarbon-utilizing microorganisms and hydrocarbon biodegradation potentials in Alaskan continental shelf areas. Appl. Environ. Microbiol. *35*:897–905.

11 WALKER, J.D., and R.R. COLWELL. 1976. Measuring the potential activity of hydrocarbon-degrading bacteria. Appl. Environ. Microbiol. *31*:189–197.

12 KAPPELER, TH., and K. WUHRMANN. 1978. Microbial degradation of the water-soluble fraction of gas oil. Water Res. *12*:327–334.

13 KAPPELER, TH., and K. WUHRMANN. 1978. Microbial degradation of the water-soluble fraction of gas oil. II. Bioassays with pure strains. Water Res. *12*:335–342.

14 FEDORAK, P.M., and D.W.S. WESTLAKE. 1981. Microbial degradation of aromatics and saturates in Prudhoe Bay crude oil as determined by glass capillary gas chromatography. Can. J. Microbiol. *27*:432–443.

15 FEDORAK, P.M., and D.W.S. WESTLAKE. 1981. Degradation of aromatics and saturates in oil by soil enrichments. Water, Air, Soil Pollut. *16*:367–375.

16 WESTLAKE, D.W.S., and F.D. COOK. 1980. Petroleum biodegradation potential of northern Puget Sound and Strait of Juan de Fuca environments. Publication No. EPA-600/7-80-133. United States Environmental Protection Agency, Washington, D.C.

17 FEDORAK and WESTLAKE, 1981, Microbial degradation of aromatics and saturates in Prudhoe Bay crude oil as determined by glass capillary gas chromatography, op cit.

18 FEDORAK and WESTLAKE, 1981. Degradation of aromatics and saturates in oil by soil enrichments, op cit.

19 WESTLAKE and COOK, op cit.

INDEX